Naturalists' Handbooks 22

# Animals under logs and stones

**2nd edition**

C. PHILIP WHEATER, HELEN J. READ
& CHARLOTTE E. WHEATER

Pelagic Publishing

Second edition published in 2023 by
Pelagic Publishing
20–22 Wenlock Road
London N1 7GU, UK

www.pelagicpublishing.com



First edition published in 1996 by
The Richmond Publishing Co. Ltd

*Naturalists' Handbooks*
Series editor: William D.J. Kirk

British Library Cataloguing in Publication Data
A catalogue record for this book is
available from the British Library

ISBN 978-1-78427-417-7 Pbk
ISBN 978-1-78427-418-4 ePub
ISBN 978-1-78427-419-1 PDF

https://doi.org/10.53061/UXAG4304

Cover images, clockwise from top: *Neobisium carcinoides* (© Frank Ashwood), *Platyarthrus hoffmanseggii* (© Frank Ashwood), *Eurynebria complanata* (© Nigel Cattlin/naturepl.com), *Blaniulus guttulatus* (© Frank Ashwood).

Printed in the UK by Short Run Press Ltd

# Contents

# Editor's preface

A remarkable diversity of fascinating animals can be found in the habitat under logs and stones, including small mammals, reptiles, amphibians and a very wide range of invertebrates. This *Naturalists' Handbook* gives an introduction to the natural history, biology and ecology of these animals and provides advice on how to study them, together with identification keys to the many animal groups. The first edition of *Animals under logs and stones* was published 27 years ago, in 1996. This second edition is considerably expanded, with many more colour illustrations and many more identification keys. It offers an updated and more detailed coverage of the subject. *Animals under logs and stones* complements several other titles in this series: *Common ground beetles* (No. 8), *Ants* (No. 24), *Studying invertebrates* (No. 28) and *Amphibians and reptiles* (No. 31).

This habitat is particularly suitable for ecological investigation because it can be created easily and conveniently by placing objects such as bricks or tiles on the ground. Factors such as the size, position and spacing can be varied easily. The book includes suggestions of topics that can be investigated in this way. These created habitats can also be used for observing lifecycles, detecting species for species records, monitoring species for conservation or simply discovering the world of hidden animals that usually go unnoticed.

William D.J. Kirk
December 2022

# About Naturalists' Handbooks

*Naturalists' Handbooks* encourage and enable those interested in natural history to undertake field study, make accurate identifications and produce original contributions to research. A typical reader may be studying natural history at sixth-form or undergraduate level, carrying out species/habitat surveys as an ecological consultant, undertaking academic research or just developing a deeper understanding of natural history.

# About the authors

**Phil Wheater** is Professor Emeritus in Environmental and Geographical Sciences at Manchester Metropolitan University. After a long university career, he has semi-retired to follow his interests in the ecology and management of human-influenced environments, and invertebrate conservation and management. He is a keen advocate of field ecology and continues to teach university students, including on field courses.

**Dr Helen Read** is the Conservation Officer for the City of London Corporation at Burnham Beeches, a National Nature Reserve in Bucks. She is an active member of the British Myriapod and Isopod Group having been Secretary and Bulletin Editor for many years and has previously served on the Council of the British Arachnological Society and been its Secretary. She is currently updating the Synopsis of the British Fauna volume on millipedes.

**Charlotte Wheater** is an artist who has produced illustrations for a range of outputs, including websites, research reports, infographics, and books.

# Acknowledgements

This book emerged from teaching notes and identification keys used at Manchester Metropolitan University to support students studying field ecology. We are grateful for the copious feedback from students and colleagues at the Manchester Metropolitan and Manchester Victoria Universities and members of the British Myriapod Group and British Isopod Study Group who contributed to the success of the first edition. A large number of people gave freely of their time and advice to help us to produce that edition, including Geoff Allen, Gordon Blower, Sarah Corbet, Rod Cullen, John Dallingwater, Henry Disney, Glyn Evans, Peter Hammond, Steve Hopkin, Richard Preece, Dennis Unwin and Jonathan Wright.

This second edition includes totally revised keys reflecting the major taxonomic changes that have taken place over the last 25 years. We would like to thank all those experts who helped in the production of this edition by supplying information (often unpublished) and advice, as well as constructive criticism of our text and keys and commenting on the accuracy and accessibility of the work. We are extremely grateful to the series editor, William Kirk, for his support and attention to detail which has greatly improved this volume. In addition, we would particularly like to thank the following who commented and provided advice and information on particular sections of the book:

- Penny Cook and Mark Langan commented on the text for Chapters 1, 2, 3, 5 and the introduction for Chapter 4, while Vikki Bengtsson commented on the section on the decomposition of wood;
- Martin Albertini reviewed the general invertebrate (A), hexapod (M) and insect larvae (V) keys, and Mike Dobson commented on amphibious leeches;
- Hugh Jones reviewed the flatworm key (B) and gave us access to unpublished information and records;
- Kevin Butt reviewed the earthworm key (C);
- Christian Owen and Ben Rowson reviewed the slug (D) and snail (E) keys;
- Matthew Shepherd contributed to the production of the mite key (F), and Felicity Crotty reviewed the final version – both kindly gave us access to their draft key;
- Mike Davidson reviewed the harvestman key (G) and Paul Richards gave us access to his tabular key;

- Francis Farr-Cox and Gerald Legg reviewed the pseudoscorpion key (H);
- James Bell and Paul Selden reviewed the spider key (I);
- Steve Gregory reviewed the woodlouse key (J);
- Tony Barber reviewed the centipede key (L);
- Peter Shaw reviewed the springtail key (N);
- Judith Marshall reviewed the earwig (O), cockroach (P) and orthopteroid (Q) keys;
- Gary Skinner reviewed the ant key (R);
- Chris Terrell-Nield reviewed the beetle family (S), ground beetle (T) and rove beetle (U) keys;
- Martin Hartup reviewed the amphibian (W) and reptile (X) keys;
- Will Trewhella and Stephanie Wray reviewed the small mammal key (Y);
- Jaime Martin and Charlotte Rideout commented on some of the keys.

All images associated with the keys (Chapter 4) were drawn by Charlotte Wheater. Unless otherwise stated, all other images are by Phil Wheater and Helen Read. Those marked CC are used without modification under appropriate licences from Creative Commons:

- Figure 3.11 – 'Nematode' by KSRE Photo is licensed under CC BY 2.0 – https://creativecommons.org/licenses/by-sa/2.0/legalcode
- Figure 3.12 – 'Horsehair Worm' by Alastair Rae is licensed under CC BY-SA 2.0 – https://creativecommons.org/licenses/by-sa/2.0/legalcode
- Figure 3.16 – 'File:Enchytraeidae2.jpg' by D. Sikes is licensed under CC BY-SA 2.0 – https://creativecommons.org/licenses/by-sa/2.0/legalcode
- Figure 3.26 – 'Crane Fly Larva' by treegrow is licensed under CC BY 2.0 – https://creativecommons.org/licenses/by/2.0/legalcode
- Figure 3.33 – 'Rote Samtmilbe – Red velvet mite' by gbohne is licensed under CC BY-SA 2.0 – https://creativecommons.org/licenses/by-sa/2.0/legalcode
- Figure 3.38 – 'File:European Yellow-Tailed Scorpion (Euroscorpius flavicaudis)-Sheerness UK-August 2013 2013-08-28 16-07.jpg' by User:Dikhou is licensed under CC BY-SA 3.0 – https://creativecommons.org/licenses/by-sa/3.0/legalcode
- Figure 3.61 – 'File:Festival of Proturans Part I (8643068105).jpg' by Andy Murray is licensed under CC BY-SA 2.0 – https://creativecommons.org/licenses/by-sa/2.0/legalcode
- Figure 3.66 – 'File:Nemobius sylvestris (Gryllidae) (Wood cricket) – (male imago), Molenhoek, the Netherlands.

We are grateful to the following for supplying additional images:

- Marc Baldwin (MB) – Figs 3.45, 3.50, 3.86, 3.87, 3.89;
- John Davy-Bowker (JDB) – Fig. 3.15;
- Steven J. Falk (SJF) – Fig. 3.65.
- Martin Hartup (MH) – Fig. 3.88;
- David Hawkins (DH) – Fig. 3.19;
- Steve P. Hopkin (SPH), courtesy of Ailsa Hopkin – Fig. 3.59;
- Hugh Jones (HDJ) – Figs 3.14 and 5.3;
- Paul Richards (JPR) – Fig. 3.35;
- Barry Webb (BW) – Fig. 3.31.

# 1 Introduction

Logs and stones, as well as other materials on the ground surface, offer refuges for a fascinating array of animals. While decaying wood is a major feature of many woodlands, logs and stones (and their equivalents) can also be found in many other habitats such as grasslands, heathlands, wasteland in urban areas, river and canal banks, and private gardens. Not only are these often highly accessible microhabitats for study, they also provide an interesting interface between the damp depths of the soil and the drier and less protected open ground surface. The communities of organisms that live beneath these features are often little noticed and even less studied, yet they support a wide range of interesting animals, which can provide substantial scope for ecological research. Such animals may occur as individuals, or gather in relatively large aggregations. In addition, it is often possible to identify differing behaviours, from mating to egg-laying, together with some fascinating predator–prey relationships. Studying these organisms gives an insight into an amazing microcosm of life covering a range of animals from those that use the microhabitats as shelters to others living most of their lives in this relatively hidden world.

## 1.1 Cryptozoa

The animals living in such habitats have been termed 'cryptozoa' (from the Greek *kryptos* meaning hidden, and *zōon* meaning animal). Savory (1971) suggests that this best describes those invertebrate animals living in the cryptosphere – the vegetation litter zone which includes those under logs and stones. The term is therefore used to cover those animals living in areas of high humidity and the absence of light.

Some of the animals living beneath logs and stones are relatively large and easy to find, although they may not always be seen in large numbers. They exhibit a wide range of life forms, from tiny invertebrates to rather large amphibians and reptiles, from relatively slow slugs or snails to very fast centipedes, from animals that may live much of their lives under logs and stones to those that simply exploit the environments some of the time, from generalist to specialist feeders, and from herbivores to carnivores. Communities found under logs and stones tend to be dominated by invertebrates such as millipedes, centipedes, woodlice, beetles and spiders, as well as occasional vertebrates such as small mammals, amphibians and reptiles. In some cases (for instance with various flies) it is the larval form that is found while the adults live elsewhere. In other groups, such as ground and rove beetles, both larvae and adults may be found together.

These microhabitats can act as refuges for surface-active animals, protecting against drying out or freezing. The environments under which the cryptozoa live are influenced by a range of factors, including the broader habitat type in which they are found, the type of material (e.g. log, stone, or artificial material), and whether the refuges are exposed or sheltered from climatic impacts such as sunlight and direct water impacts (rain and runoff). These microenvironments may provide food in the form of decaying plant material, fungal hyphae and spores, and even other animals. They are sometimes used by organisms for specific behaviours, for example as sites for mating aggregations, egg-laying, and hibernation in winter or aestivation during the summer. Alternatively, some animals spend their entire life in this habitat. As a result, the communities of animals under logs and stones vary across the year, and since some species can be found during every season, this adds to the potential such habitats have for ecological investigation.

Animals may exploit a range of refuges, including those made of artificial materials. Logs and stones are natural structures, but equivalent artificial environments are created out of corrugated iron sheets and roofing tiles, rubbish such as carpets, cloth and polythene, and building materials including bricks and breeze blocks (Figs 1.1–1.6).

Gilbert (1989) recorded a wide range of animals living among brick rubble that had been left for varying periods of time. Spiders were the most frequent early colonisers, with centipedes, millipedes and woodlice being abundant after 4–6 years. The communities after 12–15 years were rich in a full range of animal groups, adding molluscs, earthworms and beetles to the mix. Cryptozoa will also shelter beneath garden pots (plant pots), planters and other containers, compost bins, etc. Such communities will be richer when pots rest on the soil surface rather than on other hard surfaces such as paving (although even here some animals can often be found). Many of these artificial habitats may be even richer than more natural structures. Indeed, monitoring programmes can exploit artificial refuges to study a range of species including amphibians and reptiles. Habitats in urban areas and those sites that undergo regular management (e.g. some agricultural sites) are more likely to be disturbed. This may mean that refuges in such sites have the potential

**Fig. 1.1** Decaying log

**Fig. 1.2** Collapsed stone wall

**Fig. 1.3** Building waste

**Fig. 1.4** Corrugated iron

**Fig. 1.5** Invertebrate habitat – 'Bug hotel'

**Fig. 1.6** Reptile refuge

to be more important as a buffer against physical disturbance, and impacts from pollution, than they would be in a relatively undisturbed woodland, for example. Of course, in some more disturbed areas, the range of species found will be limited to more tolerant groups.

The animal communities covered by this book are those encountered under logs and stones resting on the ground in terrestrial situations. We do not include those on the seashore or in very wet places, although where relevant we have flagged some species of particular interest in these types of habitat. Very different animal communities may be found within decaying wood itself and under bark. This book does not include those animals that live within decaying wood, although some that feed on it are covered. For further details of the ecology and conservation of some of the (saproxylic) animals found associated with decaying wood, see Kirby & Drake (1993) and Stokland *et al.* (2012).

**saproxylic animals**
*'Saproxylic organisms are species which are involved in or dependent on the process of fungal decay of wood, or on the products of that decay, and which are associated with living as well as dead trees'* (Alexander, 2008). Such organisms (frequently invertebrates) are often quite rare and hence may have high conservation value.

Logs and stones that simply rest on the ground provide shelter at the surface. Those that are partially buried within the soil may harbour animals that are members of the soil fauna rather than that of the leaf litter. Most work to date has concentrated either on species classed as soil inhabitants (e.g. Wallwork, 1970, 1976; Davis *et al.*, 1992; Nielsen, 2019), or on the animals associated with habitats on or above the surface of the soil, including the vegetation layers (e.g. Curry, 1994; Leather, 2004). However, in reality there is an interchange between these communities and individual species can occur in each at different points in their lifecycle. Large piles of stones and decaying wood may provide a range of microhabitats that differ from those found under logs and stones that rest directly on the ground. These different habitats may have microclimatic differences including differential exposure to sunlight and hence a range of temperatures, as well as variation in water availability, perhaps even leading to risks of extremes such as desiccation and flooding.

Decaying wood is a valuable resource for many animals including invertebrates (Stubbs, 1972; Kirby & Drake, 1993; Humphrey & Bailey, 2012). In the past, forestry practice dictated that dead wood should be removed, largely because of the fear of disease transmission, but also sometimes for aesthetic reasons. However, it is now realised that, except in exceptional circumstances, decaying wood is a key constituent of healthy ecosystems and generally decaying wood is now increasingly left undisturbed. This includes both standing (dead trees) and fallen (branches and twigs) decaying wood. Indeed, another change to management practice has been the deliberate creation of wood piles. While forestry practice often involves the temporary siting of log piles, conservation efforts now also include the creation of more permanent piles of logs, branches and other woody debris (Figs 1.7–1.9). Read (2020) gives details of the creation

**Fig. 1.7** Forestry log pile

**Fig. 1.8** A biodiversity log pile

**Fig. 1.9** Overgrown biodiversity log pile

of a log pile designed to benefit a wide range of biodiversity. In some cases, such structures have been targeted at providing shelter for small mammals. In others, the accumulation of different sizes of logs and other woody debris can support a variety of fungi, mosses and liverworts, lichens and a wide range of invertebrates. Fry & Lonsdale (1991), Kirby (1992), Stokland *et al.* (2012) and PTES (undated) give further details of the importance of decaying wood for the conservation of invertebrates.

Stones may also be valuable, albeit to a lesser extent, since they offer refuges, although not the same food resources or complexity of habitat as does decaying wood. Gilbert (1989) found linyphiid spiders (money spiders) to be among the early colonisers of brick rubble. In all types of refuges, there may be a succession within the communities of colonising species, from early pioneers through to more settled and diverse communities. Little work has been done on the mixed communities using refuges, and much remains to be discovered about how these develop and function.

The range of animals encountered may be rather diverse and therefore seem a little challenging when first examining these habitats. Although the identification keys in this book (and those from other more specialist texts) are required for identification of many species with certainty, there are some very common and obvious species with which even a novice can quickly become familiar. Examining the ecology and behaviour of even one common species can be very rewarding and a range of interesting research topics are suggested later in this book (Chapter 5).

## 1.2 Classification

The community living under logs and stones includes representatives of many different groups of animals. Those referred to in this book are listed in Table 1.1. The keys (Chapter 4) should make it possible to name the animals found, sometimes at species level and sometimes at the level of genus, family or higher grouping. The classification of the ground beetle *Pterostichus madidus* (Fig. 1.10) is shown as an example of the hierarchical arrangement of taxonomic categories and illustrates the relationships between taxonomic levels (Table 1.2).

A species is given a unique binomial (two names: the generic name and the specific epithet) which, being Latinate, should always be italicised. When first mentioned in a text the full binomial should be used. On subsequent mentions of a species or other member of the same genus, the generic name may be abbreviated to the initial letter (e.g. *P. madidus*). In some more specialist texts, this species name is followed by the authority (i.e. the person who first named the animal) together with the date on which that name was first published. If that name is still the officially recognised name, then the authority is given as it stands. Where the name has been changed over the years, the authority and date are placed in parentheses (as in Table 1.2 for *P. madidus*). Changes in names are not arbitrary. Rather, they are approved by the scientific community on the basis of previous misidentification, confusion with similar species, or recognising that

**Fig. 1.10** *Pterostichus madidus* – common blackclock

**Table 1.2** Classification of *Pterostichus madidus* – common blackclock

Kingdom – Animalia (animals)
  Phylum – Arthropoda (arthropods)
    Class – Insecta (insects)
      Order – Coleoptera (beetles)
        Family – Carabidae (ground beetles)
          Genus – *Pterostichus* (blackclock beetles)
            Species – *Pterostichus madidus* (Fabricius 1775) (common blackclock)

**Table 1.1** Taxonomy of animals found under logs and stones

Classification based on Tilling (2014), extended and updated to take account of subsequent changes

| Phylum | Subphylum | Class | Subclass | Order | Common name |
|---|---|---|---|---|---|
| Nematoda | | | | | roundworms |
| Nematomorpha | | | | | hairworms* |
| Platyhelminthes | | Turbellaria | | | flatworms |
| Nemertea | | | | | proboscis worms* |
| Annelida | | Clitellata | Hirudinea | Arhynchobdellida | leeches* |
| | | | Oligochaeta | Haplotaxida | potworms |
| | | | | Crassclitellata | earthworms |
| Mollusca | | Gastropoda | | Pulmonata | slugs and snails (part) |
| | | | | Prosobranchia | snails (part) |
| Arthropoda | Chelicerata | Arachnida | | Acari | mites |
| | | | | Opiliones | harvestmen |
| | | | | Scorpiones | scorpions* |
| | | | | Pseudoscorpiones | pseudoscorpions |
| | | | | Araneae | spiders |
| | Crustacea | Malacostraca | | Isopoda | woodlice |
| | | | | Amphipoda | amphipods |
| | Myriapoda | Pauropoda | | | pauropods |
| | | Diplopoda | | | millipedes |
| | | Chilopoda | | | centipedes |
| | | Symphyla | | | symphylids |
| | Hexapoda | Entognatha | | Diplura | two-tailed bristletails |
| | | | | Protura | proturans |
| | | | Collembola | | springtails |
| | | Insecta | Apterygota | Archaeognatha | jumping bristletails |
| | | | | Zygentoma | silverfish* and firebrats* |
| | | | Pterygota | Dermaptera | earwigs |
| | | | | Dictyoptera | cockroaches |
| | | | | Orthoptera | crickets (grasshoppers, etc.*) |
| | | | | Hemiptera | true bugs* |
| | | | | Raphidioptera | snakeflies* |
| | | | | Lepidoptera | butterflies* and moths* |
| | | | | Diptera | true flies – larvae (adults*) |
| | | | | Hymenoptera | ants (bees* and wasps*) |
| | | | | Coleoptera | beetles |
| Chordata | | Amphibia | | Caudata | newts* |
| | | | | Anura | frogs* and toads |
| | | Reptilia | | Squamata | lizards and snakes |
| | | Mammalia | | Eulipotyphla | hedgehogs* |
| | | | | | shrews* |
| | | | | Rodentia | mice* and voles* |
| | | | | Carnivora | weasels* and stoats* |

* Groups that are infrequently found under logs and stones in the UK

species are wrongly attributed to particular genera, families, etc. This has become more frequent as DNA analysis (genetic profiling) on various groups reveals the presence of more than one species in what was previously thought to be a single species. For example, for a long time the centipede *Geophilus carpophagus* was thought to have two forms with different numbers of pairs of legs. It was only in 2001 that these were revealed to be two separate species with the shorter species now being called *G. easoni*. When species names change, synonyms for the correct specific name may exist – where these have been in wide current use, we have highlighted them within the keys. For instance, *P. madidus* (Fig. 1.10) has sometimes been referred to by a subgeneric name *Steropus*. Some groups of particularly numerous species may be allocated to a subgenus (a taxonomic level lying between genus and species). Species names can be checked using a resource such as the Catalogue of Life or the Tree of Life Project websites (see Chapter 6). The classification used for invertebrate animals in this book follows that of Tilling (2014). The use of scientific names is important in order to ensure that organisms are referred to correctly without ambiguity. See the Introduction to Chapter 4 for a discussion of the use of common names.

# 2 Environmental conditions under logs and stones

The influence of the environment under logs and stones on the animal communities found will be discussed in this chapter. There is more information about particular groups of animals in Chapter 3.

As we have seen, the habitat under logs and stones represents an interface between the soil and the world above the ground. It can form a temporary refuge for many of the animals inhabiting both environments. Animals found under logs and stones may live there all the time, or they may be occasional visitors. They may be transient inhabitants of the soil such as some adult beetles which dig into the soil when egg-laying or which overwinter in the upper soil layers. Or they may be temporary residents of the soil, such as some crane flies, moths and beetles, which live in the soil as larvae, emerging above the ground as adults. Other animals may live mainly above the soil and use logs and stones as temporary refuges from adverse climatic conditions, to escape from predation, or even to find food themselves among the other organisms that live there.

## 2.1 Microclimate

The soil layer that is in contact with the underside of logs and stones depends upon the depth to which the log or stone is buried (Fig. 2.1). Stones simply resting on the surface are in contact with the topsoil or leaf litter layer. Those that are buried more deeply make contact with lower soil levels and will often provide more moist and stable habitats. The position of the log or stone will therefore influence the community of animals found beneath it. The very localised climatic conditions such as those under logs or stones are termed microclimate. For the animals found under logs and stones, the most important aspects of microclimate are temperature and humidity. The temperature on the surface and in the top layer of soil is strongly influenced by the amount of solar radiation falling on the surface. This will vary both regularly (on a daily and seasonal basis) and irregularly (with degree of shade or cloud cover). Temperatures under logs and stones may show much less daily or seasonal variation than temperatures outside and they will be influenced less by irregular changes in ambient temperature. Smaller, thinner refuges will buffer the environment beneath them to a lesser

**Fig. 2.1** Depth of stones buried in the ground: foreground – resting on surface; background – buried

extent than will larger, thicker materials. The permanent shade afforded by a log or stone means that the humidity also tends to remain fairly constant. The humidity under a refuge depends, not only on the ambient levels, but also upon the degree of burying and the nature and moisture status of the soil. Many invertebrates, such as most woodlice, millipedes or mites, require high humidity levels. Such animals may aggregate in suitable sites to prevent desiccation, behaviours that can be observed in many common woodlice. Others, such as many species of insects, have waterproof cuticles and can tolerate drier conditions.

**capillary forces**
Where a liquid (here soil water) is held within the small spaces between soil particles. Capillarity (capillary action) is the main force holding, as well as moving, water through narrow spaces both vertically and horizontally against the force of gravity.

## 2.2 Substrate

The temperature and humidity of deeply buried items closely reflect those of the soil around them. If the underside of the object is above the ground water level (i.e. the level at which the soil or rocks are permanently saturated with water), most of the available water is held in the soil by capillary forces in small channels within the soil structure. The greater the number of small channels that there are, the better is the water holding capacity of the soil. Sandy soils consist of large particles and so have rather few large channels, which have weak capillary forces and allow much water to drain away leaving the soil quite dry. Clay soils consist of very small particles and so have many tiny channels, which hold water well. These soils can become much wetter and heavy clays may become waterlogged, especially in winter. In contrast, when they dry out, they may become hard and sometimes form surface cracks. Sandy soils warm up faster in the sun and cool down faster at night and, although they tend to be generally warmer, they may have more variable temperatures. Clay soils are cooler and experience fewer extreme fluctuations of temperature. There are, of course, many other soil types, such as loams, which show intermediate characteristics. Soil types may be identified using keys such as those in Trudgill (1989), or Kennedy (2002). A quick and easy method of classifying soils is to roll a small amount between your fingers and thumb – clay soils stick together into a sausage-shape, while sandy soils are gritty and will not hold together. Compaction of the soils around (and underneath) refuges can decrease the degree to which water can percolate and hence increase moisture levels at the surface.

The nature of the substrate under the log or stone also influences the communities present. The microhabitat and its microclimate depend on the type of soil, the depth of the top layer, and the presence or absence of decaying plant and animal material. The upper regions of the soil are usually richer in organic matter than are the lower levels (Fig. 2.2). Because they usually have large quantities of decaying plant material, such as leaf litter, woodland soils often have strong organic profiles with one of two types of humus layers at the surface. Mor humus, often found on sandy soils under conifers, is rather acidic and lacking in calcium. Mull humus, found more frequently under deciduous trees, is neutral or slightly alkaline. The presence of calcium in soil promotes a rich earthworm fauna, which will mix the organic and inorganic components of the soil, reducing the degree of obvious layering. Calcium is also important for other animals, including snails and millipedes, and thus may be associated with richer invertebrate communities. Mull humus tends to contain more large invertebrates, while mor humus is often dominated by mites and springtails. The communities under logs and stones will tend to reflect the local soil fauna. These woodland soils are extreme cases and there is a range of intermediates between them. Mull and mor humus may also be seen in open habitats such as moorlands or heathlands. However, in grassland soils there are some complications, mainly due to human influence.

**humus**
Dark organic matter formed when plant and animal material decays.

**soil horizons (from the surface downwards)**
O (organic) horizon: mostly organic matter including decomposing leaves – many cryptozoa found here.
A (topsoil): mixture of minerals from the parent material with organic matter from above – where the majority of plant roots and soil animals may be found.
B (subsoil): mainly minerals from the parent material and leaching from the A horizon above but little organic material – plant roots penetrate this layer together with some (deep-living) soil animals.
C (parent material): developed from minerals from the base rock and leaching from above – few organisms penetrate this layer.

**Fig. 2.2** Soil profile showing O horizon (organic material and decomposing plant material) at the surface overlaying A horizon (topsoil), B horizon (subsoil) and C horizon (parent material)

Most grassland soils approach mull type humus, although the annual turnover of organic matter may be higher than that in woodlands due to the faster growth and quicker decay of herbaceous vegetation. Much of the organic matter in grassland soil derives from the decay of root systems in situ rather than from leaf litter. Further details of soil biology may be found in Davis *et al.* (1992) and Neilson (2019).

Different materials, such as stone or wood, provide different microhabitats. The chemical and physical composition of stones may influence the animal communities. For example, many snails require calcium for their shells and may do better under limestone rocks. Darker rocks heat up more in direct sunlight and hence may be more attractive for heat-loving species such as ants (in this case for their nests), and rocks with a relatively complex structure underneath may provide more spaces for small animals. Surface features under logs and stones may provide a range of microhabitats suitable for different animals. Cracks, crevices and splits which appear very small to us are gaping chasms to a small invertebrate (Fig. 2.3). It is sometimes difficult to appreciate this because we are used to dealing with the world at our own scale, but the range and number of such features may be important in determining the community make-up. Change in the gaps between the wood and any bark, and the degree of splitting of the surface of a log is fundamental to the process of decay and hence will alter over time.

**Fig. 2.3** Lithobiomorph centipede heading for a crack in a stone

## 2.3 Decay and decomposition

The state of decay of logs may also be important and as decomposition progresses the influences thereon may change (Figs 2.4–2.7). The breakdown of wood is largely due to fungi and invertebrates. Of the invertebrates, beetles are especially important. Speight (1989) recognises three phases of the breakdown of wood by invertebrates: colonisation, decomposition and humification. The colonisation phase involves pioneer species such as beetles. Mites and fungi are

**Fig. 2.4** Fresh logs

**Fig. 2.5** Older log beginning to split

**Fig. 2.6** Decaying log with peeling bark

**Fig. 2.7** Humification (log decaying into the soil)

also pioneer species and are often carried on the bodies of beetles. Fungi, in particular, make early entry into the wood itself, and many will be already present in the wood when it falls. There is some evidence that fungal decomposition softens the wood sufficiently to enable beetles to be able to chew it. Adult invertebrates can carry fungal spores so that when they lay their eggs in decaying wood, the fungi they bring in are also able to colonise. The fungal action, often assisted by the strong jaws of the beetle colonisers, begins the initial decomposition. The decomposition phase involves invertebrate species which cannot tunnel into the wood themselves and require the presence of the pioneers first. When this phase begins, decomposition increases in speed as a variety of beetles and fungi become established. At the same time other species move in, which are predatory or parasitic on the wood-feeding invertebrates or fungi. The humification phase occurs as the amount of undecayed wood decreases. Now many of the species associated with earlier phases decline, and so do their predators and parasites, while those from the soil may increase with the availability of more organic material.

Fungi are a crucial element in the decay process and fundamental to the cryptozoan community. It is now thought that certain fungi (known as endophytic fungi) are present within live wood (as well as dead wood) on the tree, so that when it dies or falls, they are instantly ready to exploit the change in conditions. Living wood is saturated with water, but once wood is dead the air can penetrate it and this allows the fungi to grow and start to break down the wood into its constituents of hemicellulose, cellulose and lignin. How fungi break down wood depends on many different factors, but the initial stages can be loosely categorized as either brown rot, white rot or soft rot. In brown rot, the hemicellulose and cellulose are broken down first and the lignin remains, which is red-brown in colour (hence the name) and tends to be fragile and crumbly. In white rot, either the lignin is broken down first (selective white rot) or the lignin, hemicellulose and cellulose are broken down at the same time (simultaneous white rot). Both types result in a soft white wood. Soft rot, a third type, degrades the cellulose and hemicellulose, while some soft rots also decompose the lignin. In the early stages of decomposition, the combination of tree species and the succession of fungal species determines the type of decay. As time passes, and the decay proceeds, the resulting material and associated species become more similar. The final mix of decayed wood and invertebrate frass (droppings) slowly becomes incorporated into the soil. Where a mix of decayed wood and invertebrate frass is found within hollow trees it is sometimes referred to as wood mould (a better name would be saproxylic humus); this is an extremely valuable habitat for rare invertebrates.

The invertebrate decomposer community is unable to digest wood directly until fungi have softened the wood and destroyed the inhibitory chemicals within it. Fungi also improve its nutritional quality by altering the ratio of carbon to other mineral nutrients such as nitrogen and phosphorus: carbon is lost as carbon dioxide, whereas nitrogen and phosphorus remain. Fungal hyphae, the extensive feeding stage of the fungus, also cross the transition zone between the soil and the above-ground leaf litter and logs. While they are involved in the process of decay of organic matter of all forms (including leaf litter, twigs, decaying roots within both woodland and grassland soils and even the carcases of dead animals), fungi also provide food for the invertebrates that feed on the hyphae. It is thought that many decomposers such as some woodlice and millipedes, while appearing to eat the softened wood may only be digesting the fungi and other microorganisms living on it. Fungal mycelium (networks of hyphae) and spores are rich in nutrients. Fungi may also benefit from the activities of invertebrates which carry spores between different pieces of decaying wood. Invertebrates that tunnel into wood aerate the site facilitating fungal growth, although selective grazing on different species of fungi will alter the balance between different fungal species. Some fungi kill certain invertebrate species and feed on them. For further information about the role of fungi in decomposition see Spooner & Roberts (2005), Watkinson *et al.* (2015), and Boddy (2021).

The conditions created by the decomposition process are also important for the communities under logs. The wood, initially hard, perhaps with some peeling bark, becomes softer and moist as decomposition progresses. Since the wood just underneath the bark decays first, the gap formed under the bark may offer good retreats for cryptozoa. The species of tree will have an impact here in determining the type of gap that is formed between the bark and the wood. Tight, thin bark will have smaller species living beneath it than coarser, looser bark. Finally, the decomposed material is incorporated into the adjacent soil. The phases of decomposition may all be present at the same time in different parts of a log resting on, or partially within, the soil. The parts of the wood that are deeper in the soil often break down first, since they do not dry out as frequently as the surface of the wood. Many species of woodlice, millipedes and springtails are associated with the decomposition and humification phases, as they feed on the decaying wood as well as on the fungi present at this time. Predators, such as spiders, centipedes, ground beetles and rove beetles and some mites, also exploit these phases, feeding on the other animals. Most of the species associated with the early phases of decay are animals feeding within decaying wood and under bark, and so are beyond the scope of this book, but some of the under-log predators will feed, when they can, on the detritivore species. Decaying wood habitats are transitory and, although it may take some years for total decomposition to occur, the resource will eventually decline. Different tree species will differ in terms of the time it takes for full decomposition: oak may take many decades, while birch will take only a few years. Species such as birch have an interesting decomposition pattern with the wood often decaying before the bark, so leaving tubes of bark with rotten wood in the middle on the woodland floor. As the characteristics of the habitat alter, so too do the communities of animals and plants living within them. Changes in moisture content, the nutritional value of the wood, degree of shelter and other factors may all be important, to different extents to different organisms. Invertebrate frass under the bark of decaying wood and on the soil under decaying wood is a sign of an active invertebrate population and

**detritivore**
These are organisms that consume detritus (dead and decaying plants and animals and their waste products).

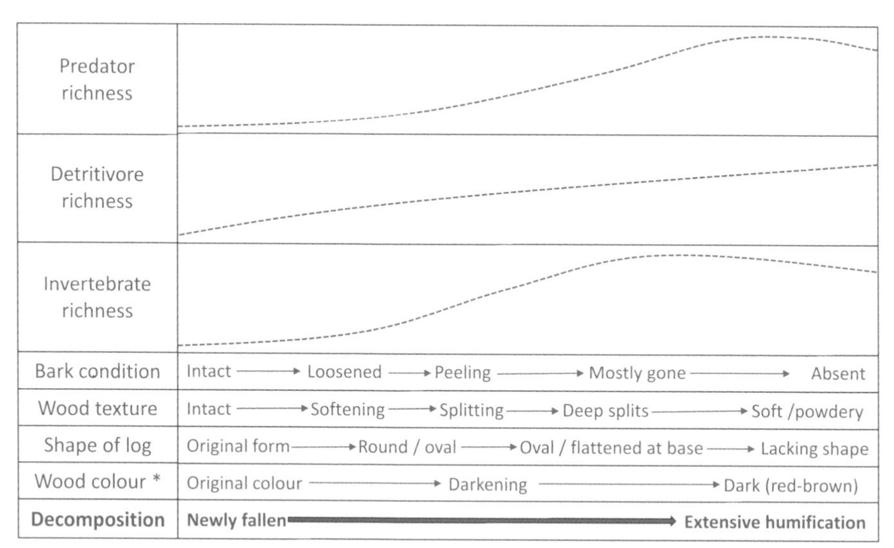


**Fig. 2.8** Generalised overview of changes during the decomposition of logs (see over)

an important step in the decomposition process, contributing to changes in the conditions under the log. As the environment changes, both in terms of the material ageing and the community developing, different successional stages may be seen (e.g. Fig. 2.8). Successional change is characteristic of such transitory habitats. This type of succession, where the primary source of energy is from dead organic material, is called heterotrophic succession and is mainly driven by decomposer species. Swift *et al.* (1979) give a more detailed description of the decomposition process of woody material.

**succession**
Ecological succession describes the sequence of biological communities over time. These changes may be driven by environmental changes and/or be due to modifications caused by the organisms that are living within the habitat at any one time. Different stages within a successional series are called seres and these may be typified by particular communities of organisms.

## 2.4 Refuges as islands

The size of a log or stone and its distance from the nearest similar log or stone may also influence the animal communities found associated with it. A patch of habitat under a log or stone is an 'island' in the sense that it is isolated in a 'sea' of habitat of a different kind. This may be particularly the case when the surrounding habitat is very different from that beneath the refuge, for example, in a relatively short-cropped sward in a grassland. In contrast, a continuous blanket of leaf litter in a woodland may provide better transit zones between logs. The larger the island, the greater is the probability of the arrival of new species (because more randomly moving species will find it), while the probability of extinction is smaller (because large populations are less likely to die out than small ones). Larger islands may also support a greater diversity of habitats, and more diverse environments generally support a wider range of animals. The relationship between species number ($S$) and island area ($A$) for a series of islands is often found to fit the equation:

$$S = CA^z$$

Where $C$ and $z$ are constants that are characteristic of particular types of islands.

This relationship implies that the number of species increases quickly with increased island area, eventually flattening out as the community approaches the total number of potential arrivals (Fig. 2.9). Putting this a different way:

$$\log(S) = \log(C) + z\log(A)$$

This means that a graph of $\log(S)$ against $\log(A)$ gives a straight line with a slope of $z$ and a $y$-axis intercept of $\log(C)$ (Fig. 2.10). The value of $z$ is a characteristic of particular types of island that can be useful, for example, in planning the size and arrangement of nature reserves. Where these refuges act as islands that are relatively isolated from each other (i.e. without much of a shared fauna), the value of $z$

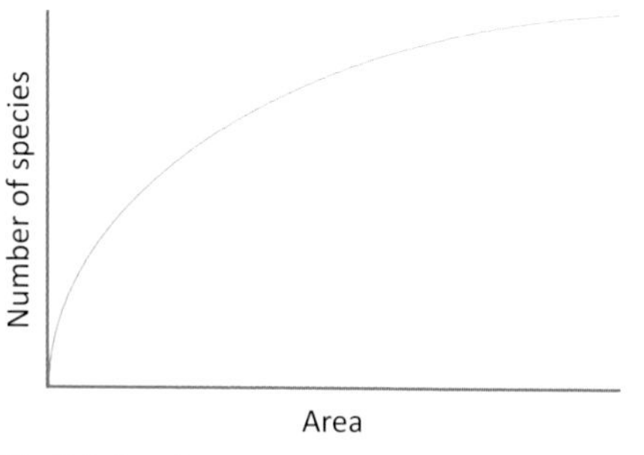

**Fig. 2.9** Species–area curve

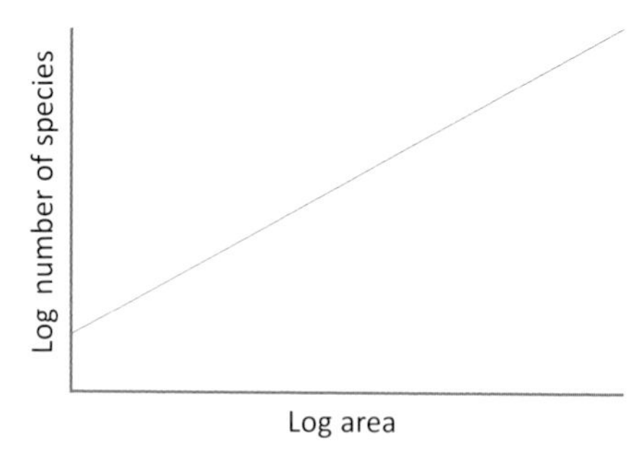

**Fig. 2.10** Log species–area curve

will be higher than when they are effectively small subsets of a wider interconnected environment (i.e. where species move freely between them).

The Theory of Island Biogeography deals with the species composition of communities of plants and animals in island habitats, making predictions about the relationship between the number of species and the distance that the island is from the nearest source of colonists (see Gormon, 1979; Berry, 2009). The number of species on an island reflects the dynamic balance between the rate of arrival and establishment of new species and the rate of extinction, or loss of species (Fig. 2.11). New species are less likely to arrive on a remote island than on an island near the mainland, or main source of colonists. Large islands, and those close to a source of colonists, are therefore more likely to have a complex fauna since they will recruit animals more quickly and these communities are more likely to persist. Furthermore, once herbivores and decomposers are present, predators are more likely to move in. A discussion of how the Theory of Island Biogeography relates to habitat islands such as trees can be found in Leather & Barbour (1999). In many cases, logs and stones may not have been in situ long enough for an equilibrium of species numbers to have been achieved. Investigations of such refuges as islands could prove to be an interesting avenue for study (see Chapter 5).

The local population densities and the dispersal abilities of different species (or groups of species) will influence which animals colonise newly deposited refuges first. While stones may not often be deposited naturally, fallen wood is constantly being dropped from trees. A combination of these factors, together with the characteristics of the wood itself (size, species, amount, stage of decay, etc.) will influence the early communities colonising the resource.

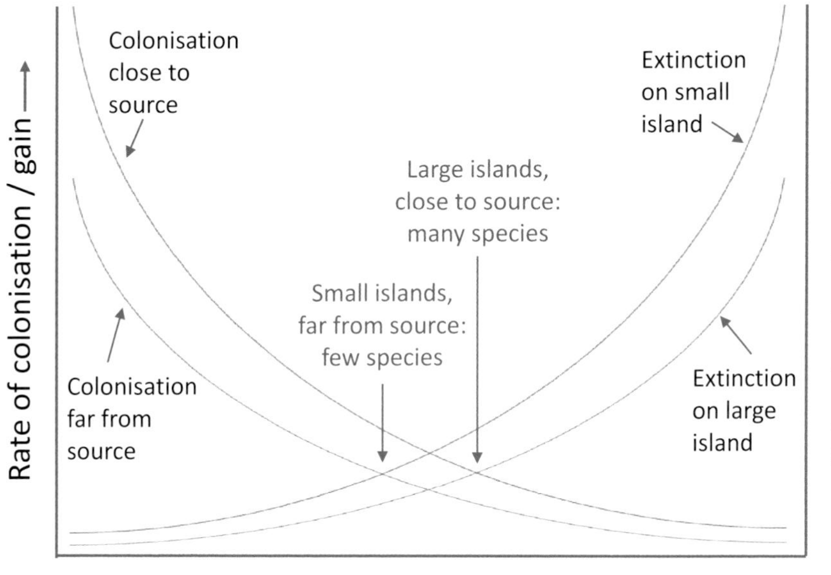


**Fig. 2.11** Equilibrium of species according to the Theory of Island Biogeography

## 2.5 Ecosystem services

The animal communities under logs and stones impact on the local and broader environment in a number of ways. In addition to increasing the biodiversity and conservation value of a site, they provide ecosystem services including nutrient cycling, soil respiration, soil development and formation (pedogenesis), soil aggregation, as well as water drainage and storage. The various mechanisms by which these processes are mitigated by cryptozoa will depend both on the environment and the communities involved. In the past, the benefits of leaving dead and decaying wood on the woodland floor were not always fully recognised. Indeed, there were concerns that the insects and fungi associated with decay might attack living trees. As it became clear that such concerns were not valid and that there was, in fact, substantial benefit to be gained in retaining decaying wood, management practices changed, even to the extent of creating different types of piles of wood to support different communities of organisms. There is substantial scope for these and other aspects of the ecology of animals under logs and stones to be researched (see Chapter 5).

# 3 The cryptozoa

## 3.1 Introduction

Most animals inhabiting the restricted spaces under logs and stones are small or flattened. However, large cavities (e.g. those under log piles) can house bigger animals including small mammals and amphibians. Ecologists have categorised soil invertebrates into size groups (e.g. Wallwork, 1970). The microfauna (animals less than 0.2 mm long) require specialised techniques for their study and are not covered by this book. Here we are mostly concerned with the mesofauna (0.2–10 mm in length) and macrofauna (10–80 mm long), and we cover some megafauna (over 80 mm long) later in this chapter. The mesofauna includes roundworms, mites, pseudoscorpions, bristletails, two-tailed bristletails, proturans, and springtails, together with many spiders, harvestmen, ants and insect larvae, as well as smaller worms, slugs, snails, woodlice, millipedes, centipedes and beetles. The macrofauna includes most worms, slugs, snails, millipedes, and centipedes, together with the larger spiders, harvestmen and insects. Vertebrates are mainly classed as megafauna. Swift *et al.* (1979) suggested that as body width may better relate to the use of available microhabitats, then width could be used rather than length to classify soil animals as follows: microfauna defined as having a body width of less than 0.1 mm; mesofauna as 0.1–2.0 mm; macrofauna 2–20 mm; and megafauna over 20 mm width. Fig. 3.1 indicates how various animals found under logs and stones may equate to these two systems.

Within the community of animals under logs and stones, those that use the habitat as their home for all or part of their lives interact with other species that use it merely as a retreat or resting site. Species usually found within the soil may migrate up to the surface on occasions. Other, surface-active species may occasionally burrow into the soil and use logs and stones as refuges or places to find prey. Some species fall into both these categories at different times of the year or during different stages of their development. For example, the millipede *Cylindroiulus punctatus* (Fig. 3.2) spends the summer breeding in logs, but during the winter moves into the soil (Geoffroy, 1981). Hence the community is very fluid with a high turnover and mobility of animals both on a daily and annual basis.

Eisenbeis & Wichard (1987) classified soil animals into three groups based on the strata in which they lived and

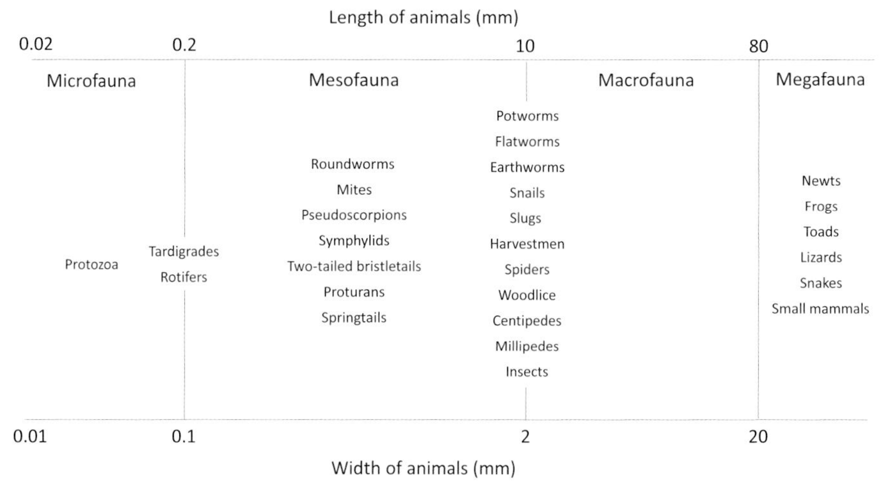


**Fig. 3.1** Sizes of animals found under logs and stones (after Wallwork, 1970 and Swift *et al.*, 1979)

reflecting on their life forms. Euedaphic animals live within the soil pore systems (i.e. the spaces between soil particles and clusters of soil particles) and sometimes extend these by forming yet more soil pores. Such animals are often small and have a worm-like body shape, and often lack pigmentation. Epedaphic soil animals live on the soil surface and in leaf litter. They have a wide range of body forms but are often flattened and strongly pigmented. Hemiedaphic animals tend

**Fig. 3.2** *Cylindroiulus punctatus* – blunt-tailed snake millipede

to dig burrows or exploit existing burrows in order to avoid unfavourable environmental conditions such as cold, heat and lack of available moisture. They include those epedaphic and atmobiotic (those living above the ground surface in vegetation) species that use burrows on a temporary basis, for example overwintering to avoid cold conditions. They are adapted to burrow and to enlarge existing gaps in the soil and may have fossorial (digging-adapted) limbs that are short and stout, sometimes with broad spines. Note that these terms are generally used for smaller arthropods, slightly different classifications are often used for larger animals (e.g. see Key C – Earthworms: epigeic – living on the surface of the ground in leaf litter and under logs and stones; endogeic – living deeper in the soil; anecic – living throughout the soil).

Groups living mainly in the soil include roundworms, earthworms and potworms, some mites, most geophilomorph centipedes, some millipedes, symphylids, two-tailed bristletails, springtails, mole crickets, some ants for some of their time (i.e. when in their nests), certain insect larvae (e.g. of flies and beetles), and some adult beetles such as rove beetles. Those active mainly on the surface include many slugs and snails, woodlice, most millipedes, lithobiomorph centipedes, spiders, harvestmen, cockroaches, earwigs, crickets, foraging ants, numerous beetles and most vertebrates.

## 3.2 Activity patterns

Many species show distinct rhythms of activity, peaking either at night or during the day (see Figs 3.3, 3.4). For example, most harvestmen species are nocturnal while many surface-active spiders are diurnal hunters. The timing of activity may differ between similar species within a group. Many ground beetles are nocturnal, for instance, but diurnal species, such as *Notiophilus* species, are not uncommon. In addition, for some species in this family, the daily rhythm of activity depends on the habitat; *Pterostichus madidus* is nocturnal in woodlands and diurnal in grasslands (Greenslade, 1963). Variability of this kind may turn out to be widespread when more species are investigated. In some groups, such as spiders and millipedes, mature males tend to be more active than the females. Examining the activity patterns of different species from different habitats would be a useful research project (see Chapter 5).

Many animals also have seasonal rhythms of activity that reflect their life histories. Activity usually peaks around the breeding season, which may be in spring, summer, or

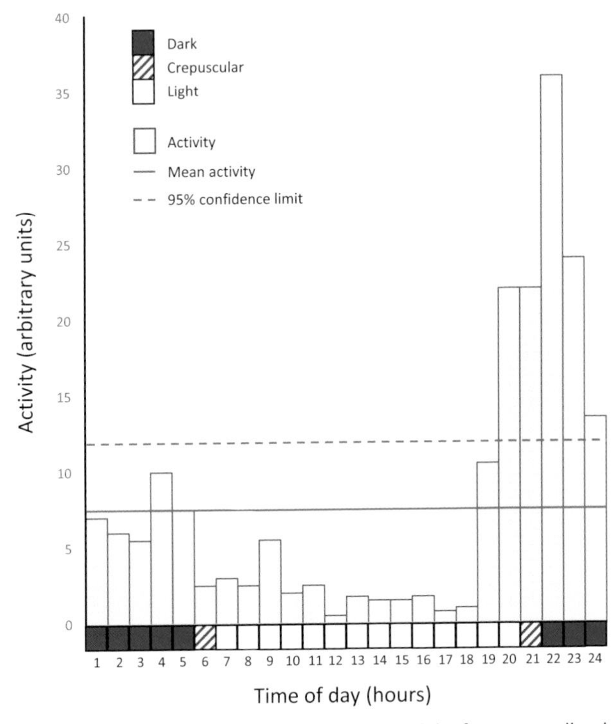


**Fig. 3.3** Daily activity levels of *Pterostichus madidus* from a woodland

early autumn. Some species have more than one breeding season, giving rise to two or more activity peaks in the year. Many animals have a period of inactivity during winter or summer at times of adverse conditions such as cold or drought. For some animals that have a lifecycle lasting a single year, the overwintering stage can be the egg (e.g. the cockroach *Capraiellus panzeri*) or larva (e.g. in many ground beetles). Other ground beetles may live and breed over several seasons and overwinter as both larvae and adults (e.g. *Carabus* species).

## 3.3 Lifecycles

Lifecycles and breeding strategies are diverse. Although many species of insect found under logs and stones have a classic endopterygote lifecycle (complete metamorphosis) from egg, to larva, to pupa and finally adult (e.g. ants and beetles), others show an exopterygote lifecycle (incomplete metamorphosis) where eggs hatch into nymphs, which look similar to the adults (although without wings), that then undergo several moults, finally emerging as adults

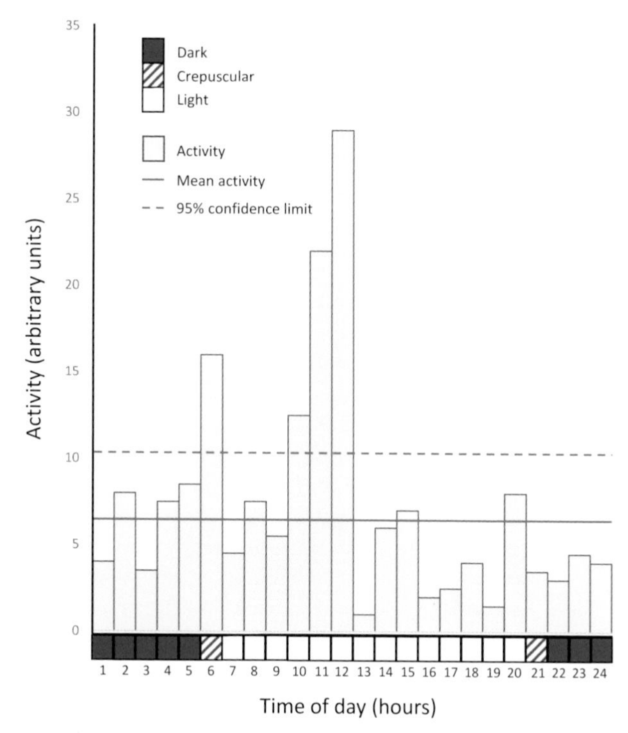


**Fig. 3.4** Daily activity levels of *Ocypus olens*

(e.g. earwigs, cockroaches and crickets). Non-insect invertebrates have an even wider range of types of lifecycle. Many emerge from eggs as miniature versions of the adults, moulting several times before becoming fully functioning adults (as in arachnids such as spiders, harvestmen and pseudoscorpions). In some such groups, additional structures are added at each juvenile moult (as in millipedes and some centipedes where body rings or segments are added as the animal matures).

While insects do not continue to moult as adults, this is not the case in other groups of invertebrates which can moult as adults (e.g. woodlice and some millipedes). In the case of a few millipedes, adult males may moult twice between their breeding seasons, firstly into a non-copulatory (intercalary) form before moulting again to become a fully functional copulatory form. Two extremes of lifecycle strategies can also be recognised: semelparity and iteroparity. In the former, animals produce many young on a single breeding occasion, dying shortly afterwards. In the latter, most animals breed on several occasions, usually producing fewer young each time. These strategies can be seen within, as well as between, invertebrate groups, and even within genera (e.g. some millipede species of the genus *Cylindroiulus* exhibit semelparity, while others are iteroparous). Parthenogenesis (where females produce young in the absence of males) also features in some invertebrate groups (including some woodlice and millipedes). There are invertebrates (such as slugs, snails and earthworms) that are hermaphroditic (possessing both male and female sexual organs). Here cross-fertilisation is the norm and self-fertilisation is prevented by the position of the sex organs (at opposite ends of the body) or maturation of the male and female organs on an individual at different times.

**parasites and parasitoids**
Parasites are organisms that live on or in another organism (the host), obtaining some benefit (such as food or shelter) from it. Parasitoids are organisms that are free-living as adults but have a parasitic larval stage. The females of certain Hymenoptera, as well as some Diptera and Coleoptera, will lay their eggs into, on or near to other invertebrates. As the subsequent larvae develop, they consume the host. In parasites, the host is not usually killed, whereas in parasitoids, the host invariably perishes. Parasites and especially parasitoids have quite specific host species.

## 3.4 Food and feeding

Animal species can be classified in terms of their feeding habits. Phytophages consume plant material including root systems, woody tissue and green material above ground. Fungivores feed on fungal hyphae and spores. Carnivores, which eat other animals, include true predators which kill and consume the prey organism, while parasites and parasitoids feed on the prey organism or host while it is still alive. Saprophages feed on dead and decaying material including dung, carrion and plant matter such as leaf litter or wood. Polyphages consume a range of food types including some or all of the above. Fig. 3.5 illustrates a simplified energy pyramid of animals under logs and stones.

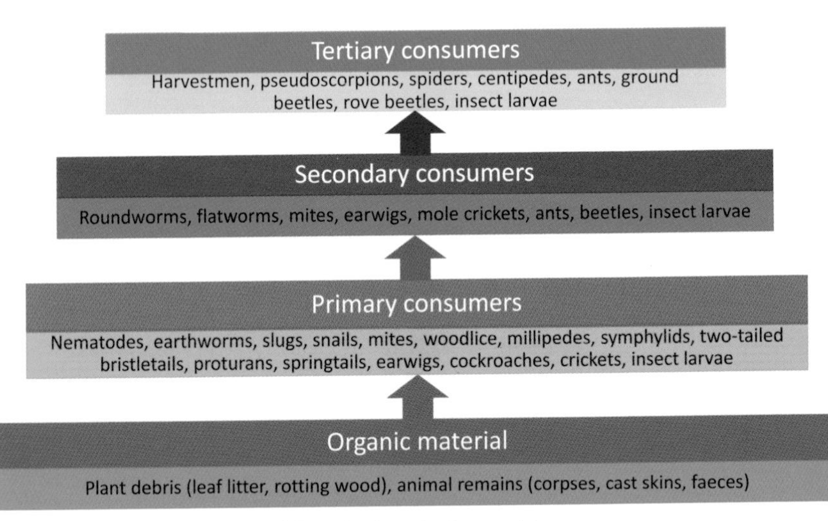


**Fig. 3.5** Simplified energy pyramid for animals under logs and stones

Table 3.1 indicates the main range of feeding types found in various cryptozoa. The true picture is actually much more complex, with some species of beetle having narrowly specialised diets, whereas other species are much more polyphagous. The morphology of the animal in question can provide clues as to the type of prey as well as how it is caught. For example, several species of ground beetles specialise in preying on springtails, which are small and fast moving and which use their furca (spring) to jump away when disturbed. These ground beetle species employ very different mechanisms to catch and subdue their prey (Figs 3.6–3.8). *Notiophilus* species have large eyes which

**Table 3.1** The diets of common animals found under logs and stones

| | Phytophages | Fungivores | Saprophages | Predators | Parasites | Polyphages |
|---|---|---|---|---|---|---|
| Roundworms | + | + | (+) | + | + | + |
| Flatworms | | | + | + | | |
| Earthworms and potworms | + | + | + | | | |
| Slugs and snails | + | + | + | + | | |
| Mites | + | + | + | + | + | + |
| Harvestmen | | | + | + | | (+) |
| Pseudoscorpions | | | | + | | |
| Spiders | | | | + | | |
| Woodlice | (+) | + | + | | | |
| Millipedes | (+) | + | + | | | |
| Centipedes | (+) | | | + | | (+) |
| Symphylids | + | | + | | | |
| Two-tailed bristletails | | | + | | | |
| Springtails | + | + | + | | | + |
| Earwigs | + | | + | + | | + |
| Cockroaches | + | | + | | | + |
| Crickets, grasshoppers, etc. | + | | + | + | | |
| Ants | + | | | + | | + |
| Beetles | + | + | + | + | | + |
| Beetle larvae | + | + | + | + | | + |
| Fly larvae | + | + | + | + | + | + |
| Amphibians | | | | + | | |
| Reptiles | | | | + | | |
| Small mammals | + | | | + | | |

+ indicates that the group often exploits the food type

(+) indicates occasional exploitation

**Fig. 3.6** *Notiophilus biguttatus* (common springtail-stalker) – head showing large eyes

**Fig. 3.7** *Leistus terminatus* (black-headed plate-jaw) – mandibles with expanded side plates

**Fig. 3.8** *Loricera pilicornis* (hair-trap ground beetle) – antennae with setal cage

allow it to see springtails from several centimetres away. Even then, they tend to concentrate on fairly slow-moving species. Species of *Leistus* have broad mandibles which are lined with setae and act as a cage, trapping springtails below. Similarly, *Loricera pilicornis* uses the setae whorls on the first few segments of its antennae as a setal cage to prevent springtails from jumping away once captured. An analogous situation occurs in some harvestmen where the spines and other protrusions from the head and palps produce a cage within which small prey can be captured and manipulated. Other adaptations include the long legs of the tiger beetle *Cicindela campestris*, which together with their large eyes and diurnal behaviour, support their active hunting technique, while the long sharp mandibles of *Cychrus caraboides* and its narrow head allow it to feed on snails within their shells (Figs 3.9–3.10).

**Fig. 3.9** *Cicindela campestris* (green tiger beetle) – showing long legs

**Fig. 3.10** *Cychrus caraboides* (buzzing snail-hunter) – narrow head and long mandibles with sharp teeth

## 3.5 The biology of the invertebrate cryptozoa

Here we give brief descriptions of the major groups of invertebrate animals that may be found under logs and stones. The classification of these groups is shown in Table 1.1 (see page 9). Invertebrate animals are those without a backbone. These include soft-bodied animals such as earthworms and slugs, as well as those whose bodies are protected by shells (e.g. snails) or hard exoskeletons (e.g. beetles).

### Roundworms (phylum Nematoda – Key A couplet 5)

Roundworms (or nematode worms, or eelworms) are relatively little-known animals and much remains to be discovered about both their phylogeny and their ecology. They are small (usually 0.1–2.0 mm long) and so are often overlooked. Nematodes are unsegmented and slightly tapered at each end (Fig. 3.11), and may make eel-like thrashing movements (hence their alternative common name). Some experts consider that there may be up to 1 million species worldwide. In soil they are concentrated in the top few centimetres where they may reach very high densities. Up to 25 million have been found per $m^2$ in oak woods (Wallwork, 1970). Data from a number of studies in European grasslands shows densities of between 1 million and 55 million nematodes per $m^2$ (summarised by Curry, 1994). Densities are often much lower than this at the surface, as the animals require damp conditions such as those found under logs and stones. Mekonen *et al.* (2017) summarise the roles that nematodes play in soil processes as well as their potential as bioindicators of soil health.

Some species are mixed feeders, while others are phytophagous, predatory or parasitic. Predatory species

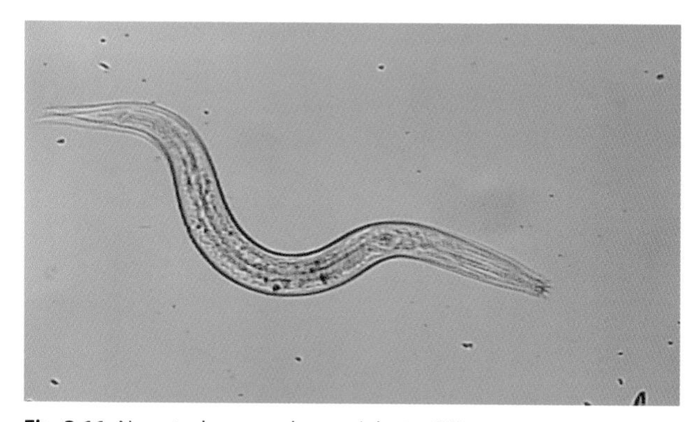

**Fig. 3.11** Nematode – roundworm (photo: CC)

are often larger than fungal or plant feeders, but are also usually less common. Many free-living soil nematodes (including many members of the order Rhabditida) feed on bacteria. Some nematodes also feed on fungi. These and the plant feeders possess stylet mouthparts which they use to pierce the fungal hyphae or plant cells. Examples of fungivores and herbivores include many members of the orders Aphelenchida and Dorylaimida respectively. Others (including members of the order Mononchida) are predators feeding on other soil microfauna. Although frequently their prey feed on dead or decaying material, soil nematodes themselves are not commonly saprophages. See Freckman (2012) for further information on nematodes in the soil.

**Hairworms (phylum Nematomorpha – Key A couplet 5)**

Hairworms (or horsehair worms, or Gordian worms) are parasitoids, with free-living adults and larvae that are parasitic on a range of arthropods including grasshoppers, cockroaches and beetles. Most species are aquatic, but in the UK there are five semi-aquatic species that live in damp soil as adults. They are longer than roundworms (often over 50 mm) and tend to be more pigmented (Fig. 3.12). The larvae, which are very small (around 0.1 mm long) infect their hosts when ingested and develop into mature adults, by which stage they fill most of the body cavity of the host. Once mature, they emerge through the host's exoskeleton forming a tangled knot in the process (hence the name Gordian worm) to mate and produce eggs which hatch into pre-parasitic larvae. See Schmidt-Rhaesa (1997) for further information.

**Fig. 3.12** Nematomorpha (hairworm) emerging from a grasshopper (photo: CC)

### Flatworms (phylum Platyhelminthes – Key B)

The phylum Platyhelminthes (flatworms) includes parasitic groups such as tapeworms (Cestoda) and flukes (Trematoda) as well as the free-living flatworms (Turbellaria). Until recently, free-living flatworms (also called planarians) were best known for their aquatic species. However, the introduction of a number of high profile antipodean terrestrial species, as well as the work of Jones and colleagues have increased knowledge of the range, ecology and distribution of terrestrial species (e.g. Jones, 1998, 2005). There are now thought to be around 22 species in the UK (of which six are probably native), although more species may be found in the coming years. Many of the introductions have arrived in plant material and soils associated with garden plants and the nursery trade, especially from Australia and New Zealand. Two of these introduced species have created some concern because of their predation on earthworms and the fact that they seem to be spreading quite widely across the UK (e.g. Murchie & Gordon, 2013).

Terrestrial flatworms are smooth and shiny and, while most larger species have flattened bodies, smaller species are more rounded in cross-section (Fig. 3.13). With the exception of *Bipalium kewense* which has a halfmoon-shaped head, flatworms' bodies taper towards the head end, which is often held off the ground when moving. Although they have eyes (either on the head end or distributed along the sides of the body), these may be difficult to see in strongly pigmented species (some flatworms can be quite colourful). Terrestrial flatworms are predators or scavengers on soil invertebrates.

**Fig. 3.13** *Marionfyfea adventor* – flatworm

**pharynx**
In flatworms the pharynx is a muscular tube that is used to suck food into the digestive system.

**cilia**
In flatworms these are fine hair-like projections that can produce a beating motion. They are used to assist with locomotion.

They have only one opening to the digestive cavity, through which they both take in food and expel waste material. The mouth is on the underside of the animal and is difficult to see except when feeding occurs. Then, the pharynx is protruded and either pierces or wraps around the prey. Terrestrial flatworms move smoothly using cilia on the underside and by muscular contractions that can extend the body by up to four times its resting length, leaving a trail of mucus behind it. See Jones (1998, 2005) for further information.

### Proboscis worms (phylum Nemertea – Key A couplet 6)

Proboscis (or ribbon) worms are usually thought of as aquatic animals with most UK species being found in freshwater or marine environments, however there are a few terrestrial species that can be found in damp habitats. All three such British species have been introduced, most probably from Australia and New Zealand. They have a flattened oval or round shape in cross-section (Fig. 3.14) and a separate mouth and anus (unlike flatworms). The eyes are at the head end, and in some species are in clusters. They glide smoothly on cilia underneath the body on a trail of mucus. Their name (proboscis worm) is associated with their habit of extending a long proboscis tipped with a sharp stylet from a hole just above the mouth, which they use to capture their prey of

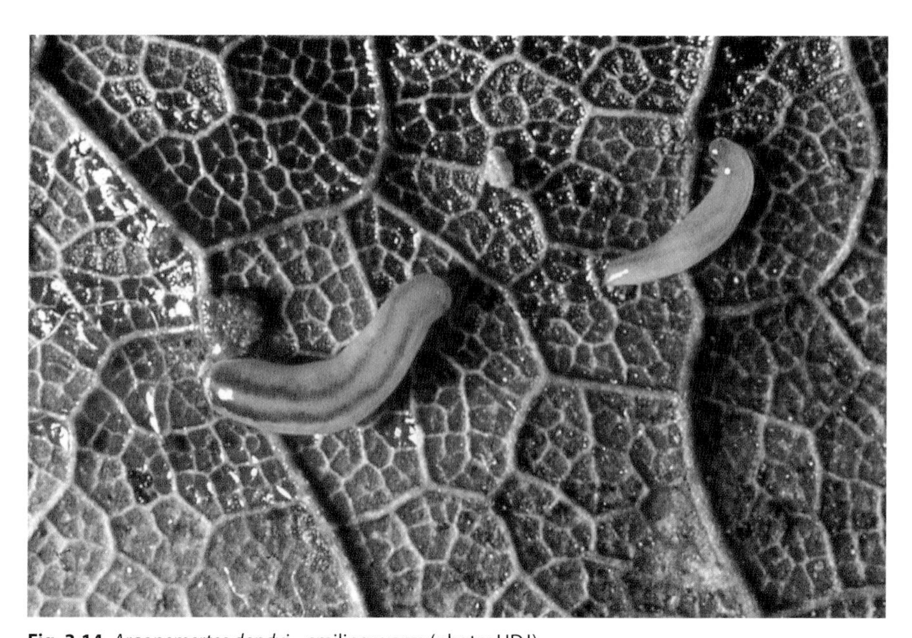

**Fig. 3.14** *Argonemertes dendyi* – smiling worm (photo: HDJ)

small invertebrates. They also use the stylet as a defence and escape mechanism by embedding it into the substrate when disturbed so as to pull themselves rapidly forward in order to escape. See Gibson (1994) and Moore *et al.* (2001) for further information.

### Leeches, potworms and earthworms (phylum Annelida – Key A couplet 4 and Key C)

Although leeches are aquatic animals, a few semi-aquatic species from the order Arhynchobdellida can live in damp terrestrial habitats. Four out of the 17 recorded UK species can sometimes be found under stones near to fresh water (one species in particular may even be found some distance from the water's edge). This amphibious habit can be employed when searching for food, or in order to overwinter, sometimes in groups. These species may scavenge on carrion. Three of the four species also feed on smaller invertebrate prey (swallowing them whole), while the fourth (*Hirudo medicinalis* – medicinal leech) has what might be thought of as a typical habit of feeding on the blood of vertebrates, including humans when available. Several more aquatic species, such as *Erpobdella octoculata* – eight-eyed leech (Fig. 3.15), may also forage on the land at night if food is scarce. Leeches have segmented bodies with a sucker at each end which are used in locomotion and, in the case of the anterior sucker, feeding. See Elliott & Dobson (2015) and Key A (couplet 4) for further information.

**Fig. 3.15** *Erpobdella octoculata* – eight-eyed leech (photo: JDB)

**Fig. 3.16** Enchytraeidae – potworms (photo: CC)

Potworms are small, unpigmented worms from the family Enchytraeidae, which live in highly organic soils (Fig. 3.16). They can be found in quite high densities. Typically 10–20 mm long, they resemble paler, smaller earthworms from which they can be differentiated by the position of the clitellum (or saddle – see Key C). Potworms can be found in high densities, especially in highly organic soils such as those in deciduous forests and peat bogs; they are less abundant in grasslands. They consume soil particles, feeding on the organic matter (fungal hyphae, plant material and microorganisms). Species of potworms are difficult to separate from each other without close examination of the internal organs (see O'Connor, 1967, and Schmelz & Collado, 2010 for further details).

Earthworms are the most familiar group of terrestrial annelid worms (e.g. Figs 3.17, 3.18), the majority of which belong to the family Lumbricidae (although there are two rare species, which are single UK representatives from two other families). The Lumbricidae are mainly saprophages, feeding on decomposing plant material. Some species prefer food rich in nitrogen and sugars, and avoid those high in the polyphenols (such as tannins) found in some tree leaves (e.g. in beech and oak). Earthworms ingest mineral soil in large quantities, and consequently have an important role in soil formation (as do potworms). They help to break up

**tannins**
Tannins are water soluble polyphenolic chemicals, which many trees produce to help to prevent insects and other invertebrates from eating their leaves.

**Fig. 3.17** *Aporrectodea longa* – blackhead worm

**Fig. 3.18** *Lumbricus terrestris* (lob worm) emerging from burrow

organic material and form humus, as well as mixing the soil, by burrowing and by ingesting organic and inorganic particles and transporting them up and down the soil profile – an aspect of their biology that was studied by Darwin (1881). Most of the earthworms under logs and stones are litter-dwelling species; they are often found partially within burrows. Although all earthworms are hermaphrodite and, therefore, have male and female reproductive organs on the same animal, mating does take place. Reproduction usually begins with the transfer of spermatozoa between two animals. Copulation may last up to four hours, the animals lying head to tail, touching at the clitellum (the saddle), often assisted in staying together by the genital pads or tubercula pubertatis. See Sheppard *et al.* (2014), Sherlock (2018) and the Earthworm Society of Britain website (see Chapter 6) for further information on earthworm biology.

**Slugs and snails (phylum Mollusca – Keys D and E)**

Slugs and snails live in damp places where they are not in danger of drying up. Most slugs and snails are nocturnal and many live near the soil surface, often using logs and stones as refuges. Snails are, in general, more frequent on lime-rich than on acid soils because they require calcium carbonate for shell production (Figs 3.19–3.21). Although slugs do not have shells in which they can shelter from predators and dry conditions, they are better able to burrow into the soil

**Fig. 3.19** *Cornu aspersum* – common garden snail (photo: DH)

**Fig. 3.20** *Cepaea hortensis* (white-lipped snail) showing white lip to mouth

**Fig. 3.21** *Cepaea nemoralis* (brown-lipped snail) showing brown lip to mouth

or cracks in stones or wood (Figs 3.22–3.24). Most species of slugs and snails are generalist feeders on decaying plant material, algae, lichens and fungi. Some molluscs, especially many slugs, feed on living higher-plant tissue, and a few of these are pests on crops. Many species of slugs and snails will feed on carrion and a few are wholly or partially predatory. Some predators, such as Zonitidae and *Vitrina pellucida*, feed on other snails, while others, such as *Testacella* species, prey on earthworms.

**Fig. 3.22** *Arion ater* (large black slug) – typical colour form

**Fig. 3.23** *Arion rufus* (large red slug) – typical colour form

**Fig. 3.24** *Limacus maculatus* – green cellar slug

Molluscs are hermaphrodite; each individual possessing both male and female reproductive organs. Despite this, mating still occurs, following a sequence of courtship behaviour in which the animals circle each other, touching frequently. The eggs, which are often spherical and gelatinous, are usually laid in the summer or autumn, often in clumps in the soil or in cracks under logs and stones where they can be quite commonly found (Fig. 3.25). The young look like

**Fig. 3.25** Snail eggs

miniature versions of the adult and usually reach maturity after about a year, although some species may take up to 4 years. An important cause of mortality is predation by birds, small mammals, adult and larval beetles and the larvae of some flies (especially the Sciomyzidae). For further information see Kerney & Cameron (1979), Rowson *et al.* (2014), Cameron (2016) and the Conchological Society of Great Britain and Ireland, and the Malacological Society of London websites (see Chapter 6).

### Arthropods (phylum Arthropoda)

Many of the most obvious invertebrates found under logs and stones are arthropods. These may be species such as woodlice, which spend much of their time (sometimes in groups) under refuges, or nocturnal predators such as harvestmen, which shelter during the day. Those arthropods that live in the soil may emerge under logs and stones to feed or reproduce.

The arthropods include the subphyla Chelicerata (class Arachnida: mites, harvestmen, scorpions, pseudoscorpions and spiders), Crustacea (order Isopoda: woodlice), Myriapoda (pauropods, millipedes, centipedes and symphylids) and the Hexapoda. The latter is a large subphylum that includes the classes Entognatha (three groups of wingless, six-legged insect-like animals), and Insecta (the insects – sometimes called the Ectognatha). These two classes were previously considered to be one group – the Insecta.

The Entognatha contains one subclass (the Collembola – springtails), and two orders (the Diplura – two-tailed bristletails, and the Protura – proturans or coneheads). All three were previously included within the Insecta in the subclass Apterygota, which also included the three-tailed bristletails (order Thysanura) – now included in the Insecta as two separate orders: the Archaeognatha (jumping bristletails) and the Zygentoma (silverfish and firebrats). See Barnard (2011) for further information.

The insects are generally six legged, with their bodies divided into head, thorax and abdomen. Although adult insects often have wings, some groups, particularly those found under logs and stones, do not. In some orders, such as fleas, the adult wings are thought to have been lost during evolution. In others, such as jumping bristletails, and silverfish and firebrats, winglessness is the primitive condition. Some groups that are typically winged include species where the wings are either reduced in size, or where the wing-cases (elytra) are fused, preventing the wings from being used for flight (e.g. in some ground beetles).

See Barnard (2011), Marshall (2017), Brock (2019) and the websites of the Amateur Entomologists' Society, Buglife, and the Royal Entomological Society (see Chapter 6) for further information.

Some of the wingless insects found under logs and stones are the immature stages of winged insects, members of the subclass Pterygota. In those belonging to the superorder Exopterygota, which includes earwigs, cockroaches, grasshoppers, true bugs and snakeflies, the juveniles (nymphs) look like small versions of the adults, with external wing buds. Such nymphs grow into adults through a series of moults, during which they increase in body size and wing length, and finally achieve sexual maturity. This process is termed incomplete metamorphosis in contrast with the complete metamorphosis which occurs in the other superorder of pterygote insects, the superorder Endopterygota. In this group, which includes flies and beetles, the larvae look very different from the adults; they are often grub-like and lack any external signs of wings. The larvae develop through a series of moults and then enter a resting stage, the pupa, in which the internal transformation from a larva to an adult takes place. Most insect larvae found under logs and stones are those of true flies (Diptera) and beetles (Coleoptera). See Chu & Cutkomp (1992) and Marshall (2017) for further information.

It is the larvae, rather than adults, of flies (Diptera) that are commonly found under logs and stones (Figs 3.26, 3.27).

**Fig. 3.26** Tipulidae (crane fly) larva (photo: CC)

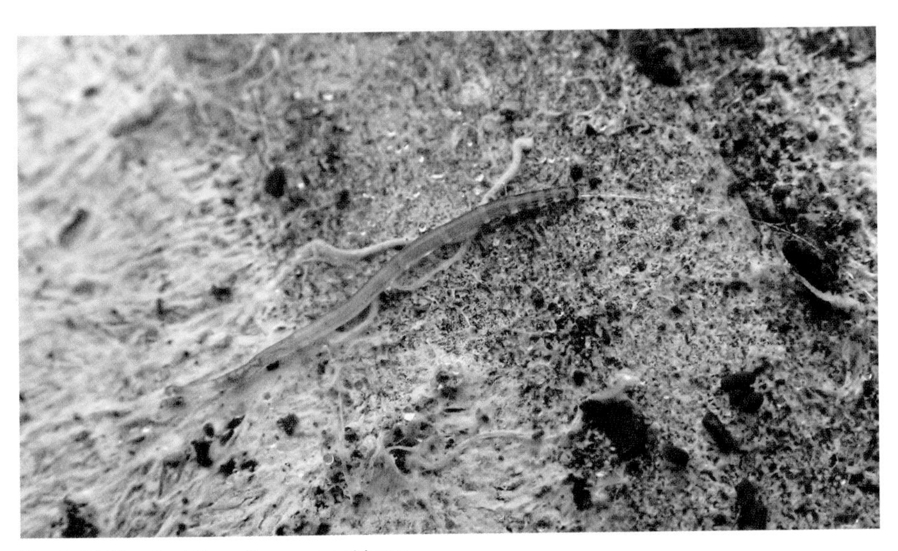

**Fig. 3.27** Mycetophilidae (fungus gnat) larva

The most frequently found in these habitats belong to the suborder Nematocera, which includes crane flies (Tipulidae), midges and fungus gnats. Most of these larvae feed on decomposing material and they are mainly found in damp areas especially where there is abundant organic matter. The more predatory larvae of another suborder, the Brachycera, are also found in damp areas rich in organic material. The larvae of some species of a third suborder, the Cyclorrhapha, feed on carrion or dung and are restricted to places where such food occurs, although phytophagous species of this group may be present in agricultural land in association with plant roots. Often the species of fly larvae found under logs and stones are also present in the surrounding soil. In many moist areas, dipteran larvae are present throughout the year. Some species have only one generation per year, while others have two or even three. See Smith (1989) and Marshall (2012) for further information on fly larvae.

The larvae of several families of beetles can also often be found under logs and stones. Click beetle larvae (Elateridae) are waxy, elongated animals, commonly called wireworms, which feed on living or decaying plant material or, sometimes, animal prey (Fig. 3.28). Larvae of dung beetles such as Scarabaeidae may be found when dung is present near to a log or stone. Other scarab beetle larvae may be found in the soil (Fig. 3.29). In other cases, rotting wood can be a food source for some beetle larvae, including stag beetles, which

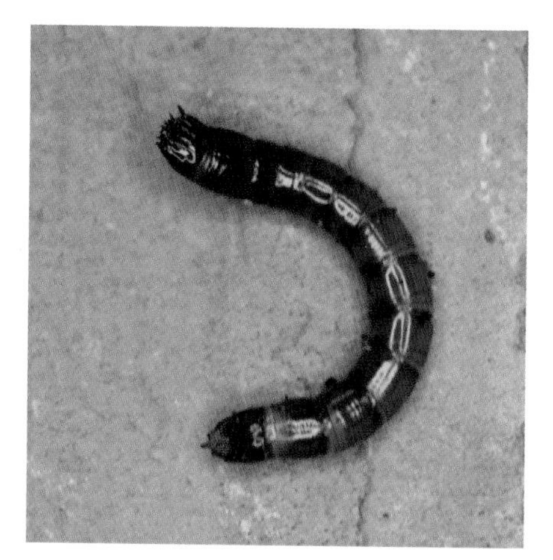

Fig. 3.28 Elateridae (click beetle) larva

Fig. 3.29 Scarabaeidae larva

Fig. 3.30 Staphylinidae (rove beetle) larva

can be found under log piles. Other beetle larvae are active predators, using the refuge both for shelter and also to hunt prey. These include the larvae of both ground beetles and rove beetles (Fig. 3.30). See Marshall (2018) and Hammond *et al.* (2019) for further information on beetle larvae.

### *Mites (class Arachnida: subclass Acari – Key F)*

Mites are extremely numerous in soils and leaf litter and are commonly found under logs and stones. As they are very small (usually less than 1.5 mm, and often under 1 mm, in length) they are often overlooked and only the larger ones are readily noticed (Figs 3.31–3.34). Despite their small size, they are important components of the ecosystem. Many mite

**Fig. 3.31** Pycnonoticae – *Damaeus* species (photo: BW)

**Fig. 3.32** Prostigmata – *Linopodes* species

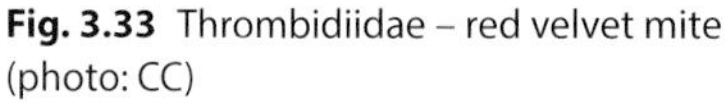

**Fig. 3.33** Thrombidiidae – red velvet mite (photo: CC)

**Fig. 3.34** Mesostigmata – *Pergamasus* species

species are parasitic on vertebrates or invertebrates, but here we are concerned only with free-living groups. Oribatid mites, which are particularly common under logs and stones, are saprophages and fungus feeders. Mesostigmatid mites of the genus *Pergamasus* are predators, feeding on nematodes and the eggs of other small invertebrates.

While adult mites have eight legs, in most species there is a six-legged larval stage followed by up to three nymphal stages. In some species there is a specially adapted stage (often the middle nymphal one) called a hypopus, which is phoretic (often on beetles). These nymphal stages may look quite different from the adults. This hitchhiking stage does not occur in the Oribatida, presumably because their habitat (predominantly leaf litter) is relatively constant and reliable. Phoresis may be a useful behaviour when the food source is relatively ephemeral.

**phoresis**
Behaviour exhibited by invertebrates, such as some mites and pseudoscorpions, where they attach onto a (usually larger) more mobile animal in order to 'hitchhike' as a dispersal mechanism.

The most obvious members of the suborder Prostigmata belong to the family Trombidiidae some of which appear velvety because they are covered in fine hairs. Some species are large (well over 1 mm), broadly strawberry shaped and bright red. The suborder Oribatida usually have tough cuticles and are dark brown. They are variably shaped and range from 0.02–1.3 mm long. Among the most common forms of soil oribatids are the box or armadillo mites, such as *Steganacarus* species, which are totally enclosed in a tough brown cuticle and can tuck their heads and legs in so that they look more like stones or seeds than animals. Others are roughly disc-shaped (for example, *Euzetes* species). Some members of the Poronoticae (a subcohort of the Oribatida – for example, *Euzetes globulus* and species in the family Galumnidae) have lobe-like flaps (pteromorphs) extending from the sides of the body that can either be fixed and immoveable or hinged so that they can move. Pteromorphs are protective features, which, together with other flaps and flanges, protect the vulnerable parts of the mite. Another common genus is *Damaeus*, species of which are up to 1.3 mm long, and are black with a spherical body and long spindly and knobbly legs. Within the order Mesostigmata, members of genera such as *Hypoaspis* and *Sejina* have several chestnut brown shields on otherwise pale bodies. *Pergamasus* species are larger (up to 1.5 mm), chestnut brown, and are active, predatory mites. In the male, the second pair of legs is enlarged and modified for grasping the female during mating. See Evans *et al.* (1961), Coleman *et al.* (2017) and Shepherd & Crotty (2020) for further information.

***Harvestmen (class Arachnida: order Opiliones – Key G)***
Harvestmen, like other arachnids, have eight legs, and are typified by having long legs relative to the size of the body (Figs 3.35, 3.36). They need moist surroundings to prevent desiccation and often shelter under logs and stones. They are most frequently found in autumn when the majority of species mature and many become inactive during the winter. They are mainly nocturnal hunters though some species eat dead invertebrates and others have been recorded feeding

**Fig. 3.35** *Trogulus tricarinatus* (photo: JPR)

**Fig. 3.36** *Nemastoma bimaculatum* (two-spotted harvestman) – missing a leg

on bird droppings, the gills of fungi and vegetable material. In some species (e.g. *Lophopilio palpinalis*) the pedipalps have a number of inwardly and downwardly facing structures and spines (Fig. 3.37) which, together with a trident of tubercles (bumps) on the front of the head, are thought to provide a cage to assist in the capture of their prey (often

**Fig. 3.37** Head of *Lophopilio palpinalis* showing spiny structures on the pedipalps

fast-moving animals such as springtails) – this is a similar strategy to some carabid beetles that specialise on springtails as prey. Harvestmen usually live for one year with the female slightly outliving the male, but not surviving long after egg-laying. Most eggs are laid in the autumn and hatch the following spring, although some species lay eggs at intervals during the year and these hatch after about two weeks. The young often moult within hours of hatching, and continue to moult at intervals of about 10 days, reaching maturity after 7–8 moults. Harvestmen seem to prefer to rest on walls and in angles and cracks and are often found in aggregations, sometimes comprising huge numbers of animals. As a defensive mechanism they may secrete drops, or even a fine jet, of nauseous fluid. They also cast limbs if handled even slightly roughly and, unlike spiders, they do not regenerate these limbs so that individuals with fewer than eight legs are not uncommon (e.g. Fig. 3.36). When identifying such animals check for limb bases which should still be present even if the limb itself is missing. See Hillyard (2005), Davidson (2019), and the Opiliones pages of the British Arachnological Society website (see Chapter 6) for further information.

### *Scorpions (class Arachnida: order Scorpiones – Key A couplet 10)*

Scorpions have eight legs, a pair of pincers and a segmented tail that ends in a stinger. There is only one species of scorpion found outdoors in Britain: *Euscorpius flavicaudis* (Fig. 3.38).

**Fig. 3.38** *Euscorpius flavicaudis* – European yellow-tailed scorpion (photo: CC)

Originally from north-west Africa and southern Europe, this species was introduced originally into the docks at the Isle of Sheppey in Kent possibly over 200 years ago. It has now been recorded from two other dockland sites in the south of England probably confirming its introduction via shipping. It is a relatively small scorpion (35–45 mm long), with a black body, and yellowish legs and tail. The animals often hide in cracks in masonry waiting for prey to pass by. They feed on a variety of arthropods, including woodlice and small insects. Although possessing a venomous sting, it is not thought to cause humans much harm (although see Vaucel *et al.* 2020 for a reported case involving a 10-year-old boy). See Benton (1991) for further information.

### *Pseudoscorpions (class Arachnida: order Pseudoscorpiones – Key H)*

Pseudoscorpions are tiny but aggressive predators, which use large pincers to catch their prey. Unlike true scorpions, they do not have a tail or stinger. Although some species reach high densities in the surrounding leaf litter, pseudoscorpions can often be overlooked under logs and stones, which they may use as occasional refuges or when hunting prey. When found, pseudoscorpions are more likely to be seen on the underside of logs and stones when they are turned over, rather than on the soil surface. By far the two most common species found under logs and stones are *Chthonius ischnocheles* and *Neobisium carcinoides* (Figs 3.39, 3.40). Most

**Fig. 3.39** *Chthonius ischnocheles* – common chthoniid

**Fig. 3.40** *Neobisium carcinoides* – moss neobisiid

species require damp microhabitats and several species move into the soil for the winter, and migrate back to the surface in the spring. Nymphs and adults may build silken chambers in which to hibernate, and these silken chambers are also used for breeding. The females of many species construct silken chambers in which they produce eggs and where the young hatch. The male leaves a packet of sperm on a stalk for the female to collect. There are three nymphal stages, with the young often being paler than the adults, as well as smaller. Like mites, pseudoscorpions may disperse by phoresis, attaching to flies or harvestmen. This phoretic behaviour is facultative; it is not an essential part of the lifecycle. It seems to be especially associated with species living in short-lived habitats, including under logs, but more often in birds' nests or compost heaps, and may allow the rapid colonisation of new habitats and, hence, survival of a population. See Legg & Jones (1988), Legg (2019) and the pseudoscorpion pages of the British Arachnological Society website (listed in Chapter 6) for further information.

***Spiders (class Arachnida: order Araneae – Key I)***

There is a diverse range of spiders that can be found associated with logs and stones (Figs 3.41–3.43). Spiders are predators and can be divided very broadly into two groups depending on the way they catch their prey; some use webs, whereas others stalk their prey. All spiders spin silk which is used

**Fig. 3.41** *Amaurobius* species – laceweb spider

**Fig. 3.42** *Clubiona terrestris* – a sac spider

**Fig. 3.43** *Pardosa saltans* (Lycosidae: wolf spider) with egg sac

when moulting and for coating the egg mass, as well as (in some species) for prey capture. Webs are more common among species living in open habitats where flying insects can be intercepted, but there are also some species that build webs in subterranean cavities or partially protruding from logs or stones. Often these webs include a retreat in which the spider sits and from which it emerges to retrieve captures entangled in the body of the web.

The type of web gives a reliable clue to the spider in residence (e.g. Figs 3.44, 3.45). Some species (such as *Nesticus cellulanus*) produce webs under logs and stones comprising a loose framework of fine threads. Others produce much more distinctive webs, some of which have refuges built into them where the spider can hide, retreat when disturbed or lay its eggs. A tangled bluish-white lace web found under or on the side of a log belongs to *Amaurobius* species, which are responsible for similar webs along window ledges in houses. These spiders have an extra silk-spinning organ, a cribellum, which is a plate in front of the spinners. This produces very fine silk threads which are combed out by a row of special hairs, the calamistrum, on the fourth pair of legs, giving the silk a woolly texture. A common under-stone dweller is *Coelotes* which spins a tube-shaped retreat with a silk collar

**Fig. 3.44** *Pisaura mirabilis* web – nurseryweb spider

**Fig. 3.45** *Agelena labyrinthica* (labyrinth spider) in funnel web (photo: MB)

spun around the entrance and catches mainly beetles that walk into the web. After paralysing them by injecting poison, the spider carries the prey back to the retreat to feed. *Hahnia* species build small sheet webs without retreats close to the ground. The spider rests on the upper surface of the web waiting for prey. The rare ladybird spider (*Eresus sandaliatus*) and the scarce purseweb spider (*Atypus affinis*) live in silk-lined burrows. In the latter species, these extend above ground as tubes, which are often camouflaged by debris. The money spiders, Linyphiidae, make hammock webs, which are strung horizontally in low vegetation. They are seen most easily on dewy or frosty mornings, sometimes among leaf litter close to or almost under logs and stones. Ground-running spiders such as clubionids and gnaphosids are frequently found resting in small, tight-fitting tube-like webs. These are used not for prey capture but as a retreat during the day, the spider actively hunting for its prey during the night. See Eberhard (2020) for more information about spider webs.

Sometimes under a stone there is a small round package, usually pale-coloured, perhaps spherical or shaped like a flying saucer, close to the spider. This is an egg case made of silk. Wolf spiders, Lycosidae, carry their egg cases attached to the spinners (see Fig. 3.43), and the spiderlings may also be carried on the abdomen of the mother after hatching. Spiders hatch as miniature versions of the adults and moult many times before reaching maturity after about a year. The interval between moults increases as the spider grows. The number of moults, usually 5–10, depends on the ultimate size of the spider. At the final moult the sexual organs and palps are fully developed. The male uses the palps to transfer sperm to the female. He charges them with sperm from the genital opening on the underside of his abdomen and uses them to deliver the sperm to the genital opening on the underside of the female. Mating may be hazardous if the male is treated as potential prey by the female, thus in some wolf spiders, for example, the males signal to the females before approaching. Different species mature at different times of the year, but late spring is a peak time for many and wolf spiders are particularly obvious then. Most species live for a year, but the lifecycle has been followed in only a few and so details of diet and habits are poorly known in many of the smaller ground-living species. Note that spiders do not spontaneously regenerate their limbs but may recover them at the next moult, hence finding spiders with missing legs is not uncommon. As with harvestmen, when identifying such animals check for limb bases which should still be present even if the limb itself is missing. See Foelix (2010), Bee *et al.* (2017), Harvey *et al.* (2017), and the spider pages of the British Arachnological Society website (see Chapter 6) for further information.

### *Woodlice and amphipods (class Crustacea: orders Isopoda and Amphipoda – Key A couplet 13 and Key J)*

The terrestrial Crustacea include the Amphipoda as well as the Isopoda (woodlice). Like the familiar freshwater shrimp (*Gammarus* species) most amphipods are aquatic, but there are a few more terrestrial species (landhoppers) that have been introduced into the UK. Most of these are restricted to hot-houses, but one species, *Arcitalitrus dorrieni*, can be found in damp (mainly deciduous) woodlands and gardens across the UK, especially in the west of England, and most frequently in the south-west of England and south Wales. Amphipods are compressed from side to side (rather than being flattened from top to bottom as in woodlice) and have two pairs of antennae. *A. dorrieni* is up to 15 mm long and darkly coloured (Fig. 3.46). The species is a decomposer, breaking down leaf litter, and is generally found in places where the leaf litter is very deep. See Gregory (2016) for further information.

**Fig. 3.46** *Arcitalitrus dorrieni* – landhopper

Woodlice are extremely common under all sorts of refuges and can often be found in large aggregations in seemingly unsuitable places around houses and gardens in urban areas (Figs 3.47–3.49). They can be very common in woodlands where they help to break down leaf litter. They feed mainly on dead and decaying vegetable matter which they excrete as smaller fragments. They may also feed on fungal hyphae growing on the leaf litter and, very occasionally, they also eat living plant material. Two specialist woodlouse predators may also be found under logs and stones; these are the two *Dysdera* species (woodlouse spiders – Fig. 3.50) which are distinctive because of the orange-red cephalothorax (the front part of the body), smooth cream-coloured abdomen (hind part of the body) and the very large jaws with which they can tackle woodlice.

During mating, sperm is transferred directly from the male to the female after a fairly simple courtship involving the male tapping the female with his antennae. The eggs develop inside a fluid-filled brood pouch on the underside of the female. The young hatch from the eggs, but remain in the brood pouch for several days, the fluid preventing them from drying out. When they emerge they have only six pairs of legs; within 24 hours they moult and gain their

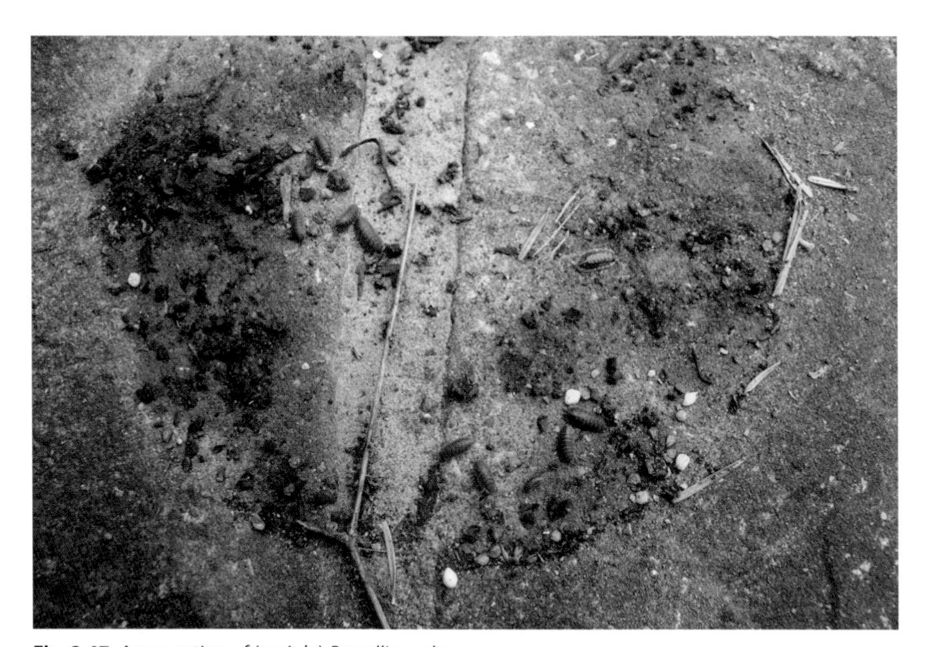

**Fig. 3.47** Aggregation of (mainly) *Porcellio scaber*

**Fig. 3.48** *Porcellio scaber* (left) and *Philoscia muscorum* (right)

**Fig. 3.49** *Armadillidium vulgare* – pill woodlouse

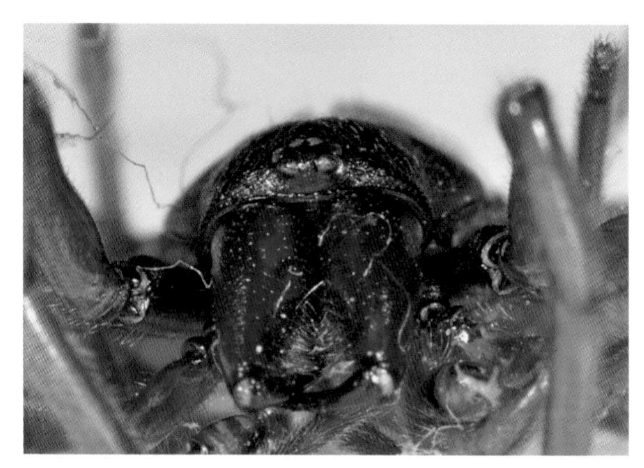

**Fig. 3.50** *Dysdera crocata* – woodlouse spider (photo: MB)

final, seventh pair of legs. Further moults take place at regular intervals throughout life. The hind part of the cast skin is shed a few days before the front end. The woodlouse may eat its cast skin, although these can often be found intact in places where woodlice occur and look like almost perfect halves of white woodlice. Woodlice usually reach maturity in a year and may live two to three years, or more. See Sutton (1980), Harding & Sutton (1985), Lee (2015) and the woodlouse pages of the British Myriapod and Isopod Group (BMIG) website (see Chapter 6) for further information.

### *Millipedes (Class Diplopoda – Key K)*

Millipedes are primarily saprophages, consuming large quantities of leaf litter and dead wood and contributing significantly to the breakdown of vegetable matter (Figs 3.51–3.54). Some species will feed on live plants, including vegetables, young seedlings and fruit, while some feed on fungi, and a few species will also occasionally scavenge from carcases. Many live in the leaf litter and humus layers, and they often take refuge, particularly during dry spells, under objects such as logs to retain moisture. One group of species making use of refuges are the very active animals, such as *Tachypodoiulus niger*, which hide during the day but spend the night wandering, often walking up trees and walls. In contrast, other species, such as *Cylindroiulus punctatus* (Fig. 3.2), are nearly always found in association with dead wood, especially that resting on the ground. The undersides of such logs and the small gaps between the log and its bark are particularly rich habitats for millipedes.

**Fig. 3.51** *Tachypodoiulus niger* (white-legged snake millipede) mating

**Fig. 3.52** *Blaniulus guttulatus* – spotted snake millipede

**Fig. 3.53** *Polydesmus* species – flat-backed millipede

**Fig. 3.54** *Glomeris marginata* – pill millipede

Nests of eggs are laid usually in the spring. The egg hatches into a legless larva which immediately moults into a six-legged animal. At subsequent moults more body rings are added with a corresponding increase in the number of legs (two pairs per ring). The flat-backed millipedes rarely have more than 20 rings, but the snake millipedes may achieve 40 or more. Millipedes mature after 1–2 years, but some species continue to grow and moult after this, breeding each year. Millipedes sometimes live to a great age, for example, the pill millipede, *Glomeris marginata*, has been recorded as living 11 years. See Blower (1985), Hopkin & Read (1992), Lee (2015) and the millipede pages of the BMIG website (see Chapter 6) for further information.

***Centipedes (class Chilopoda – Key L)***

Centipedes are active and powerful predators, which capture a variety of prey using their large poison jaws, although some have been recorded as feeding on plant material. The poison jaws are not part of the mouthparts of the animal but actually modified front legs, able to inject venom into the prey. There are four orders commonly found in Britain. When a log or stone is turned over the most obvious are the Lithobiomorpha (stone centipedes, e.g. Fig. 3.55). These centipedes are relatively short with 15 pairs of legs. They often

**Fig. 3.55** *Lithobius variegatus* – variegated lithobius

run very fast; when a stone is lifted many of the larger species disappear over the top in a flash. They hunt actively at night and use logs and stones as damp retreats during the day. In contrast, the Geophilomorpha (soil centipedes) are long and thin and have many stout legs (over 30 pairs – Fig. 3.56). They

**Fig. 3.56** Geophilomorph centipede

are blind and are generally found crawling through the soil and leaf litter, but frequently rest under logs partly buried in the soil. The less common Scolopendromorpha (which include giant centipedes and *Cryptops* species) are mainly a tropical group and are intermediate between lithobiomorph and geophilomorph centipedes in terms of leg number (21 pairs) and habits. There are only six species of these found in Britain, of which only three (*Cryptops* species – Fig. 3.57) are found outdoors, especially in human-influenced sites. All of the final order, Scutigeromorpha (house centipedes), recorded from the UK are introduced, often associated with imported plants, fruit or vegetables, and have only occasionally been found away from buildings.

Mating in centipedes may involve elaborate courtships, but direct transfer of sperm does not take place. The sperm are usually enclosed in a package or spermatophore which is deposited by the male on a web for the female to pick up. Eggs are usually laid in the spring and early summer. In some species of Geophilomorpha the female guards and cleans the eggs and the newly hatched young by coiling round them. The young animals have an almost complete complement of legs, although they gain coxal pores and other structures

**coxal pores**
These are thought to be involved in osmoregulation (regulation of water balance).

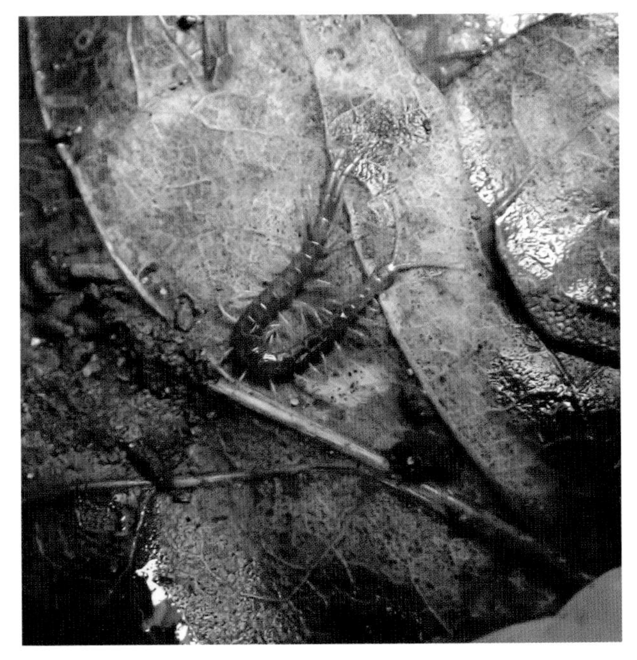

**Fig. 3.57** *Cryptops* species

at each moult. Egg-laying in the Lithobiomorpha is rather different: the female lays eggs one by one and coats each with soil particles forming a shell and then abandons them. Although at first the larvae lack the full complement of legs, they are active and able to feed on their own. Centipedes may take about 2–3 years to mature. Some continue to moult after this, living to 5–6 years of age. Some centipedes are able to produce substances that are sticky and are effective in deterring predators. See Lewis (1974), Barber (2009), Lee (2015) and the centipede pages of the BMIG website (see Chapter 6) for further information.

***Symphylids* (class *Symphyla* – Key A couplet 16)**

The symphylids are small, pale, blind animals that look rather like white lithobiomorph centipedes (Fig. 3.58). They feed mainly on fragments of dead vegetation in the soil, but will also eat the roots of plants, thus earning themselves a reputation as minor horticultural pests in some situations.

**Fig. 3.58** Symphylid

**Fig. 3.59** Pauropod (photo: SPH)

As in the centipedes, sperm transfer is indirect; copulation does not take place. From the genital opening on the fourth segment, a male deposits between 150 and 450 spermatophores (packets of sperm) on tall stalks. A female will eat about 18 of these each day. Some are consumed, but others are stored in a pouch in the mouth. When the female lays her eggs she takes them from the opening on her fourth segment, places them in her mouth and transfers the sperm to them. See Hopkin & Roberts (1988) and the symphylid pages of the BMIG website (see Chapter 6) for further information.

***Pauropoda (class Pauropoda – Key A couplet 16)***

The pauropods are a little-studied group of myriapods that are small (0.5–1.5 mm long), blind animals, which are often overlooked and are rarely found (Fig. 3.59). They have forked antennae and 9–11 pairs of legs. While most have elongate whitish bodies, one species (*Trachypauropus britannicus*) is stout, darker in colour and with more sclerotised plates (harder body). They have been recorded as feeding on fungal hyphae and decaying material. See Barber *et al.* (1992), Scheller (2008) and the pauropod pages of the BMIG website (see Chapter 6) for further information.

***Two-tailed bristletails (class Entognatha: order Diplura – Key M couplet 11)***

These are small (under 5 mm long), white, wingless animals represented, in Britain, by one genus, *Campodea* (Fig. 3.60). As their common name implies, these animals have two tails (or cerci) at the end of the abdomen, which in *Campodea*

**Fig. 3.60** Dipluran (*Campodea* species) – two-tailed bristletail

species are long and filamentous. Most two-tailed (also called two-pronged) bristletails feed on decaying plant material. They are not uncommon, although they are usually found at quite low densities. See Delaney (1954) and Barnard (2011) for further information.

***Proturans (class Entognatha: order Protura – Key M couplet 9)***

These small (under 2 mm long), pale animals have no antennae or eyes on the head and no long cerci ('tails') at the rear (Fig. 3.61). The first pair of legs are used as sense organs and are held in front of the animal as they move; they appear to play little part in locomotion. Proturans feed on decaying plant matter and fungal hyphae. They are not abundant and because of their small size are often missed. See Pass & Szucsich (2011) and Galli *et al.* (2019) for further information.

**Fig. 3.61** Proturan (photo: CC)

### *Springtails (class Entognatha: subclass Collembola – Key N)*

One reason that springtails (Figs 3.62, 3.63) are well known is because some species can jump; in general, those species that are more often found on the soil surface are better at jumping than those living under the soil. The spring (or furca or furcula) originates at the hind end of the animal and is usually held folded forwards, under the body. When

**Fig. 3.62** Sminthurid springtail

**Fig. 3.63** *Orchesella villosa* – springtail

the animal is disturbed, for instance by a potential predator, the spring is released suddenly catapulting the springtail into the air in a forward somersault. Another characteristic feature of springtails is the ventral tube, on the body segment immediately behind the legs, which absorbs water and ions from the substrate.

Some soil dwelling springtails have false ocelli, which are not eyes (as true ocelli are) but produce a sticky secretion that deters predators. Most springtails feed on decomposing plant material and fungal hyphae. They can be very abundant in soil and leaf litter and are frequently found under logs and stones. See Hopkin (1997, 2007), Dallimore & Shaw (2013), and the UK Collembola website (see Chapter 6) for further information.

*Earwigs (class Insecta: order Dermaptera – Key O)*

Only four native species of earwig are found in Britain, and only two of these are very common. They are typified by curved pincers (modified cerci) at the end of the abdomen which are used for catching prey, in defence, and in male-to-male conflicts (Fig. 3.64). They are nocturnal insects, resting by day in dark crevices and for this reason are often found under logs and stones. Although they can tolerate low temperatures, they spend much of the winter buried in the soil or in other protected sites. They are polyphagous, feeding on carrion, small invertebrates and some living plant material, such as flower petals (although their habit of boring holes into flower petals may actually be to gain

**Fig. 3.64** *Forficula auricularia* (common earwig), female (left) and male (right)

**instar**
The stage between successive moults in arthropods until sexual maturity. This can occur in the larval (or nymphal) stages, as well as in the pupal stage. In insects, adults do not undergo further moults. Although some non-insect invertebrates continue to moult after sexual maturity, these stages are not usually referred to as instars. In some invertebrates the number of instars is fixed, whereas in others the number depends upon the prevailing environmental conditions – including temperature and humidity. Note that in woodlice, millipedes and centipedes, the term stadium (plural stadia) is more often used to describe the time between moults.

access to prey within the flower itself). The female lays her eggs in the soil and looks after them during the winter. The eggs hatch in early spring and the larvae emerge from the soil when they reach the second instar. Until then, and even after emerging above ground, they are fed by the mother. Family groups seem to remain together and earwigs are often seen in clusters. Many insects are very pale just after moulting and earwigs are particularly so, sometimes appearing almost white. They reach maturity during the late summer and mate before searching for winter resting sites. See Marshall & Haes (1988), Sutton (2015) and the Orthoptera & Allied Insects website (see Chapter 6) for further information.

*Cockroaches (class Insecta: order Dictyoptera – Key P)*

In Britain, cockroaches are most often seen in buildings, however most of these are introduced species. There are three native cockroach species that live among vegetation, mainly in the southern counties of England and Wales. They are polyphagous exopterygotes and are not common members of the communities found under logs and stones. Males of all three species have full wings and are fully capable of flight (Fig. 3.65). Female *Ectobius pallidus* (tawny cockroach) also have full wings and readily fly. In *E. lapponicus* (dusky cockroach), the females have slightly reduced wings and

**Fig. 3.65** *Ectobius lapponicus* (dusky cockroach) – male (photo: SJF)

usually remain on the ground, while in *Capraiellus panzeri* (lesser cockroach) the females have vestigial wings and cannot fly. The two *Ectobius* species have a two-year lifecycle with adults emerging from late May and are present until September. Nymphs hatch from early June and overwinter, emerging the following year as adults. *C. panzeri* has a one-year lifecycle with adults maturing in July or August and being present until September or October. This species overwinters as eggs in oothecae (egg cases), hatching into nymphs in April or May. See Marshall & Haes (1988), Sutton (2015) and the Cockroach Species File Online and Orthoptera & Allied Insects websites (see Chapter 6) for further information.

### *Grasshoppers and crickets (class Insecta: order Orthoptera – Key Q)*

Among the Orthoptera (the grasshoppers and crickets), the only species found under logs and stones in the UK are members of the families Gryllidae (true crickets – Fig. 3.66) and the single species of Gryllotalpidae (*Gryllotalpa gryllotalpa* – the mole cricket – Fig. 3.67). *G. gryllotalpa* is an odd-looking stout animal with large front legs highly adapted for burrowing. The female lays several hundred eggs in an underground chamber and guards these and the newly emerged nymphs for the first couple of months. The young feed on roots and insect larvae encountered underground and reach maturity a year after hatching. Mole crickets are usually found in damp grasslands and in Britain are rare, appearing to be mainly restricted to areas in southern England.

True crickets (Gryllidae) may occasionally be found under logs and stones. There are four British species of true crickets plus one occasional introduction from continental

**Fig. 3.66** *Nemobius sylvestris* – wood cricket (photo: CC)

**Fig. 3.67** *Gryllotalpa gryllotalpa* – mole cricket (photo: CC)

Europe. Of the four British species, one rare species (the scaly cricket *Pseudomogoplistes vicentae*) is limited to shingle beaches, while another (the house cricket *Acheta domesticus*) is restricted to buildings and is only found outdoors during warm summers or on warm habitats such as rubbish tips. The other two species both live in southern England and can sometimes be found under logs and stones. The field cricket (*Gryllus campestris*) lives in areas of low vegetation where adults and older nymphs live in burrows. The wood cricket (*Nemobius sylvestris*) can be found among leaf litter in woodlands.

Other groups of Orthoptera such as bush-crickets, groundhoppers and grasshoppers are more likely to be seen on or beside, rather than under, logs and stones. See Marshall & Haes (1988), Benton (2012), Sutton (2015) and the Orthoptera & Allied Insects website (see Chapter 6) for further information.

### *Ants (class Insecta: order Hymenoptera: family Formicidae – Key R)*

The order Hymenoptera includes ants, bees, wasps, sawflies and ichneumon flies. This order largely comprises species that are strong fliers and are not usually found under logs or stones, although a number of bumblebees, solitary bees and solitary wasps nest in burrows they dig in the ground, and may be seen in the vicinity at ground level. Most ants, on the other hand, are wingless and can be found under refuges in very large numbers, especially if they have built a nest under a log or stone or other material (Fig. 3.68). Ants

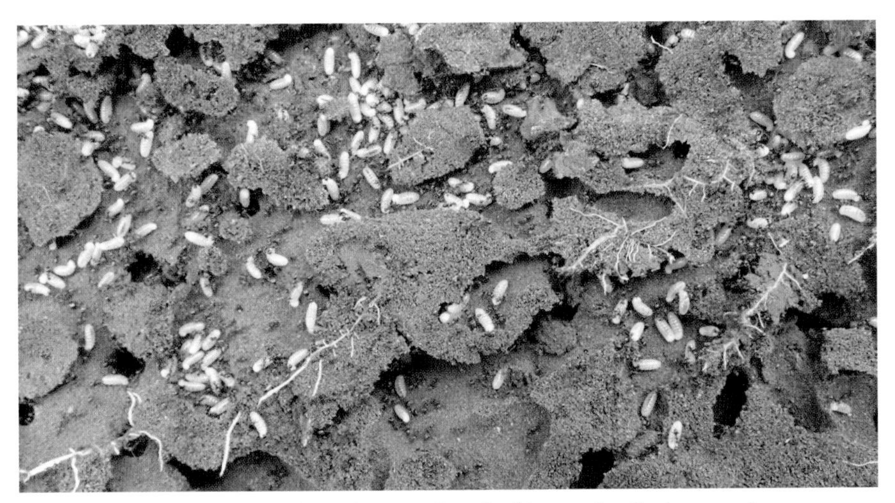

**Fig. 3.68** *Lasius niger* (black garden ant) nest under a dustbin revealing the larvae and pupae

**Fig. 3.69** *Formica rufa* (red wood ant) nest on a log

are social insects living in well-defined colonies consisting of one or more queens, workers and new reproductives (queens and males). The majority of the animals seen either in a nest or foraging elsewhere are workers, which are non-reproductive females. Ants may also be found under logs and stones searching for food to take back to their nests. In the case of wood ants, which nest in large mounds up to 2 m high (Fig. 3.69), huge numbers of workers can be seen on the woodland floor and may be found under refuges (Fig. 3.70). Most ants are polyphagous, feeding on vegetable material, carrion and other invertebrates. Some collect nectar or 'milk' aphids for the honeydew they produce. Although ants are active predators, a variety of other animals may live in ants' nests, including woodlice, millipedes, spiders, butterfly larvae and beetles. These guests do not appear to be attacked by the ants and may 'help' the ants, perhaps by secreting substances, which the ants consume, or by removing unwanted debris. Disturbance of a nest (including by lifting a covering log or stone) produces frantic movement as the workers hastily gather towards the pale-coloured larvae and pupae and carry them to safety. Sometimes the structure of the tunnels may be seen clearly. Ants have a number of defence mechanisms; depending on the species they may

**Fig. 3.70** *Formica rufa* (red wood ants) on their nest on a woodland floor

**Fig. 3.71** Flying ants gathering on a stone

attack by biting, stinging, spraying formic acid, or be more passive by playing dead, rolling into a ball and rolling away.

At certain times of the year, males and fully developed females emerge (Fig. 3.71). Both are winged and after a nuptial flight the males die and the queens lose their wings and seek nesting sites. They lay their eggs, either in an existing nest alongside the resident queen, or in a new nest site. After tending the first generation to hatching and feeding them herself, the queen concentrates on egg-laying. Future generations are then tended by existing workers. Queens live for several years and colonies can be quite long lasting. Ants within a nest will usually attack an invading ant, and individuals meeting others on trails will either attack them or avoid them depending on the species involved. However, there are situations where different species of ants will coexist within the same nest. At its most benign, this can be when one species makes its own nest within the confines of that of another species, as *Formicoxenus nitidulus* does within the nests of *Formica* species. In other cases, the situation is not so congenial and results in the workers of one species being slaves to another. For example, the blood-red slave-maker ant (*Formica sanguinea*) raids the nest of the common black ant (*F. fusca*) and brings back pupae to its own nest where they become workers for the new colony, despite being of a different species. *F. sanguinea* may also exploit *F. fusca* colonies by killing the queen and taking the nest over. See Brian (1977), Skinner & Jarman (2024), Lebas *et al.* (2019) and the Bees, Wasps & Ants Recording Society, and AntWiki websites (see Chapter 6) for further information.

### *Beetles (class Insecta: order Coleoptera – Key S)*

There are more than 4,000 British species of beetles in over 100 families, several of which are well represented among the cryptozoa. The ground beetles (Carabidae) and rove beetles (Staphylinidae) are some of the most frequent and obvious beetles found under logs and stones and will be considered in the next two sections.

Other families of beetles are found less frequently under logs and stones. These include the burying beetles, family Silphidae (Fig. 3.72), which are only usually found in this habitat in the presence of carrion. They burrow under carcases sinking them into the ground so they do not dry out. The beetles then lay their eggs in the carcases which provide food for developing larvae. The Scarabaeidae and Geotrupidae are also infrequent under logs and stones. They include the dung beetles, which are rarely found far from animal dung, the larvae of which will also feed on carrion and decomposing litter. The Hydrophilidae are small shiny beetles living in damp places. These too may be associated with dung and with decaying plant material. The weevils, family Curculionidae, are phytophagous. Those weevil species that lay their eggs in wood are occasionally encountered under logs.

In the click beetles, family Elateridae, although adults may be seen (Fig. 3.73), it is mainly the larvae that are encountered under logs and stones (Fig. 3.28). A range of other beetle larvae may also be found under logs and stones, some of which are associated with rotting wood (see Key V for more information on insect larvae). See Harde & Hammond (2000), Lane & Mann (2016), Jones (2018), Marshall (2018) and the UK Beetles and UK Beetle Recording websites (see Chapter 6) for further information.

**Fig. 3.72** Silphid beetle – *Nicrophorus humator* (black sexton beetle)

**Fig. 3.73** Elateridae (click beetle) – *Agriotes lineatus* (lined click beetle) (photo: CC)

### *Ground beetles (class Insecta: order Coleoptera: family Carabidae – Key T)*

Most of the ground beetles (Figs 3.74–3.76), especially as larvae, are predatory, although many species include other materials in the diet, and even scavenge on carcases and feed on fruit or fungi. The majority of carabid species are active on the surface, but adults of some and larvae of many are found in leaf litter or in the upper layers of the soil. Many British species are found under logs and stones during part of their lifecycle. Although many carabid beetles are nocturnal, some, for example *Notiophilus* species, are active in daylight. Different species breed at different times of the year but there are often peaks of activity in spring and autumn. Species that overwinter as larvae usually show a peak of adult activity in mid-summer. Typically it takes about a year from the egg to adult maturity. Most Carabidae are relatively long lived and some, such as *Carabus* species (Fig. 3.74), live two years or

**Fig. 3.74** *Carabus violaceus* (violet ground beetles) under bark (photo: CC)

**Fig. 3.75** *Cychrus caraboides* – buzzing snail-hunter

**Fig. 3.76** *Abax parallelepipedus* – common shoulder-blade

more as adults. Many species overwinter as adults, perhaps burrowing into the soil to re-emerge the following spring. Some build overwintering cells under bark or in crevices of fallen logs. See Thiele (1977), Forsythe (2000), Luff (2007), Telfer (2016) and the UK Beetles and UK Beetle Recording websites (see Chapter 6) for further information.

### *Rove beetles (class Insecta: order Coleoptera: family Staphylinidae – Key U)*

Many of the larger rove beetles (Fig. 3.77) have similar ecological habits to the Carabidae. They are polyphagous with a tendency towards a predatory diet, although some smaller species feed on fungi. In one genus (*Stenus* – Fig. 3.78)

**Fig. 3.77** *Ocypus olens* (Devil's coach-horse) with tail partially raised in display

**Fig. 3.78** *Stenus bimaculatus* – rove beetle (photo: CC)

part of the mouthparts has sticky plates at the end that are able to extend under fluid pressure to catch fast-moving prey including springtails. Rove beetles are often active on the surface of the ground, and some species are frequent under logs and stones, with a few species being found associated with ants' nests. Many species can fly using the hindwings which are intricately folded beneath the shortened forewings (elytra). This is the largest family of beetles in the UK, with over 1,000 species, and many of them are difficult to separate. Several species, including the large and impressive Devil's coach-horse, *Ocypus olens*, raise up the end of the abdomen and secrete an oily substance when alarmed (Fig. 3.77). See Lott (2009), Lott & Anderson (2011), Lane (2016) and the UK Beetles website (see Chapter 6) for further information.

## 3.6 The biology of the vertebrate cryptozoa

Logs and stones can also shelter vertebrates including amphibians such as toads and newts, reptiles such as lizards, slow worms and snakes, and small mammals including hedgehogs, shrews, mice and voles. These may be seeking damp habitats, protection from predators, or prey such as invertebrate members of the communities under logs and stones.

All groups of British amphibians may be found under refugia, although some more frequently than others. After their breeding season in spring and early summer, newts such as the smooth (or common) newt (*Lissotriton vulgaris* – Fig. 3.79) and common toads (*Bufo bufo* – Fig. 3.80) may be found under logs and stones, and some will also hibernate

**Fig. 3.79** *Lissotriton vulgaris* – smooth newt

**Fig. 3.80** *Bufo bufo* – common toad (photo: WC)

over winter in these habitats. Most British frogs are usually aquatic animals. However, common frogs (*Rana temporaria*) will leave the water after breeding and, although they are less likely to be seen far away from fresh water, may overwinter under logs and stones (Fig. 3.81). The introduced midwife

**Fig. 3.81** *Rana temporaria* – common frog (photo: CC)

toad (*Alytes obstetricans*), which has a restricted distribution, spends the day hidden in damp areas, including under logs and stones, usually not too far from fresh water, emerging to feed at dusk. All British amphibians feed on live prey including invertebrates such as earthworms, molluscs and arthropods, and, in the case of larger animals, occasionally on small amphibians and reptiles. Some species such as the natterjack toad (*Epidalea calamita*) and great crested newt (*Triturus cristatus*) are rare and are protected in law from being harmed or taken from the wild. The latter species may also be vulnerable to competition from, and hybridisation with, the larger introduced Italian crested newt (*T. carnifex*).

Although many species of British reptiles can be found under logs and stones, possibly the most likely lizard to find is the legless slow worm (*Anguis fragilis* – Fig. 3.82). This species is found throughout Britain in a wide range of habitats, including woods and grassland, as well as urban sites such as road and railway embankments, and gardens. Slow worms are active from March to October and, unlike some other reptiles, tend to avoid basking in the open, choosing instead to shelter beneath debris exposed to the sun. Lizards such as the viviparous (or common) lizard (*Zootoca vivipara* – Fig. 3.83) and the sand lizard (*Lacerta agilis*) usually bask in the open but may still sometimes be found under refuges. The former are the most commonly encountered legged lizards under logs and stones, and can be found in a wide range of open habitats including heaths, grassland and open woodland. Young viviparous lizards are jet black at birth, becoming more coppery as they age and

**Fig. 3.82** *Anguis fragilis* – slow worm

**Fig. 3.83** *Zootoca vivipara* – viviparous lizard (photo: WC)

acquire adult markings. The larger sand lizard prefers dry heaths and dunes. Snakes, including the grass snake (*Natrix helvetica* – Fig. 3.84), the adder (*Vipera berus* – Fig. 3.85), and the rarer smooth snake (*Coronella austriaca*) are also commonly found under refuges that are in full sun but also close to denser vegetation to which they can retreat when disturbed. Indeed, the use of artificial refuges such as corrugated iron or

**Fig. 3.84** *Natrix helvetica* – grass snake (photo: WC)

**Fig. 3.85** *Vipera berus* (adders) under a refuge

roofing felt is a standard method of surveying these animals (for further information see Beebee, 2013). The adder is the only venomous British snake, and care should be taken when encountering this species.

Lizards and slow worms feed on invertebrates, with faster moving lizards eating spiders and insects, while the slower moving slow worm feeds on slow-moving prey such as slugs. British snakes tend to eat vertebrates, with the grass snake feeding on amphibians and fish (and hence often being found near to water), while the adder prefers small mammals and young birds, and the smooth snake mainly feeds on reptiles. For further information about reptiles and amphibians see Beebee & Griffiths (2000), Beebee (2013), Inns (2018) and the websites for Amphibian and Reptile Conservation (ARC), the Amphibian and Reptile Groups of the UK (ARG UK), and Froglife (see Chapter 6).

A variety of small mammals may be seen under refuges, especially under log piles, although they are often so fast moving that it may be difficult to know quite what you have seen once they have been disturbed. Shrews, including the pygmy shrew (*Sorex minutus*), common shrew (*Sorex araneus*

**Fig. 3.86** *Sorex araneus* – common shrew (photo: MB)

– Fig. 3.86) and even the water shrew (*Neomys fodiens*) can be found under logs and larger pieces of debris, as can both bank voles (*Myodes glareolus* – Fig. 3.87) and field voles (*Microtus agrestis*). Wood mice (*Apodemus sylvaticus* – Fig. 3.88) and yellow-necked mice (*Apodemus flavicollis*) are more often seen under wood piles, as are the two smallest British carnivores, the mustelids weasels (*Mustela nivalis* – Fig. 3.89) and stoats (*Mustela erminea*), and the omnivorous hedgehog (*Erinaceus europaeus*). Tracks and signs (including nests) of

**Fig. 3.87** *Myodes glareolus* – bank vole (photo: MB)

**Fig. 3.88** *Apodemus sylvaticus* – wood mouse (photo: MH)

**Fig. 3.89** *Mustela nivalis* – weasel (photo: MB)

small mammals may also be found under logs and stones and the two species of mustelids may also hide prey such as small mammals under logs and stones. Common shrews eat insects, earthworms and slugs, while the smaller pygmy shrew concentrates on arthropods. Water shrews feed on aquatic invertebrates. Although omnivorous, the majority of a hedgehog's diet consists of insects, earthworms and slugs. Voles are plant eaters, consuming grasses, roots, berries and seeds, while mice, in addition to plants, will also consume earthworms and arthropods. For further information about small mammals see Corbet (1989), Harris & Yalden (2008), Couzens *et al.* (2021) and the species pages of the Mammal Society website (see Chapter 6).

## 3.7 Tracks and signs of cryptozoa

A range of animal products and signs may be found under logs and stones, which may give some clue as to the organisms that have been present. These include eggs, egg cases, cocoons and pupae. Examples include the shiny, round jelly-like mollusc eggs (Fig. 3.90), and earthworm eggs encased in hardened mucus cocoons. Spiders' eggs may be formed into egg cases wrapped in silk: silk is also used to create webs and retreats (see the earlier section on spiders). Other retreats such as the chambers used for moulting by millipedes (Fig. 3.91) and the silken chambers used by pseudoscorpions for hibernation and breeding may also be found. Fragments of animals may also be present, including the remains from predators' meals such as beetle wing-cases, the disarticulated

**Fig. 3.90** Mollusc eggs

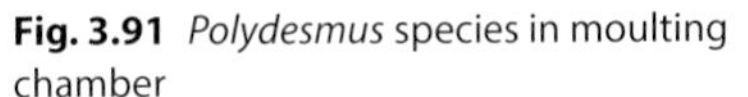

**Fig. 3.91** *Polydesmus* species in moulting chamber

**Fig. 3.92** Cast moult of a woodlouse

segments of millipedes, and exuviae, which are the discarded remains of moults, such as those from woodlice (often looking like paler, half woodlice – Fig. 3.92). Various tracks and trails may also be found including worm casts at the soil surface, mollusc mucus trails, and evidence of burrows in rotten wood and in the soil underneath a log or stone. Where vertebrates have been present there may be nesting material, evidence of runs and trails, and even recognisable droppings, some of which may have evidence of the type of prey taken by the animal. Rhyder (2021) describes the tracks and signs of vertebrates (including small mammals, amphibians and reptiles) and some invertebrates in the UK, while Eiseman & Charney (2010) give a large range of tracks and signs of invertebrates, albeit in a North American context.

# 4 Identifying the animals

**incident light source**
This is a light shining onto the specimen rather than the transmitted light shone through a semi-transparent object in other kinds of microscopy. It is important to make sure that any light used to view a specimen is strong enough and is angled to show the major features needed for identification. Modern LED lights can be useful for this since they do not produce much heat and hence do not cause specimens to dry out.

Identification usually requires a 10× or 20× hand lens or preferably a low-power stereoscopic (binocular) dissecting microscope with a good incident light source. A decent bench lamp is useful to provide a second light source on the other side. Digital microscopes also require good illumination and have the added advantage of enabling images to be captured using a computer. Although it is always preferable to identify living animals, it may be necessary to kill and preserve specimens in order to see small details. When observing fluid-preserved animals under a microscope, it is best to keep them fully immersed in fluid to avoid distortion of the image by drops of liquid, to prevent them from drying out, and to allow the appendages (such as antennae) to be free from the body. If available, tiny glass beads are useful for positioning the specimens. Well-washed river sand is a suitable alternative. Fill a solid watch glass or small petri dish approximately one third full with beads or sand, add alcohol and then the animal to be observed (Figs 4.1, 4.2).

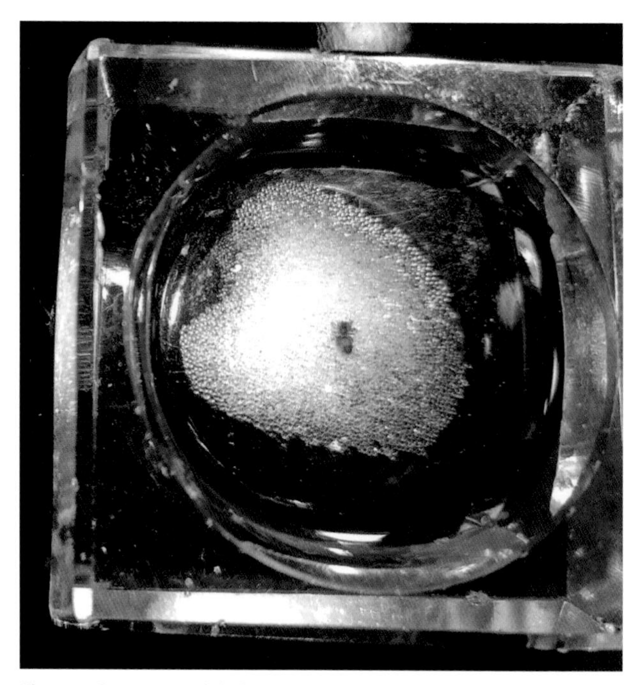

**Fig. 4.1** Specimen (*Philodromus* species) in solid watch glass

**Fig. 4.2** Close up of *Philodromus* species on glass beads

For very small animals high magnification microscopy may be required. This involves mounting the animal on a cavity slide in a drop of fluid (such as alcohol or glycerol, which doesn't evaporate as readily) and viewing it using a compound microscope with objective lenses providing ×10, ×40 or ×60 magnification. When combined with ×20 eyepieces, such microscopes give a combined magnification of up to ×1,200. Although this type of microscope usually has a built-in light source from below, for darker specimens, it may be useful to use an additional good quality light source from above. Where available, the use of phase-contrast microscopy will enable specimens to appear well illuminated against a dark background. For some small animals, species identification may require the specimen to be cleared. This involves dissolving tissues and reducing pigmentation using caustic chemicals before mounting them on microscope slides. Hopkin (2007) and Shepherd & Crotty (2020) give details of clearing for springtails and mites respectively. To measure the animal, it is usually sufficient to use a sheet of graph paper under the petri dish and estimate lengths using the squares. More accurate measurements for smaller specimens require an eyepiece micrometer (see Guthrie 1989 for further details).

When observing living animals with either a hand lens or a microscope it is often necessary to restrain them in some way. This can be done by placing them in a small polythene bag (Fig. 4.3). If the animal is too active, a ring of metal can be placed over the specimen to isolate them in a 'bubble' of polythene. Cooling animals in a refrigerator for 5–10 minutes will slow them down for a short time. Although this is often not long enough to allow detailed examination, it may help if the identification of a familiar animal is being confirmed. Various chemicals will knock animals out and, if used carefully, allow them to recover later. Ethyl acetate is one such, although you need to be careful not to leave animals in too long and actually kill them. Ethyl acetate

**Fig. 4.3** Centipede restrained in plastic bag

is an irritant and is toxic to humans and should be used with caution. Carbon dioxide is a better chemical to use if available. Small carbon dioxide injectors are commercially available as wine-bottle openers (Fig. 4.4), or bicycle-tyre inflators. These use a small cylinder of the gas, which, since it is heavier than air, can be injected into a specimen tube containing the specimen. Once the animal stops moving, it can be removed for identification, but take care that it does not recover quickly and escape.

The identification keys in this chapter are dichotomous (from the Greek *dichotomia* – meaning cut in two), that is, they use a series of numbered paired statements (couplets) to point towards the identification of the animal in question. At each point, all parts of the statement should be true before progressing to the next couplet. The number at the end of the statement indicates which couplet should be considered next. The number in brackets indicates which the previous couplet was (in case you need to backtrack). When the key presents you with an answer, check that the animal fits the descriptions. If it does not, then check back through the key and see whether you have made a mistake. Note that individual variations in animals due to developmental differences or accidental damage may sometimes cause some confusion

**Fig. 4.4** Carbon dioxide injector

regarding confirmation of identification. Damage may be obvious and such specimens should be identified with some caution. Where determination of identification is unclear, try another similar specimen (if available) or use a more specialised identification guide. Such texts are identified in the introduction to each key. The keys also reference (where available) texts and/or websites that provide species descriptions and illustrations as well as checklists and distribution maps (atlases). The latter can be helpful in assisting with the confirmation of identification by eliminating unlikely species that are restricted to particular habitats or locations.

Within these keys we have tried to use simple terminology where possible, although it is inevitable that technical terms are needed on occasions. Where this occurs we have provided an explanation within the key (and often an illustration showing the feature in question). The keys cover a wide range of different groups of organisms from very different orders of invertebrates and the terminology used by taxonomists can differ depending on the group in which they specialise. For example, simple eyes can be called ocelli (e.g. in insects) and ommatidia (e.g. in woodlice). The term ocelli can also be used for the false eyes found on the top of the head of some springtails and beetles, for example. Even within a single group such as the insects, the terminology can vary; the hardened hindwings of cockroaches, grasshoppers and crickets are termed tegmina, while in beetles they are called elytra (in earwigs, both terms have been employed). We have used the terms more commonly associated with the group in question so that readers are familiar with these terms should they wish to consult more specialist texts and keys. Most keys include an illustration labelled with some of the important characteristics used within the key. Other illustrations highlight specific features referred to in particular couplets.

Where possible in these keys we have enabled the identification of (particularly common and widespread) species. Where there are many such species and/or where they are difficult to separate without detailed microscopic examination, we have limited the identification to a higher taxonomic grouping such as order or family (see Table 4.1 for an explanation of the taxonomic hierarchy for animals). For some groups of animals, additional levels may be used – for example, in mites the level of cohort (sitting between order and family) is commonly used. Other levels can be included just above the main level using the prefix super- (e.g. superorder) or just below using the prefix sub- (e.g. subfamily). Note that

**Table 4.1** Taxonomic hierarchy

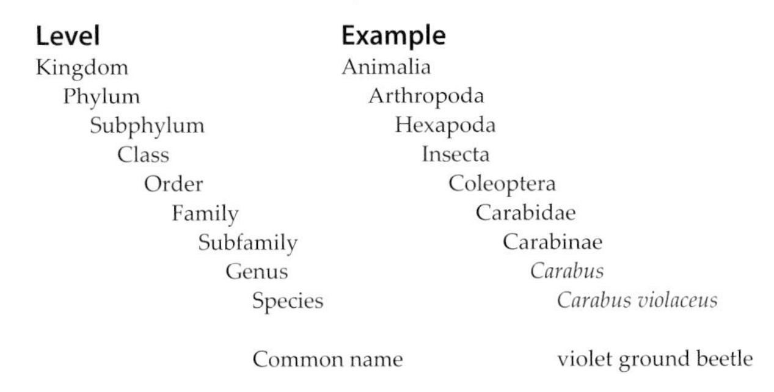

| Level | Example |
|---|---|
| Kingdom | Animalia |
| Phylum | Arthropoda |
| Subphylum | Hexapoda |
| Class | Insecta |
| Order | Coleoptera |
| Family | Carabidae |
| Subfamily | Carabinae |
| Genus | *Carabus* |
| Species | *Carabus violaceus* |
| Common name | violet ground beetle |

family names end with -dae and subfamily names with -nae. Genera and species should be italicised and names at all levels (except the specific epithet) are capitalised. Common names are generally not capitalised (but this depends on the group and the style of publication where it is recorded). Care should be taken when using common names. As can be seen from several of the keys, animals may have the same common name as other (sometimes quite different) species, some animals have multiple common names, and some (even quite common and obvious species) do not have a well-used or recognised common name. There is some debate about the usefulness of common names, especially since using the scientific name avoids all three issues. As an example, in addition to its common name of common blackclock, the ground beetle *Pterostichus madidus* is also known as the black garden beetle, rain-clock and strawberry beetle. This may mean that regional differences in nomenclature can cause confusion when discussing particular species and therefore, the agreed scientific name should always be noted. However, common names may also have their place in broadening the awareness of certain species and groups of species, especially when the species concerned are common and of interest to a wider public in terms of their obviousness, commercial attributes or conservation importance. For example, wood ants describe the major habitat for a group of ants that build large nests from pieces of vegetation (and may be found under logs and stones in woodland). More specific names may help to identify particular characteristics of individual species: for example, *Formica rufa* – red wood ant (also called the hill ant, horse ant or southern wood ant) has large reddish workers and is mainly concentrated in southern England. We have included common names for completeness and because several groups are readily acknowledged by their common name in the more popular literature (e.g. the New Zealand flatworm, *Arthurdendyus triangulatus*, which is frequently in the news for its potential as a pest species – see Section 3.5). Where there are multiple common names associated with an individual species, we have indicated the most commonly used and added alternatives within parentheses.

Until you are confident about your ability to separate specimens into broad groups, you should start with Key A to the major invertebrate groups found under logs and stones, before progressing to the appropriate specialist key(s). The guessing guide (Fig. 4.5) may also be helpful in determining which specialist key to consult (note that some rarely encountered early immature mites and Myriapoda may not key out using this guide since they have fewer pairs of legs than do later stages, including adults). Once you are familiar enough to separate specimens into major groups, you can go directly to the appropriate specialist key. The hierarchy of keys in this book is as follows:

- Key A – major invertebrate groups
  - Key B – flatworms
  - Key C – earthworms
  - Key D – slugs
  - Key E – snails
  - Key F – mites
  - Key G – harvestmen
  - Key H – pseudoscorpions
  - Key I – spiders
  - Key J – woodlice
  - Key K – millipedes
  - Key L – centipedes
  - Key M – hexapods
    - Key N – springtails
    - Key O – earwigs
    - Key P – cockroaches
    - Key Q – orthopterans (crickets and allied insects)
    - Key R – ants
    - Key S – beetles
      - Key T – ground beetles
      - Key U – rove beetles
    - Key V – insect larvae
- Key W – amphibians
- Key X – reptiles
- Key Y – small mammals

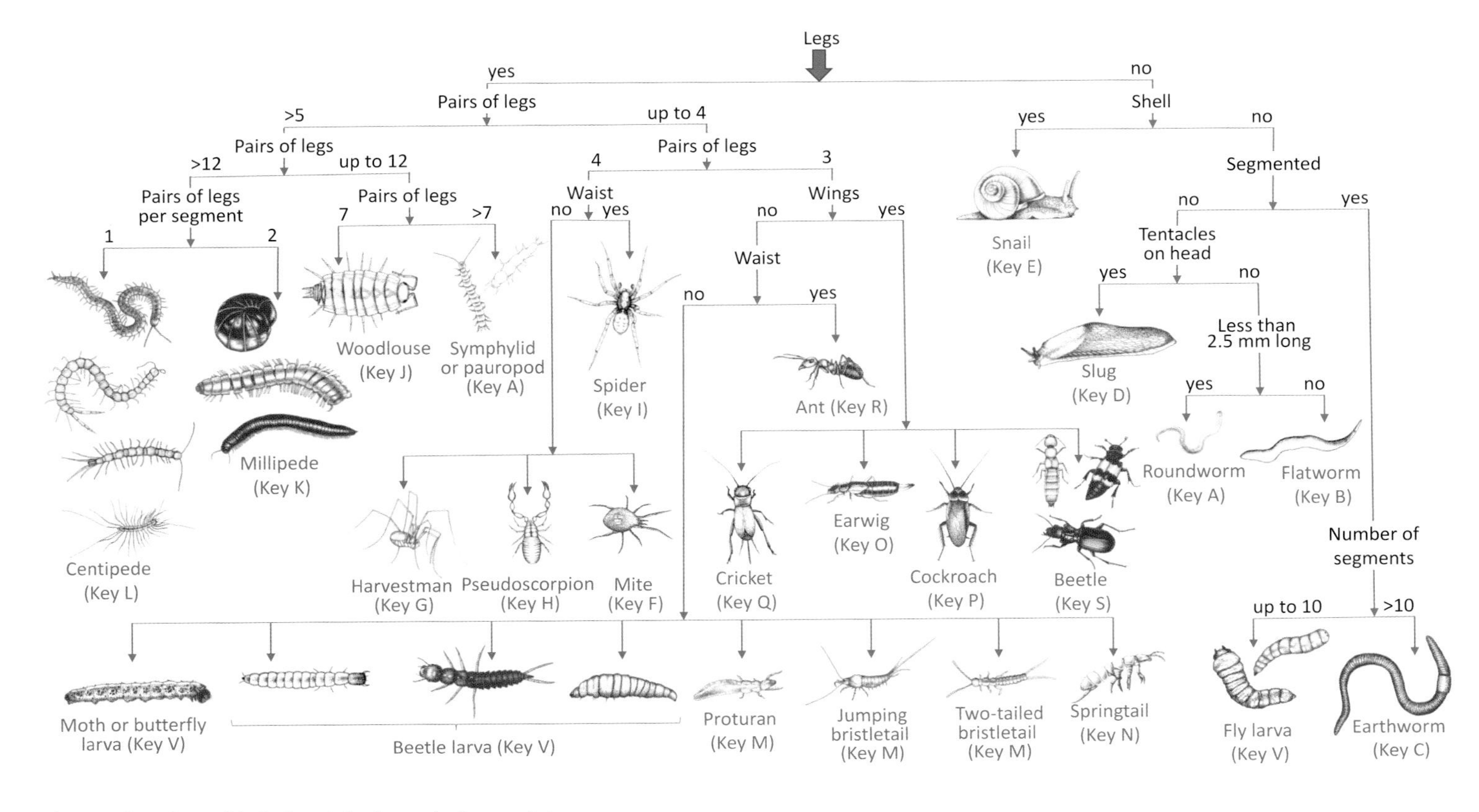


**Fig. 4.5** Guessing guide for invertebrates under logs and stones

## Key A Major invertebrate groups

This key will allow you to identify the major groups of invertebrate animals found under logs and stones. The key covers adults and (some) juveniles (larvae and nymphs) but not the inactive pupal stages of some insects. Couplets may lead you to further keys in this book to narrow down the identification (indicated by a reference to the next relevant key). Note, however, that in a book of this size it is not possible to provide the information to enable you to identify everything to species, and where relevant there is reference to more detailed/specialist keys. Where animals do not fully fit the descriptions in this key or where there is no signpost to further keys, then consult Tilling (2014) for further identification. Any sizes given are maximum adult body lengths (L) not including any appendages (i.e. antennae, pincers or tails).

1 Without jointed legs 2
– With jointed legs 8

2(1) Body composed of segments 3
– Body not segmented 5

3(2) Body with more than 20 segments (A.1 or A.2) Annelida (segmented worms) 4
– Body with fewer than 20 segments; head may be well developed (A.3 or A.4) or head pointed and tail end broader (A.5) insect larvae Key V

Some immobile pupae may key out here. These are not covered by this book – see Chu & Cutkomp (1992) for further information.

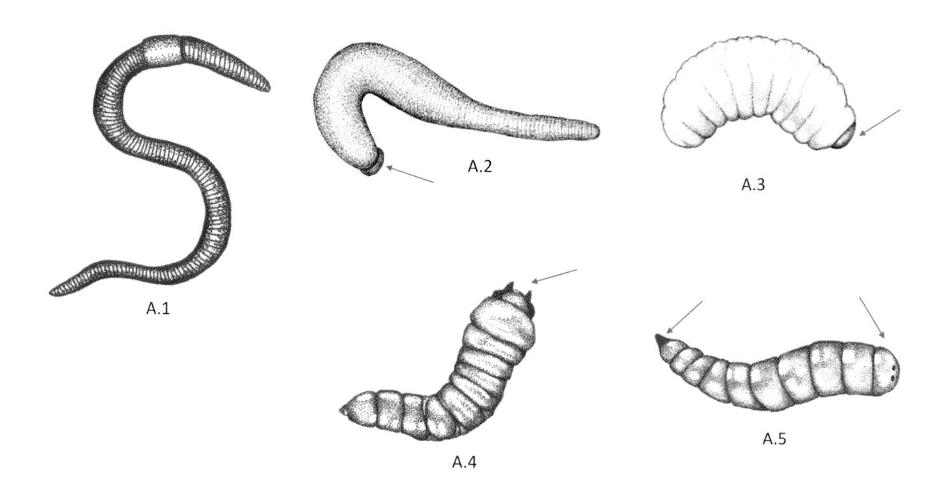

4 (3) Posterior end with sucker (A.2) – head end may appear pointed or also with a sucker; animal moves in a looping motion anchored by the suckers
Hirudinea (leeches)

There are four species of amphibious leeches in Britain that may be found out of water in wet conditions or under stones even at some distance from water (see Elliott & Dobson, 2015, and Section 3.5 for further details of the ecology and identification of leeches). Note that the sizes given here are maximum lengths and specimens found may be substantially smaller.

- *Haemopis sanguisuga* – horse leech – L:15 cm; body dark grey or greenish (may have darker flecks) with greenish yellow stripes (the pattern may not be clear unless the specimen is immersed in water); may be found under stones near to water.
- *Hirudo medicinalis* – medicinal leech – L:20 cm; body green, brown or greenish brown with thin reddish yellow stripes; may be found under stones near to water.
- *Trocheta pseudodina* – L:7.5–15 cm; body light grey or pinkish; may be found under stones near to water.
- *Trocheta subviridis* – L:15–16 cm (may be longer); body greyish green or reddish, may have two brown stripes; often in damp areas at some distance from water.

– Head and tail ends without a sucker, both pointed (A.1) Clitellata (earthworms and potworms) Key C

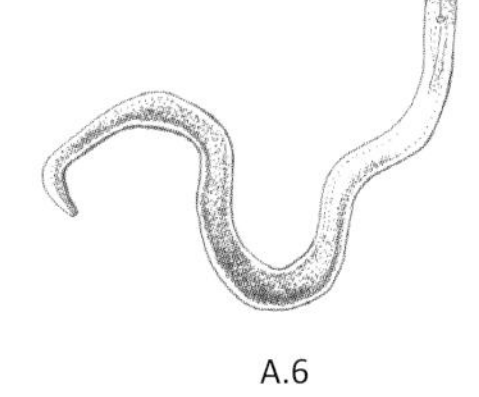

A.6

5 (2) Elongate and cylindrical (worm-like) (A.6); usually less than 2.5 mm in length
Nematoda (roundworms or eelworms)

Roundworms are small and difficult to identify (see Section 3.5 for further information).

The five British semi-terrestrial species of Nematomorpha (hair worms, or horsehair worms, or Gordian worms) may also key out here but tend to be restricted to damp soil near to open water. They are difficult to separate from nematodes but are typically longer (L:50–100 mm but can be more) and tend to be more pigmented (see Section 3.5 for further information).

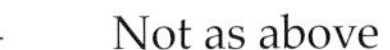

– Not as above 6

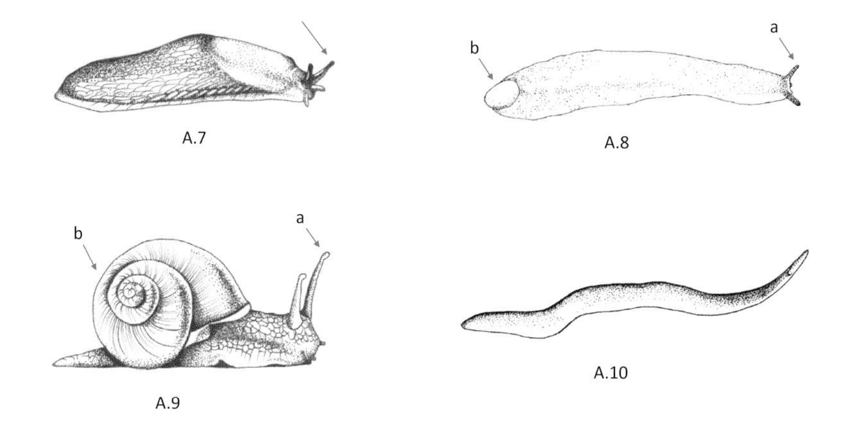


A.7

A.8

A.9

A.10

6 (5) With tentacles on head when extended (A.7, A.8a or A.9a); body slimy; may have external shell (A.8b or A.9b) Mollusca 7

– Without tentacles (A.10), may be round or flattened in cross-section Turbellaria (flatworms) Key B

Terrestrial Nemertea (proboscis or ribbon worm) species may also key out here (see Section 3.5 for further information). There are at least three introduced species, which originated from Australia and New Zealand and may have been introduced to the Isles of Scilly and the south-west of England along with plant imports. Proboscis worms can be distinguished from flatworms by their habit of pushing out a proboscis (which can be almost as long as the body) from just above their mouths when live specimens are handled. Three species have been recorded (one from the Scilly Isles and one from a few locations near Penzance in Cornwall). The most common, widespread species, and the one possibly of longest standing in the UK, is:

- *Argonemertes dendyi* – smiling worm – L:15 mm; body a flattened oval shape in cross-section, brownish pink, cream or light yellowish brown, with two darker brown stripes; recorded from across the UK including west Wales and Northern Ireland, can be found under logs in damp habitats.

7 (6) Does not have obvious shell (A.7) or shell much reduced (A.8b) and situated at the rear of the body Slugs Key D

– Has obvious shell situated towards the middle of the body (A.9b) Snails Key E

Note that semi-slugs have obvious shells (albeit too small for the animal to completely retract into it) and will key out here as snails (see Key E).

8 (1) Three pairs of jointed legs Hexapoda Key M

Note that first instar millipedes also have three pairs of jointed legs (they add more at each moult as they grow). They do not have an obvious separation between the thorax and abdomen and will usually be found with an adult millipede nearby. Later instars have more than three pairs of jointed legs even if they do not have the full adult complement (see couplet 14). Larval mites also only have three pairs of legs. These are very small and do not have a separation between the head, thorax and abdomen (see couplet 9).

– More than three pairs of jointed legs 9

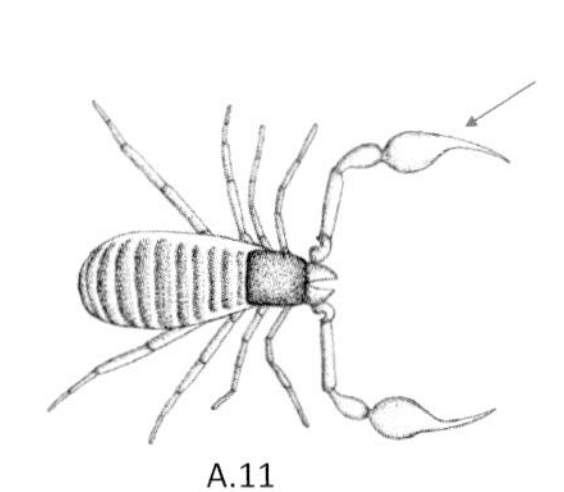

A.11

9 (8) Usually four pairs of legs; body divided into one or two distinct parts Arachnida 10

Juvenile mites have only three pairs of legs and some spiders may have long palps that make them appear as if they have five pairs of legs.

– More than five pairs of legs 13

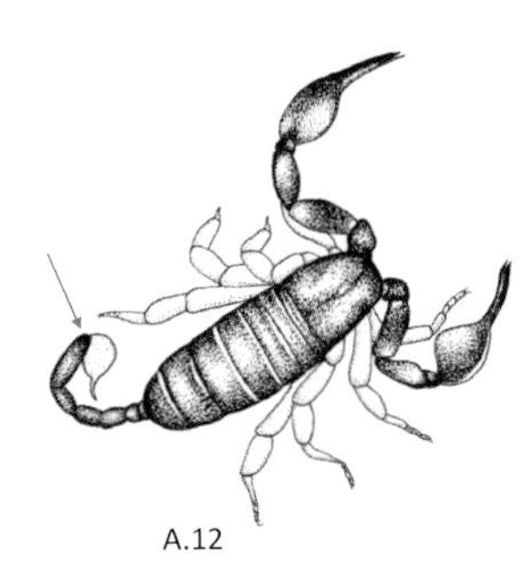

A.12

10 (9) With a pair of large palps with pincers (A.11) Pseudoscorpiones (pseudoscorpions) Key H

Scorpions also key out here but are much larger (pseudoscorpions are less than 5 mm in length) and have tapered tails ending in a sting, which is often held in a threatening position over the body (A.12) – see Section 3.5 for further information. The only scorpion present in Britain is an introduced species originally restricted to docks in Kent but which has expanded its range somewhat in recent years to London and the south-west of England. It is mostly harmless to humans.

- *Euscorpius flavicaudis* – European yellow-tailed scorpion – 35–45 mm; black body, yellow-brown legs and tail; in cracks in walls and under stones.

– Palps without pincers 11

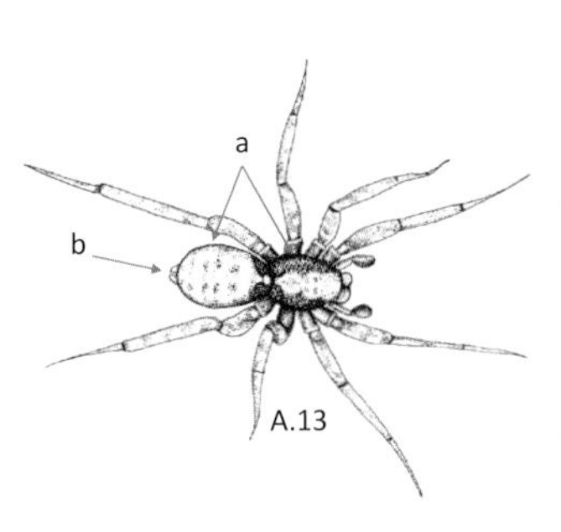


A.13

11 (10) Body divided into two distinct parts (A.13a); silk spinnerets on hind end of abdomen (A.13b) Araneae (spiders) Key I

– Body undivided, consisting of a single part; no spinnerets 12

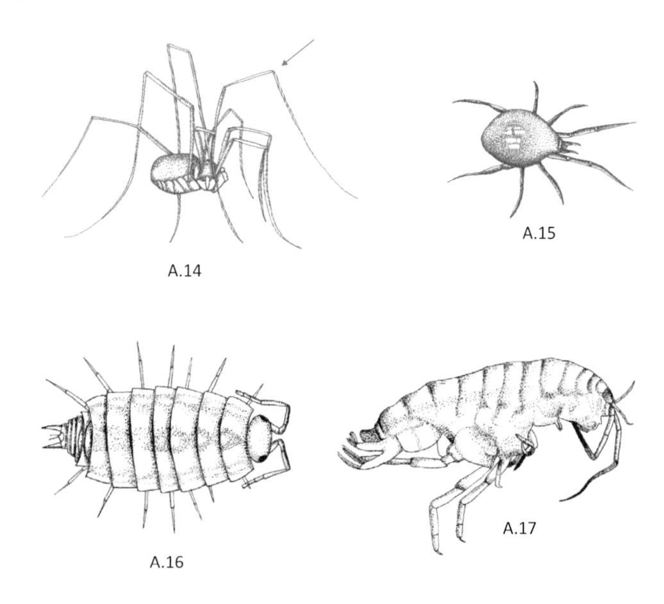
A.14
A.15
A.16
A.17

12 (11) Second pair of legs the longest (A.14); usually at least one pair of legs more than 1.5 times the length of the body; body often over 2 mm in length and segmented (easiest to see from below)
Opiliones (harvestmen) Key G

Second pair of legs not normally the longest; legs usually shorter than body; body small, usually under 2 mm in length and round or slightly elongate with four pairs of legs (A.15), or, if sausage shaped, with three pairs of legs Acari (mites) Key F

13 (9) Body less than three times as long as wide (A.16); always one pair of legs per segment and a maximum of seven pairs in all Isopoda (woodlice) Key J

One species of Amphipoda (A.17) also keys out here (see Section 3.5 for further information):

- *Arcitalitrus dorrieni* – landhopper (or woodhopper, or lawn shrimp) – L:15 mm; dark in colour (often quite orange, although this pigmentation can fade in preserved specimens), flattened from side to side rather than from top to bottom (as in the Isopoda) and with two pairs of antennae (a smaller pair above the longer pair); found in wet woods (often in quite deep leaf litter) and in human-disturbed areas, especially near the coast in the south and west of Britain. There may be some confusion between this species and the introduced species of semi-terrestrial talitrid, *Cryptorchestia cavimana* – bankhopper – L:22 mm. The latter is usually an aquatic species but can be found in habitats near to water. *A. dorrieni* can be distinguished from *C. cavimana* by having the 1st (shorter dorsal) antenna being relatively long, ending part way along the third segment of the 2nd antenna (in *C. cavimana* the 1st antenna is shorter and ends before the end of the 2nd segment of the 2nd antenna). Note that this relative 1st antennal length in *A. dorrieni* is a feature shared with several introduced species found in hot-houses. See Gregory (2016) for further details of differences between these species.

– Body very elongated, more than three times as long as wide (may be about three times as long as wide but if so with more than seven pairs of legs, usually two pairs per 'segment'/body ring) Myriapoda 14

Note that in millipedes (Diplopoda – see couplet 14) what appear to be segments are in fact two fused segments and should strictly be referred to as body rings.

14 (13) Two pairs of legs per body ring (A.18) (see A.19, A.20, A.21 or A.22) Diplopoda (millipedes) Key K

– One pair of legs per segment 15

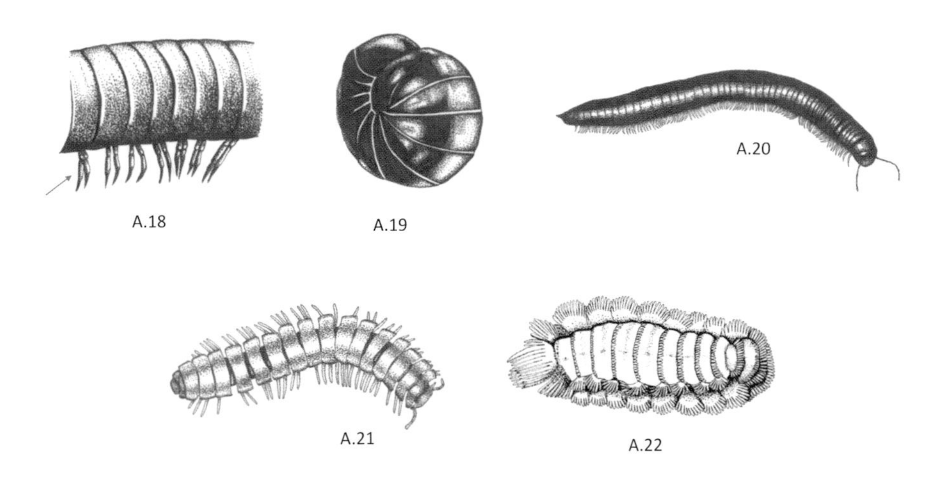

A.18 A.19 A.20 A.21 A.22

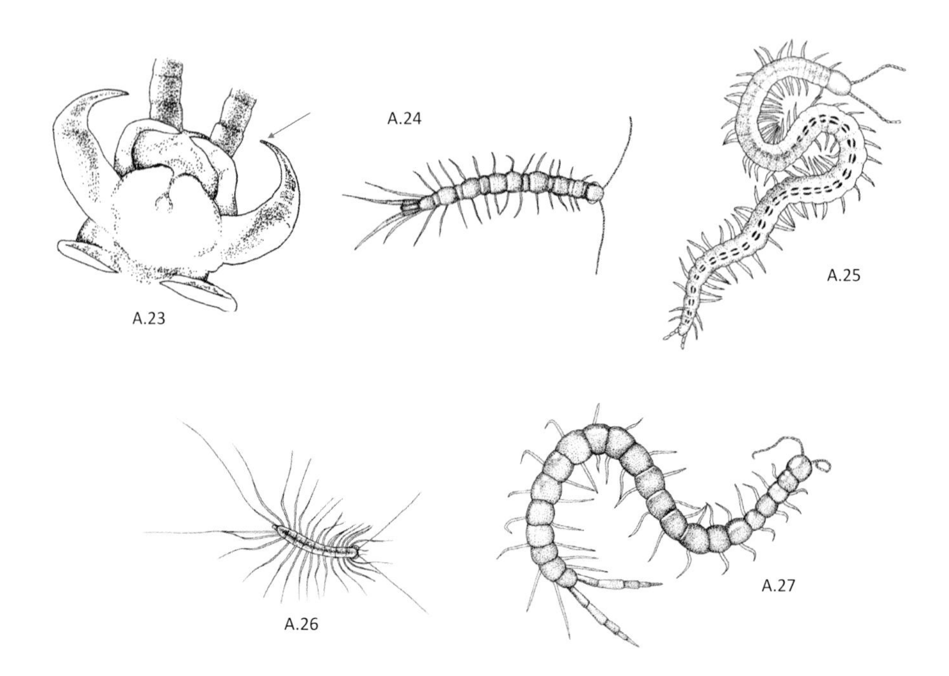


15 (14) Front legs modified into set of large poison claws (can be more easily seen from below – A.23) (see A.24, A.25, A.26 or A.27) Chilopoda (centipedes) Key L

– Front legs not modified into set of large poison claws; small (<10 mm) and pale coloured 16

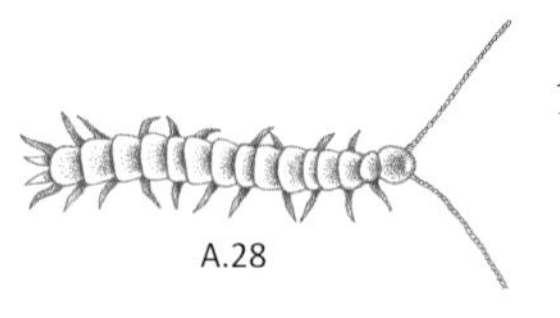


16 (15) Body whitish, 1–8 mm long with 7–12 pairs of legs (12 pairs in adults) (A.28) Symphyla (symphylids)

These are also known as symphylans or garden centipedes or pseudocentipedes. Note that the name 'garden centipede' is also one of the common names ascribed to the centipede *Lithobius forficatus*. There are 14 species of symphylids in Britain and Ireland from two families each with three genera, of which nine species have been described as reasonably common in the UK (Edwards, 1959). A study by Hopkin & Roberts (1988) confirmed five of these species as common and identified one (*Scutigerella causeyae*) as the species that would be most likely to be found outside of agricultural habitats. This is also the largest British species. However, these species are difficult to separate and Hopkin & Roberts also suggested that there was possible taxonomic confusion between this species and another of Edwards' common species (*Scutigerella lineatus*). Since these papers, this group has been little studied so the taxonomy remains unresolved. See Section 3.5 for further information.

A.29

– Not as above other taxa

Pauropoda will key out here (A.29). They are small (<2 mm long) myriapods, with branched antennae (arrowed). There are two families: the Pauropodidae which are whitish and have rather elongated and slender bodies, containing five genera and 22 species, most (14) being *Allopauropus* species; and the Eurypauropodidae containing a single yellowish, and relatively short and squat species (*Trachypauropus britannicus*). These species have been little studied and there are no current key works available. See Barber *et al.* (1992) and Section 3.5 for further information.

## Key B Flatworms

Terrestrial flatworms have been relatively little studied and there are currently no up-to-date keys for the identification of the UK fauna. There are around 22 species in the UK of which only six are likely to be native. Several introduced species have only been found in hot-houses (two species), or on introduced plants (one species), or are viewed unlikely to be established (one species with records from a single site in Oxfordshire, one species due to doubtful records and another which may be extinct). All UK land flatworms belong to the family Geoplanidae. This key enables the identification of mature specimens of some of the more obvious common terrestrial flatworms (Turbellaria: Planaria). Where the description does not match the specimen, consult Jones (1998, 2005). Sizes are for typical extended lengths (L) and widths (W) for mature specimens, but be aware that flatworms expand and contract while moving and so lengths may be rather plastic. Habitat information reflects the extent of current records rather than comprehensive habitat preferences. See Section 3.5 for further information on flatworms.

1 Head tapers to a point or a blunt rounded end (B.1) 2

– Head moon-shaped (B.2) *Bipalium kewense*

- *B. kewense* – hammerhead flatworm – L:50–350 mm, W:3–5 mm; body buff coloured with five purple stripes; an introduced species found in glasshouses and a few gardens.

2 (1) Body yellow (no patterns of flecks, lines or stripes) *Microplana scharffi*

- *M. scharffi* – L:20–50 mm, W:1–2 mm; body bright yellow (may be yellow-grey after feeding on slugs or pinkish after feeding on earthworms), underside paler; widespread records.

– Body not yellow 3

3 (2) Body orange (no patterns of flecks, lines or stripes) *Australoplana sanguinea*

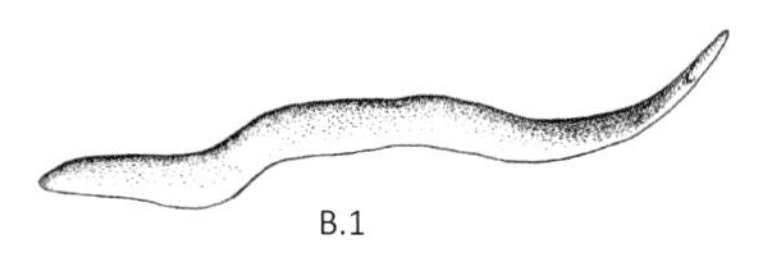
B.1

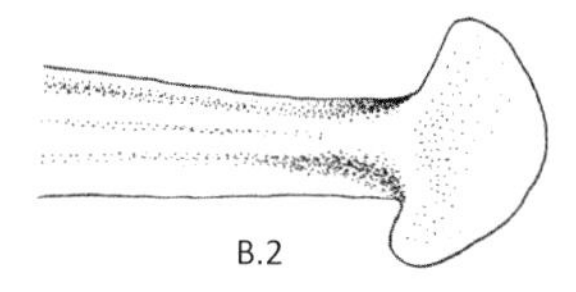
B.2

- *A. sanguinea* – Australian flatworm – L:20–80 mm, W:2–5 mm; body orange (purple-orange after feeding); abundant in sites in England and Wales.

– Body not orange 4

4 (3) Body whitish with broad brown central band *Artioposthia exulans*

- *A. exulans* – L:10–20 mm, W:1–2 mm; body greyish white with a broad brown band centrally and several brown lines along the sides, underside greyish white with two broad brown bands; records restricted to Cornwall.

– Body not whitish with broad central brown band 5

5 (4) Body dark with a strongly contrasting yellow or cream central stripe (note that some individuals of *Obama nungara* may have a faint paler midline – see couplet 6a) *Caenoplana* species

- *Caenoplana coerulea* – blue garden flatworm – L:20–80 mm, W:1–2 mm; body dark purple-brown with dark cream midline, underside cobalt blue; records from Cornwall and Liverpool.
- *C. variegata* – yellow-striped flatworm – L:50–150 mm, W:2–5 mm; body dark brown or black with broad yellow midline within which are two narrow brown lines, underside pale with a paler midline; some records from Scotland, Wales and Ireland, mainly from England, especially the south.

– Body not dark, or if dark without a strongly contrasting yellow or cream central stripe 6

6 (5) Body at rest over 30 mm long two species

- *Arthurdendyus triangulatus* – New Zealand flatworm – L:50–200 mm, W:5–10 mm; body dark brown with pale margins and underside both with brown spots; many records across the UK.
- *Obama nungara* – Brazilian flatworm – L:100 mm, W:10 mm; body dark brown with numerous black streaks (sometimes with a pale midline), underside pale; recorded from a few locations across Britain and Ireland.

– Body at rest no more than 30 mm long 7

7 (6) Body with patterns or stripes on the upper surface 8

– Body without any patterns or stripes on the upper surface four species

- *Microplana fuscomaculosa* – L:12–15 mm, W:1–2 mm; body beige-brown speckled with mahogany-brown spots; single record from a wood in Oxfordshire.

- *M. kwiskea* – L:10–20 mm, W:2 mm; body pale to mid brown (khaki) with a brown underside; scattered records across England.
- *M. terrestris* – L:10–20 mm, W:1–2 mm; body brown, dark grey or black, underside pale; common and widespread.
- *Australopacifica atrata* – L:10–30 mm, W:1–2 mm; body shiny black, underside pale with three dark lines; records from Wales and southern England.

8 (7) Body dark with blue lines or flecks — two species

- *Australopacifica coxii* – L:10–20 mm, W:1 mm; body black with two brilliant blue lines and flecks, underside with five lines; records from Isles of Scilly and Cornwall.
- *Marionfyfea adventor* – L:5–10 mm, W:1.0–1.5 mm; body patchy brown with pale blue flecks; recorded across England, Wales and Ireland.

– Body without any blue colouration on upper surface — three species

- *Kontikia andersoni* – L:10–25 mm, W:1 mm; body pale brown with three rows of dark brown spots; records from Ireland, the Isle of Man, Cornwall, Guernsey and the Isles of Scilly.
- *K. ventrolineata* – L:10–20 mm, W:1–2 mm; body black with two pale grey lines above and four darker lines underneath; scattered records mainly in south-west England and the Channel Islands.
- *Rhynchodemus sylvaticus* – snake-headed flatworm – L:10–30 mm, W:1–2 mm; body dark grey to black with two black lines and a central dark patch, pale underside, there can be a slight narrowing and some banding (looking a little like segmentation) behind the eyes; widespread records.

## Key C Earthworms

There are two groups of Annelida (segmented worms) that can be found under logs and stones: the Enchytraeidae (potworms) and the Crassiclitellata (earthworms). Potworms, which are not covered by this key, have a clitellum (saddle) restricted to segments 12 or 13: see Schmelz & Collado (2010, 2015) for identification and a checklist for this group. Adult earthworms are most easily distinguished from juveniles by the presence of a clitellum part-way down the body (closer to the head end than to the tail – see C.1 for the general morphology of earthworms). The common species of earthworms in the UK found in these habitats are members of the family Lumbricidae and are typified by the clitellum starting after segment 16 (when counting from the head end); two members of other families are rarely found in the UK. There are 28 species of Lumbricidae in the UK and Ireland, including one very rare species (found only in Ireland) which has the clitellum starting on segment 19; the rest have the clitellum beginning from at least segment 21. Adult Lumbricidae have a male pore on each side of segment 13, 14, or 15. This can be an obvious swelling around a slit, although in some species it can be very small and hard to see. This key identifies adults of those common earthworm species that have a clitellum starting on or after segment 21; if you find an earthworm with a clitellum starting before segment 20, see Sherlock (2018) for identification. Any specimens that do not exactly fit the descriptions and illustrations should be identified using the keys in Sherlock (2018) or Krediet (2020).

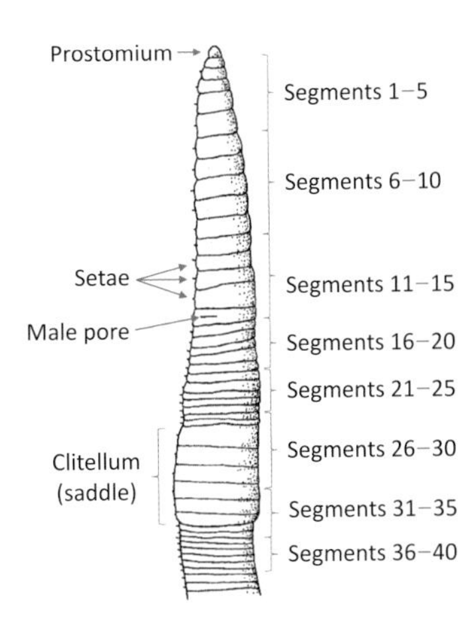


C.1

Note that the earlier keys in Sims & Gerard (1999) are also useful, although since this text was published, some names have been changed and one species has now been split into two species. Although there is currently no atlas to British earthworms, distribution data can be found at the NBNatlas website, while additional information and identification support can be found at the Earthworm Society of Britain (see Chapter 6). See Section 3.5 for further information on earthworms and potworms.

The lengths (L) given here are for adults at normal extension. Several important features are used in the keys and the species descriptions, which need a little explanation. These include the start and end segments of the clitellum, which are fairly specific to individual species (e.g. c29/30–32/33 means that the clitellum starts on either segment 29 or 30 and extends to either segment 32 or 33). The tubercula pubertatis (tp) forms swellings on the clitellum of many species in the shape of bumps, ridges or bands, which may have a role in copulation. In a similar way to the position of the clitellum, the position of the tubercula pubertatis tends not to vary within a species (e.g. tp30–32/33 means the tp starts on segment 30 and extends to segment 32 or 33). Colours can vary within a species and preserved specimens can become washed out so be aware that this may only present evidence towards confirming a species identification. Similarly, the habitat choice, including the species' ecological preferences (i.e. composter, epigeic, endogeic or anecic – see marginal note) may help with confirmation rather than acting as a reliable diagnostic tool.

**ecological groupings**
Earthworms can be classified on the basis of their ecological preferences: ***composters*** live in compost heaps or other sites with a high organic content; ***epigeics*** live on the surface of the ground in leaf litter and under logs and stones; ***endogeics*** live in the soil in horizontal burrows, only rarely coming to the surface; ***anecics*** live in the soil in vertical burrows, coming to the surface to feed on leaves which they drag down into their burrows.

1 Lobe on top of first segment extends to the rear edge of the first segment – tanylobic (C.2 from side); setae (bristles) behind clitellum closely paired (C.3) 2

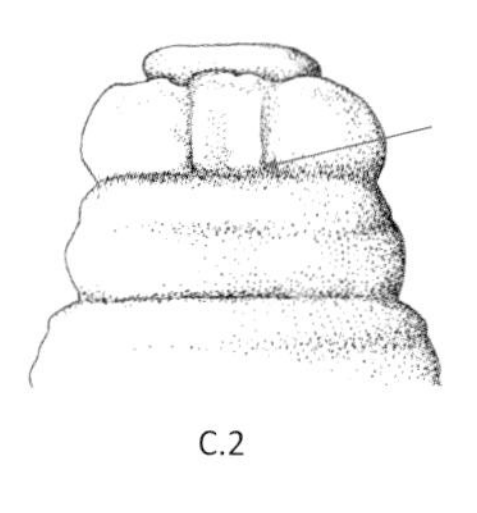
C.2

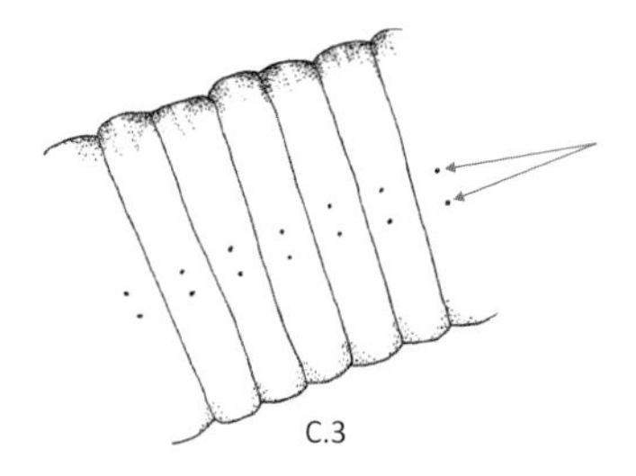
C.3

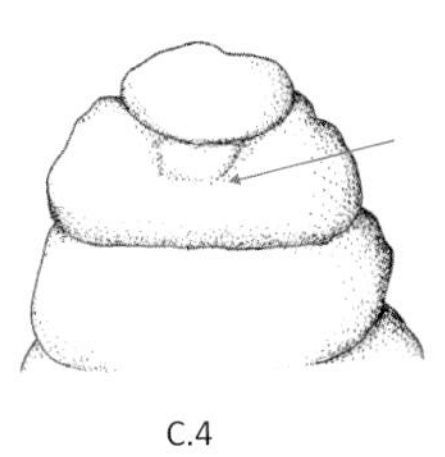

C.4

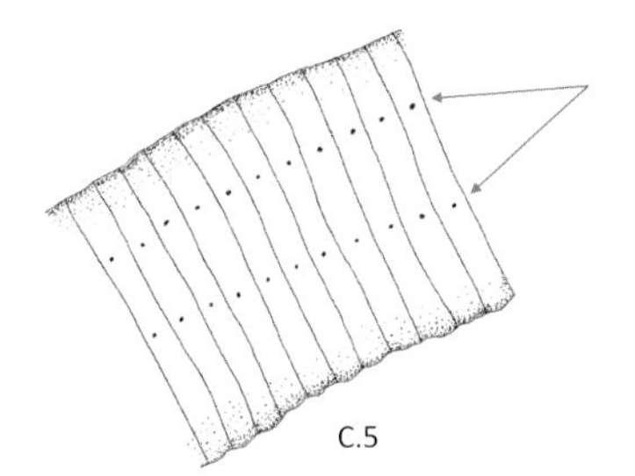

C.5

– Lobe on top of first segment does not extend to the rear edge of the first segment – epilobic (C.4 from side); setae (bristles) behind clitellum may be either closely paired (C.3) or widely spaced (C.5) 3

2 (1) Tubercula pubertatis present (a swelling forming bands or ridges on the clitellum) *Lumbricus* species

There are five British species of which two are rare. Common species are:

- *L. castaneus* – chestnut worm – L:30–85 mm; deep red; c27/28–33/34, tp29–32; common, anecic, can be in rotting logs.
- *L. rubellus* – redhead worm (or red marshworm) – L:15–150 mm; purple-red; c26/27–32, tp27/28–32; very common, epigeic (larger individuals may be anecic), recorded from all habitats.
- *L. terrestris* – lob worm (or common earthworm) – L:90–150 mm; deep purple-red; c31/32–37, tp32/33–36/37; common, anecic, mainly in grassland.

– Tubercula pubertatis absent *Bimastos eiseni*

- *B. eiseni* – small-tailed worm – L:30–64 mm; dull red; c25/27–32/33, tp absent; common, epigeic, can be found under bark.

3 (1) Setae behind clitellum closely paired (C.3) 4

– Setae behind clitellum widely spaced (C.5) 9

4 (3) Male pore (C.1) a slit-like opening on segment 13 or 14 *Eiseniella tetraedra*

- *E. tetraedra* – squaretail worm – L:20–80 mm; pale/green-brown; c22/23–26/27, tp22/23–26/27; common, endogeic, waterlogged soils.

– Male pore on segment 15 (this may be slit-like – sometimes surrounded by a swelling on adjacent segments, or be a small pit, which may be difficult to see) 5

5 (4) Tubercula pubertatis present (seen as swellings forming ridges, bands, humps or raised discs on the clitellum) 6

– Tubercula pubertatis absent *Murchieona muldalis*

- *M. muldalis* – glass worm – L:10–60 mm; pink-grey; c26/27–33/34, tp absent; rare, endogeic, woodland and grassland.

6 (5) Clitellum starts at segment 21 or 22 *Helodrilus oculatus*

- *H. oculatus* – L:35–80 mm; pale occasionally with black markings; c21/22–32, tp29–30; very rare, endogenic, waterlogged soil.

– Clitellum starts after segment 22 7

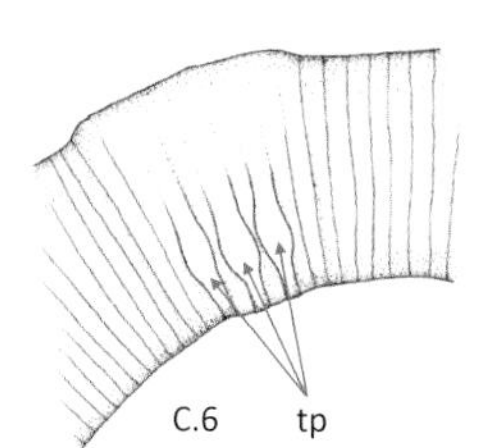


7 (6) Tubercula pubertatis consists of 3 sucker-like discs (C.6) *Allolobophora chlorotica*

- *A. chlorotica* – green worm – L:30–80 mm; two morphs – pink and green; c28/29–37, tp31 33 35; very common, endogeic, in most habitats.

– Tubercula pubertatis absent or consisting of bands or (at most 2) humps – not 3 discs 8

8 (7) Tubercula pubertatis consists of a thin band on segments 27/28 to 30/31; worm usually appearing stripy when moving *Eisenia fetida*

- *E. fetida* – brandling (or tiger) worm – L:26–130 mm; dark dorsally (stripy when moving); c24/27–32/34, tp27/28–30/32; common, epigeic, in areas of high organic content. Some non-stripy forms may be a separate species, *E. andrei* (red tiger worm), that is otherwise indistinguishable from *E. fetida* – indeed some taxonomists record *E. fetida* as an aggregated species as a result.

– Tubercula pubertatis consists of 2 humps or if as a band this is not on segments 27/28 to 30–31; not stripy when moving *Aporrectodea* species

There are six British species of which two are rare (with a further very rare species being found only from Ireland). Common species are:

- *A. caliginosa* – grey worm – L:40–180 mm; pink/grey; c25/29–34/36, tp31–33; very common, endogeic, in most habitats.
- *A. icterica* – mottled worm – L:50–140 mm; grey (can be yellowish or with a brownish tinge); c33/34–42/43, tp35/36–42/43; uncommon but widespread, endogeic, in grasslands and orchards.
- *A. longa* – blackhead worm – L:90–170 mm; dark (black) head; c27/28–35/36, tp31/32–34; common, anecic, grasslands and gardens.

- *A. rosea* – rosy-tipped worm (or pink soilworm) – L:20–110 mm; grey/pink (clitellum often orange); c23/26–32/33, tp29–30/31; common, endogeic, in most habitats.

9 (3) Tubercula pubertatis (seen as swellings forming ridges or bands on the clitellum) starts at segment 33/34 *Satchellius mammalis*

- *S. mammalis* – Celtic worm – L:24–41 mm; deep red; c31–36, tp33–34; uncommon, epigeic, woodlands with high organic content.

– Tubercula pubertatis starts before segment 33 10

10 (9) Tubercula pubertatis extending across the whole of the clitellum *Octolasion* species

There are two British species, which are common but endogeic and rarely come to the soil surface:

- *O. cyaneum* – blue-grey worm – L:65–140 mm; no pigmentation (or light blue-grey), may have a yellow tail; c29–34, tp29/30–33/34; common, endogeic, mainly wet areas.
- *O. lacteum* – white worm – L:25–160 mm; greyish, may have a yellow tail; c30–35, tp30–35; common, endogeic, very moist areas.

– Tubercula pubertatis absent or not extending across the whole of the clitellum 11

11 (10) Tubercula pubertatis starting on segment 28 or 29 *Bimastos rubidus*

- *B rubidus* – European barkworm – L:20–100 mm; dorsal surface deep red; c25/26/27–31/32/33, tp28/29–30; common, epigeic, under logs and stones and sometimes in wood ant nests.

– Tubercula pubertatis absent or starting on segment 30 or later *Dendrobaena* species

There are five British species of which two are rare. The three commoner species under logs and stones are:

- *D. attemsi* – L:26–70 mm; dark red dorsally or pale; c28/29–34, tp30–32; epigeic, under bark, leaf litter and decaying wood, uncommon.
- *D. octaedra* – octagonal-tail worm – L:20–60 mm; dark red dorsally; c27/28/29–33/34, tp31–33; epigeic, under litter, bark and stones in broad-leaved woods, uncommon but can be locally abundant.
- *D. veneta* – European nightcrawler – L:14–45 mm; deep red (stripy when stretched); c26/27–32/33, tp29/30–31/32; epigeic, on well-drained soils, common.

## Key D Slugs

The sizes given here are for the extended length of the adult (L), and the height (H) and width (W) of the shell in shelled species. Many British slugs can be found under logs and stones, although there are species that are subterranean and others that climb into vegetation. This key will assist in the identification of slugs to genus or species following the checklist by Anderson & Rowson (2020). Separating some slug species can be difficult and requires dissection and where a specimen does not closely match the description consult Rowson *et al.* (2014). General features of slugs are shown in D.1. Further information and images of many British species can be found on the Conchological Society of Great Britain and Ireland, MolluscIreland, and National Museum Wales websites (see Chapter 6). Kerney (1999) and Rowson *et al.* (2014) provide distribution maps for the British species. See Section 3.5 for further information on slugs.

1 Slugs with very small external shell situated at the rear of the body (D.2) Testacellidae

There are four British species, all of which live in the soil but may come to the surface in wet weather or be found under deeply sunk stones or other debris or moss:

- *Testacella haliotidea* – ear-shelled slug (or earshell slug, or (common) shelled slug or carnivorous slug) – L:70–100 mm, shell slightly convex (H:7–8 mm, W:5–6 mm); body grey or creamy white to pale yellow with little (or light brown) speckling, centre of back light brown; gardens and horticultural sites.
- *T. maugei* – Atlantic shelled slug (or Maugé's (shelled) slug, or dead man's fingers) – L:60–100 mm, shell convex (H:12–16 mm, W:6–7 mm); body usually greyish or brownish with dense darker speckling; gardens in south-west England.
- *T. scutulum* – orange shelled slug (or golden shelled slug, or shield slug) – L:70–100 mm, shell

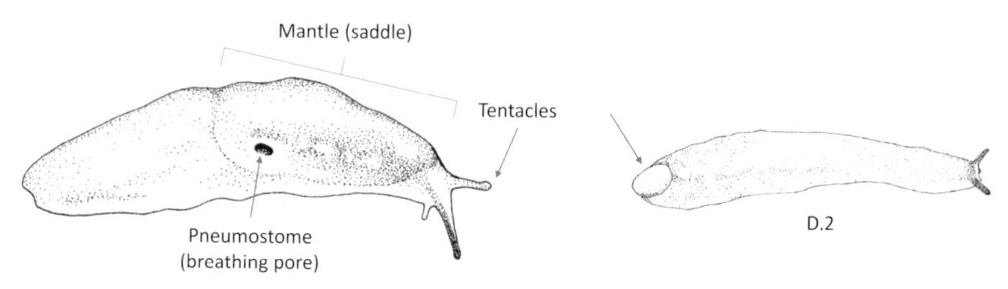

D.1

D.2

flat (H:6–7 mm, W:4 mm); body pale yellow to orange-yellow with dense darker speckling; parks and gardens. Another species (temporarily named *T.* species '*tenuipenis*') is indistinguishable from *T. scutulum* without dissection, and seems to be the more common of the two species.

A recent introduction belonging to the family Oxychilidae (see also couplet 2) also has a shell that is too small for adults to retract into: *Daudebardia rufa* – Alsatian semi-slug: L:17–20 mm; body grey above and whitish underneath; recorded from two woodlands in Wales.

Note that species of the semi-slug family Vitrinidae may key out here – they have larger shells than the Testacellidae (albeit that are still too small for the whole of the animal to retract into), and whereas in the latter the shell is at the tail of the slug (D.2), in the former, it is more central on the body (see couplet 7 in the snail key E).

– No external shell (D.1) 2

2 (1) Mantle ('saddle' on back) obvious and located towards the head end of the slug (D.1) 3

– Mantle ('saddle' on back) very small, not obvious and located at the tail end of the slug (D.3) *Selenochlamys ysbryda*

An introduced species belonging to the family Oxychilidae (see also couplet 1):

- *S. ysbryda* – ghost slug – L:50–75 mm; body white and translucent; subterranean and rarely found at the surface, in gardens and allotments.

3 (2) Pneumostome (respiratory opening) at front of mantle ('saddle' on back) (D.4) Arionidae 4

There are 16 species belonging to two genera, one of which consists of four subgenera (see couplets 4–6).

– Pneumostome at back of mantle (D.5) 7

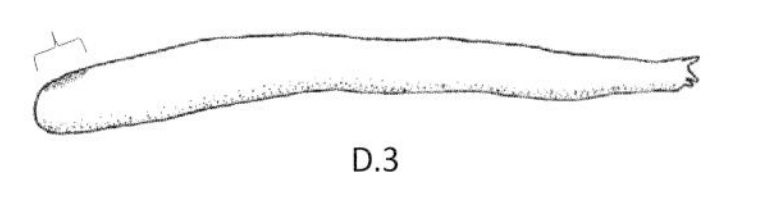

D.3

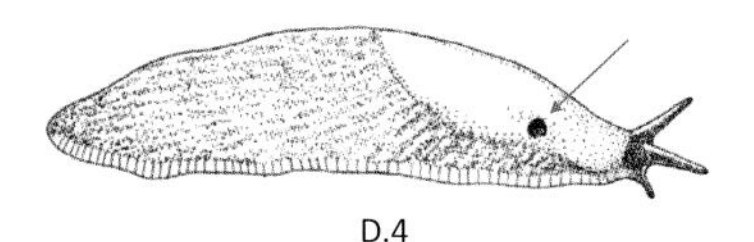

D.4

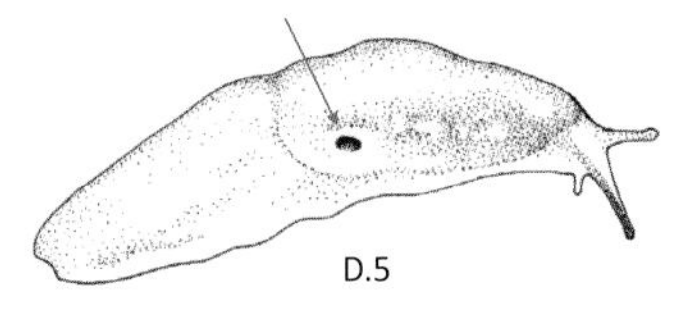

D.5

4 (3) Sole of foot yellow-orange *Arion* (*Kobeltia*) species

There are six species in this subgenus which are all small slugs (up to 40 mm), of which two are rather rare (both have lateral bands – coloured stripes running along the sides of the body, which are high up away from the foot edge (in contrast to the other species which either have faint or no bands, or where such bands are close to the foot edge). Some of the more common species are rather difficult to separate and may require dissection to confirm identification:

- *A.* (*Kobeltia*) *distinctus* – brown soil slug (or common garden slug, or darkface arion, or Mabille's orange-soled slug, or April slug) – L:25–40 mm; body often dark grey-brown, often with a suffused golden-yellow speckling over the back; widespread and common, at the soil surface and under refuges in woodlands and gardens.
- *A.* (*Kobeltia*) *hortensis* – blue-black soil slug (or southern garden slug, or garden arion, or Férussac's orange-soled slug) – L:20–35 mm; often dark blue-grey to black; in both species there are lateral bands low down (almost to the edge of the foot); at the soil surface and under refuges in woodlands and gardens.
- *A.* (*Kobeltia*) *intermedius* – hedgehog slug (or hedgehog arion, or glade slug) L:15–20 mm; variable in colour from white to orange to yellow-brown to blue-grey; lateral bands if present high on body and faint (often absent), contracts into dome shape with a prickly appearance; widespread, including in woods and wetlands.
- *A.* (*Kobeltia*) *owenii* – tawny soil slug (or Inishowen slug) – L:25–40 mm; body tawny yellow-brown to rich brown colour; widespread and common, found in woodland, scrub, gardens and roadsides.

– Sole of foot not yellow-orange 5

5 (4) Body mucus yellow-orange, often colouring the body, especially the head (stains white paper) *Arion* (*Mesarion*) species

There are three species in this subgenus, two of which are rare and restricted to only a few sites. The common species is:

- *A.* (*Mesarion*) *subfuscus* – dusky slug (or dusky arion) – L:50–70 mm; body various colours from orange-yellow to brown to grey to purple-black, with dark lateral bands high on the body sides (the two rarer species have these bands low on the body sides); widespread in both natural and disturbed habitats.

– Body mucus colourless 6

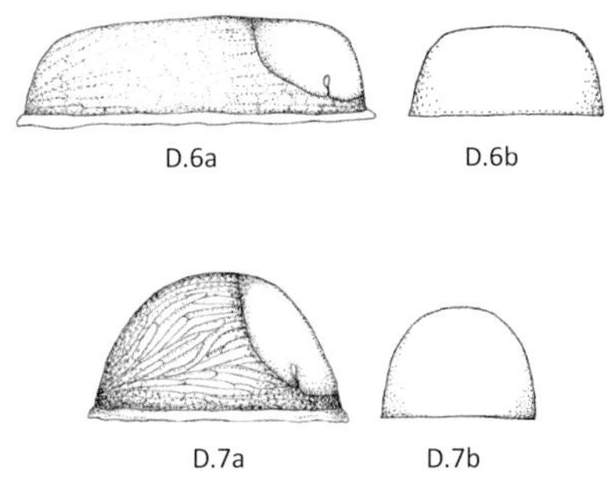

D.6a D.6b

D.7a D.7b

6 (5) Contracted body with a flattened top (like a flattened dome in cross-section) (D.6a – from the side, D.6b – in cross-section); relatively smooth appearance *Arion* (*Carinarion*) species

There are two species in this subgenus, one of which can be divided into two subspecies:

- *A.* (*Carinarion*) *circumscriptus* – L:30–40 mm – *A.* (*Carinarion*) *circumscriptus circumscriptus* – spotted false-keeled slug (or Bourguignat's slug, or forest slug, or brown-banded arion) – body grey to blue-grey or brown, often with dark blotches on the mantle; and *A.* (*Carinarion*) *circumscriptus silvaticus* – silver false-keeled slug (or heath slug, or forest arion) – body grey to blue-grey or brown, without dark blotches on the mantle. Both species are widespread in a range of habitats including woodlands and disturbed sites. Separation of these subspecies may require dissection.
- *A.* (*Carinarion*) *fasciatus* – rusty false-keeled slug (or banded slug, or orange-banded arion, or Bourguignat's slug) – L:35–45 mm; body grey to yellow-brown, usually with orange or yellow colouration below lateral bands and a pale coloured foot (not bright white as the other species often have); found in very disturbed habitats such as roadsides and waste ground, as well as woodlands and scrub.

– Contracted body with a rounded top (like a rounded dome, nearly hemispherical, in cross-section) (D.7a – from the side, D.7b – in cross-section); relatively rough appearance *Arion* (*Arion*) species

There are five species in this subgenus, which are very variable in colouration and can, therefore, be difficult to separate without dissection. Typical colour types can sometimes be separated:

- *A.* (*Arion*) *ater* – large (or great) black slug (or black arion) – L:60–140 mm; body red, brown, grey or black usually with darker fringe to foot and a sole that is lighter or the same as the body sides; widespread in almost all habitat types. This species frequently rocks from side to side when disturbed.

- *A.* (*Arion*) *flagellus* – green-soled slug (or Spanish stealth slug, or Durham slug) – L:60–100 mm; body yellow to red to brown to grey to black, often with a greenish tinge (which may also occur on the sole), surface of body with a coarse appearance with (usually) fewer than 9–10 coarse tubercles (wart-like raised projections) between the lateral bands; widespread and common, disturbed habitats including urban sites as well as woodlands and uplands.
- *A.* (*Arion*) *rufus* – large (or great) red slug (or chocolate arion) – L:60–140 mm; body yellow to red to brown to grey (occasionally black), with the edge of the foot often orange or orange-red; widespread in almost all habitat types. This species may rock from side to side when disturbed.
- *A.* (*Arion*) *vulgaris* – vulgar slug (or Spanish slug, or Iberian slug, or (false) Lusitanian slug, or plague slug) – L:60–140 mm; body yellow to red to brown to grey (occasionally black), often with a sooty grey sole; disturbed habitats including roadsides, gardens and field margins, as well as woodland edges. This species is easily confused with some forms of *A.* (*Arion*) *rufus*, and *A.* (*Arion*) species 'Davies' (Stella Davies's slug – L:100 mm), which is very similar to *A.* (*Arion*) *vulgaris* and cannot be distinguished without dissection.

A species from another genus of Arionidae also keys out here – *Geomalacus maculosus* – Kerry slug – L:60–90 mm; body dark with pale spots, rolls into a ball when disturbed; only found in south-west Ireland.

7(3) Body with long keel (ridge) running from the tail to the rear of the mantle (D.8) 8

– Body with short keel running from the tail but not extending to close to the mantle (D.9) 10

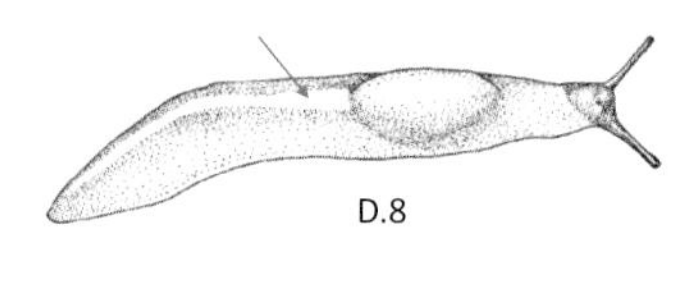
D.8

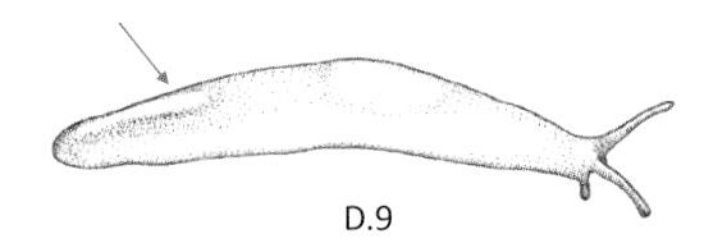
D.9

8 (7) Keel lighter in colour than rest of back
Milacidae (part)

Four species of one genus from this family key out here, of which two are rare (see also couplet 9). The more common species are:

- *Tandonia budapestensis* – Budapest (keeled) slug – L:50–70 mm; body grey to dark grey-brown (mantle often brown), sole with pale sides and a darker central band and with colourless mucus; in disturbed habitats including woodland edges and paths. The rarer *T.* cf. *cristata* (Crimean keeled slug – L:25–35 mm) has a uniformly pale sole without a central darker stripe seen in *T. budapestensis* and has a network-like pattern of darker pigmentation in the grooves between the tubercles (the raised projections on the back).
- *T. sowerbyi* – Sowerby's (keeled) slug (or keeled slug) – L:60–75 mm; body yellow-grey to dark grey-brown with a pale sole and yellow-orange mucus; lowland habitats including wood edges and grassland, also in disturbed sites. The rarer, *T. rustica* (spotted keel slug, or rustic slug – L:50–70 mm) is more lightly speckled with larger dark spots than *T. sowerbyi*.

– Keel same colour as, or darker than, rest of back 9

9 (8) Body slender and worm-like, white or pale grey
Boettgerillidae

There is one species in this family:

- *Boettgerilla pallens* – worm slug – L:35–55 mm; body white to pale blue-grey (or lilac-grey); subterranean in disturbed habitats including grassland, also in woodland.

– Body not slender and worm-like, usually not white or pale grey Milacidae (part)

One species from this family keys out here (see also couplet 8):

- *Milax gagates* – smooth jet slug (or jet slug, or greenhouse slug, or black keeled slug, or small black slug) – L:45–55 mm; body grey to brown or black; subterranean in gardens, waste ground and coastal grassland.

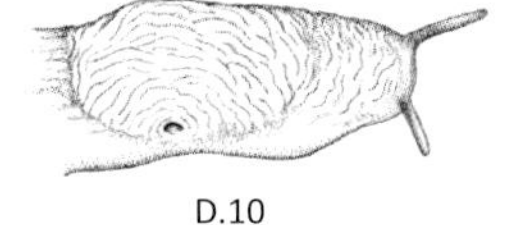

D.10

10 (7) Concentric ridges on mantle centred on a point just above the pneumostome (D.10) Agriolimacidae

There are five species in this family:

- *Deroceras agreste* – Arctic field slug (or northern field slug) – L:25–45 mm; body light brownish cream to light brownish yellow with no darker flecks or markings (mantle shorter than the distance from the rear of the mantle to the end of the tail), usually produces milky mucus when handled; widespread in the north mainly in open and upland habitats.

- *D. invadens* – tramp slug (brown field slug, or longneck field slug, or Caruana's slug, or Sicilian slug) – L:25–35 mm; body pale chestnut, mid grey to very dark brown (mantle shorter than the distance from the rear of the mantle to the end of the tail), pneumostome often with a paler area around it; an introduced species found under debris in disturbed habitats including urban sites, roadsides, gardens and farms. *D. panormitanum* (Sicilian slug – L:25–35 mm) is known from only two British sites (both in south Wales) and can only be separated from *D. invadens* by dissection.
- *D. laeve* – marsh slug (or meadow slug, or brown slug, or smooth slug) – L:15–25 mm; body pale chestnut, mid grey to very dark brown with widely spaced ridges on the mantle, which is the same length or longer than the distance from the rear of the mantle to the end of the tail; widespread in moist habitats including wet woodland, damp grassland and wetlands.
- *D. reticulatum* – netted field slug (or (grey) field slug, or grey garden slug, or reticulate slug, or milky slug) – L:35–50 mm; body white to light brownish yellow to dark brown or grey (mantle shorter than the distance from the rear of the mantle to the end of the tail), usually produces milky mucus when handled; widespread in open habitats, may be found in wood edges.

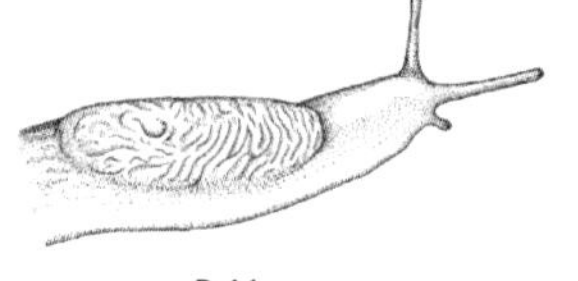

D.11

– Concentric ridges on mantle centred on the mid-line (D.11) Limacidae 11

There are nine species in this family of keelback slugs.

11 (10) Under 50 mm long when extended *Malacolimax tenellus*

- *M. tenellus* – lemon slug (or tender slug, or slender slug) – L:35–50 mm; body yellow (can be orange or brown), soft and flaccid, with yellowish mucus; found in older (coniferous and deciduous) woodland – an ancient woodland indicator species.

– Over 50 mm long when extended 12

12 (11) Body firm and muscular two genera

There are two genera containing five species, one of which is a rare introduction. The more common species are:

- *Limacus flavus* – yellow cellar slug (or yellow slug, or yellow garden slug, or house slug, or cellar slug) – L:80–130 mm; body yellow-green to yellow-brown with a pale central stripe extending to the end of the tail (this is not the case in *L. maculatus*), tentacles greyish, keel short; widespread, associated with buildings, gardens and urban sites.

- *Limacus maculatus* – green cellar slug (or Irish yellow slug) – L:80–130 mm; body pale ochre, yellow-green, or blackish, any pale central stripe does not extend to the end of the tail, tentacles greyish, keel short; widespread, in human-disturbed areas and woodlands, sometimes in large groups of individuals.
- *Limax cinereoniger* – ash-black slug (or ashy-grey slug) – L:100–150 mm; body pale brown to grey or brown with no dark markings on mantle (any darker markings only on the tail), tentacles greyish, keel long, sole often with darker stripes at the sides (not seen in *L. maximus*); mainly in older woodland (deciduous and coniferous) – an ancient woodland indicator species.
- *Limax maximus* – leopard slug (tiger slug, or great grey slug, or giant/spotted garden slug, or cellar slug) – L:100–150 mm; body pale brown to grey or brown usually with darker blotches or spots (including on the mantle), tentacles brownish, keel short; broadleaved woodland, scrub, hedges, parks and gardens.

– Body soft and relatively gelatinous and flaccid 13

13 (12) Keel paler than body *Lehmannia marginata*

- *L. marginata* – tree slug – L:60–90 mm; body pale grey to brown, often with a central pale stripe extending beyond the keel; especially found on tree trunks in woodland but also in more open rocky areas.

– Keel same colour as body *Ambigolimax* species

There are two species in this genus:

- *A. nyctelius* – Balkan three-band slug (or vine slug, or Bourguignat's slug, or striped garden/field slug) – L:50–80 mm; body pale grey to brown (often slightly pinkish), usually with distinct dark blotches, spots or bands; in disturbed habitats including gardens.
- *A. valentianus* – Iberian three-band slug (or greenhouse slug, or three-band garden slug, or Valencia slug, or Spanish slug) – L:50–80 mm; body pale grey to brown (often slightly pinkish), usually with faint dark blotches, spots or bands; mainly in disturbed habitats including gardens, also in woodlands under logs.

## Key E Snails

The sizes given here are for the shells of adult animals (height – H, and width – W). Individuals, especially juveniles, will be smaller than these measurements. Since some of the characteristics used in the identification of snails are only wholly developed in the adults, and others may be lost in weathered shells, it is only possible to be certain of the identification of specimens that are fresh and adult. For many species it is not necessary to have a living specimen and a fresh shell will often suffice. On occasions, however, diagnostic characteristics include body colour and pattern, especially in relation to the mantle, which will usually require the animal to be extended out of its shell. Shell colour can become bleached when older and especially in dead snails and it is important to recognise this when looking at the colour in any description (see Fig. 4.6). It is also worth noting that juvenile snails are smaller than the adults and frequently have much thinner and more transparent edges to the mouth of the shell. Note that there are some slugs with shells (albeit very small and situated at the tail end of the body) that will key out in Key D (couplet 1), and semi-slugs with larger shells (situated towards the middle of the body) that key out as snails in this key (couplet 7).

This key to snails will enable the identification of those families of snail found in the British Isles, following the

**Fig. 4.6** Weathering in common garden snail shells (*Cornu aspersum*)

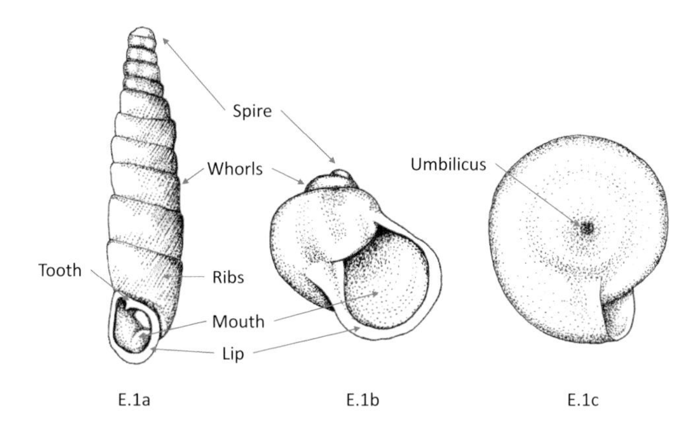


E.1a E.1b E.1c

checklist by Anderson & Rowson (2020). Where appropriate, genera or species likely to be found under logs and stones can also be identified. See E.1a–c for the general morphology of snails. The figures in Cameron (2008) and Naggs *et al.* (2014) may help in the confirmation of a species. Where a specimen does not closely match the description consult Cameron (2008), who also gives details of where dissection may be required. Further information and images of many British species can be found on the Conchological Society of Great Britain and Ireland, and MolluscIreland websites (see Chapter 6). Kerney (1999) provides distribution maps for many British species. See Section 3.5 for further information on snails.

| | | |
|---|---|---|
| 1 | Shell clearly taller than broad (E.1a) | 2 |
| – | Shell not taller, and often shorter, than broad (E.1b) | 7 |
| 2 (1) | Shell coiled to left (opening on observer's left when shell is upright and viewed from front) (E.2) | 3 |
| – | Shell coiled to right (opening on observer's right when shell is upright and viewed from front) (E.3) | 4 |

**coiling in snails**

In most species of snails the shells coil to the right (dextral), while in a few species belonging to the families Clausiliidae and Vertiginidae the shells coil to the left (sinistral). Rarely (less than 1:100,000), individuals can be found that have the opposite coiling to the norm in that species.

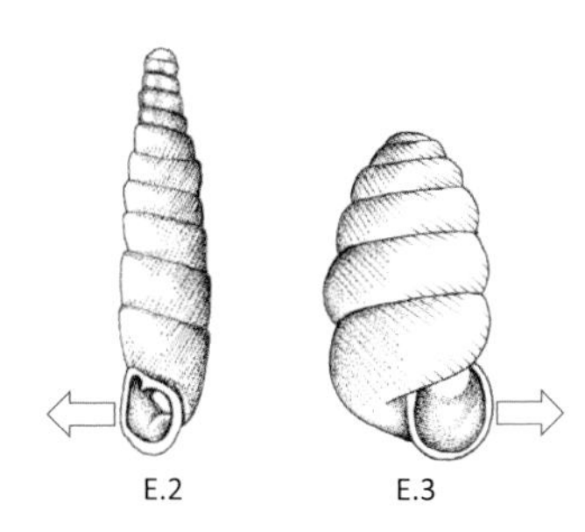

E.2 E.3

3 (2) Under 3 mm high, mouth with many teeth
Vertiginidae (whorl snails) (part)

Two fairly rare species of *Vertigo* key out here (see also couplet 4). The most common species under stones is:

- *V. pusilla* – wall (or wry-necked) whorl snail – H:2.0 mm, W:1.1 mm; shell pale yellowish brown and glossy with 6–9 teeth in the mouth of the shell that almost fill the aperture; sometimes found under stones.

– Over 3 mm high, mouth with few or no teeth
Clausiliidae (door snails)

Eight species from six genera, of which one is an introduced species, one is very rare and two species (*Balea* species – tree snails) have round mouths containing no teeth and are less often found at ground level. The most common species under logs and stones are:

- *Clausilia bidentata* – two-toothed door snail – H:9–12 mm, W:2.2–2.7 mm; shell dark reddish brown to blackish; damp areas in woods and among rocks, climbs up vertical surfaces in dry weather.
- *Clausilia dubia* – craven door snail – H:11–16 mm, W:2.7–3.2 mm; shell dark reddish brown to blackish; damp, shaded areas among rocks (rarer in woods) in ground litter but also climbs vertical surfaces.
- *Cochlodina laminata* – plaited door snail – H:15–17 mm, W:4 mm; shell reddish or yellowish brown (can be albino/unpigmented), glossy and translucent with faint growth lines across the whorls; found in woods and scrub among leaf litter and woody debris.
- *Macrogastra rolphii* – Rolph's door snail – H:11–14 mm, W:3.4–3.6 mm; shell reddish brown; found in deciduous woods in leaf litter and under logs and stones.

4 (2) Less than 3 mm high three families

Snails from three families key out here. These can be separated on the basis of their shell shape:

Aciculidae (needle snails) – shell tapers gently, no teeth in mouth, one species:

- *Acicula fusca* – point snail – H:2.2–2.5, W:0.8 mm; pale brown, shell very glossy and translucent; damp areas including leaf litter in woods.

Carychiidae (thorn snails) – shell strongly conical, three teeth in mouth, two species which may require dissection to confirm identification:

- *Carychium minimum* – short-toothed herald snail (or herald thorn snail, or minute sedge snail) – H:1.8–2.0 mm, W:1.0 mm, shell height usually less than twice the width; widespread, in wetlands and damp woodland.

- *C. tridentatum* – long-toothed herald snail (or dentate thorn snail, or slender herald snail) – H:2.0 mm, W:1.0 mm, height usually more than twice the width; widespread, in woods and hedges.

Vertiginidae (whorl snails) – (part, see also couplet 3) shell cylindrical or ovoid, 13 species from three genera key out here:

*Columella* – two species which have semi-cylindrical shells, no teeth at the mouth and a thin lip:

- *Columella aspera* – rough whorl – H:2.0–2.5 mm, W:1.3–1.4 mm; shell opaque, dark brown; widespread including dry woodland and poor grassland.
- *C. edentula* – toothless chrysalis snail (or toothless column snail) – H:2.5–3.0 mm, W:1.3–1.5 mm; shell translucent, pale yellowish brown; damp and calcareous habitats in marshes, woods and grassland.

*Truncatellina* – two species which have an almost cylindrical shell (0.9 mm wide) with no teeth at the mouth and a thickened lip, both of which are very rare and found in dry grassy habitats.

*Vertigo* – nine species (six of which are rare or very rare) which have ovoid (or slightly conical) shells, 1.0–1.5 mm wide, with teeth at the mouth (except in one of the very rare species) and a thickened lip; the three commoner species are:

- *Vertigo antivertigo* – marsh whorl snail – H:2.0–2.2 mm, W:1.2 mm; shell dark brown, quite smooth, mouth with 6–10 prominent teeth; under flood debris in wet areas.
- *V. pygmaea* – common whorl snail (or crested vertigo snail) – H:1.7–2.3 mm, W:1–1.12 mm; shell pale brown, mouth with 4–7 prominent teeth; dry calcareous grassland and wetlands.
- *V. substriata* – striated whorl snail – H:1.7–1.9 mm, W:1.1 mm, pale brown with regular striations, mouth with thin lip containing 5 or 6 teeth; damp habitats.

– Over 3 mm high 5

5 (4) Height of mouth less than half height of shell (at most 40% of height) 6

– Height of mouth nearly or more than half height of shell (more than 40% of height) three families

Three families key out here which are rarely found under logs and stones:

Ellobiidae (hollow-shelled snails) – three species from two genera; shell up to 12 mm high, with more than five whorls and mouth with teeth; found near the tidal strand line and the edge of saltmarshes.

Pomatiidae (round-mouthed snails) – one species – *Pomatias elegans* – round-mouthed snail – shell large (H:13–16 mm, W:9.0–11.5 mm) and thick with heavy spiralling and convex whorls; found on loose calcareous soils.

Succineidae (amber snails) – five species from four genera; shell with less than five whorls and mouth without teeth; found in wet places, e.g. on bare mud, floating vegetation or freshwater margins. *Oxyloma elegans* (Pfeiffer's amber snail – H:9–12 mm, occasionally up to 20 mm; shell with a rough appearance) and *Succinea putris* (large amber snail – H:10–17 mm, can be up to 24 mm; shell smooth) are the most common species.

6 (5) Shell glossy and transparent; more than 4 mm high — Cochlicopidae (pillar snails)

Three British species from two genera:

- *Azeca goodalli* – three-toothed moss snail – H:5.3–6.8 mm, W:2.4–2.7 mm; shell often reddish brown (can be albino/unpigmented), very smooth shell with very shallow indentations between whorls and mouth with many teeth; woods and scrub, often in rocky areas.
- *Cochlicopa* cf. *lubrica* – slippery moss snail (or glossy pillar snail) – H:5.0–7.5 mm, W:2.4–3.0 mm, shell pale to dark brown (may be white), deepish indentations between whorls, mouth with no teeth; damp woods, grasslands and marshes.
- *C.* cf. *lubricella* – least slippery moss snail (or lesser moss snail, or thin pillar snail) – H:4.5–7.0 mm, W:2.0–2.5 mm; shell paler and less glossy than *C. lubrica*, indentations between whorls shallower, more cylindrical, mouth with no teeth; drier habitats.

The two species of *Cochlicopa* are difficult to separate and there is some debate about the taxonomy of this genus.

The single British species from the family Ferussaciidae (ground snails) keys out here. This is a much smaller species, especially in terms of its width:

- *Cecilioides acicula* – blind snail (or blind agate snail, or blind awlsnail) – H:4.5–5.5 mm, W:1.2 mm; shell glossy and transparent (becomes white and opaque after death); subterranean sometimes found dead in debris and ants' nests.

– Shell opaque (of various colours) — five families

Five families key out here:

Chondrinidae (chrysalis snails) – one British species with at least seven whorls and mouth with at least seven teeth:

- *Abida secale* – large chrysalis snail (or juniper chrysalis snail) – H:6–8 mm (sometimes taller), W:2.2–2.6 mm; shell pale brown, mouth edge very thick and white; limestone rocks.

Cochlicellidae (pointed snails) – shell strongly conical with untoothed and unthickened mouth – two *Cochlicella* species (one a recent introduction) key out here, both are coastal species. The more common species is:

- *C. acuta* – pointed (or conical) snail – H:10–20 mm (or more), W:4–7 mm; shell white or pale brown often with darker bands or patches, very conical shell; found on sandy and calcareous soils at the coast.

Enidae (bulin snails) – two species – mouth toothless with white lip:

- *Ena montana* – mountain bulin – H:14–17 mm, W:6–7 mm; shell brown with 7–8 whorls and a granular appearance; ancient woodlands and hedges, uncommon, only in southern England.
- *Merdigera obscura* – lesser bulin – H:8.5–9 mm, W:3.7 mm; shell brown with up to 7 whorls which have a fine lined appearance; woods and hedges, widespread and common.

Lauriidae (chrysalis snails) – three species (one of which is very rare) from two genera. The two common species are:

- *Lauria cylindracea* – common chrysalis snail – H:3.0–4.4 mm, W:1.8 mm; shell pale brown, translucent and glossy; widespread, in woods, grassland and gardens, especially among rocks and on walls.
- *Leiostyla anglica* – English chrysalis snail – H:3.0–3.7 mm, W:1.9 mm; shell dark reddish brown, not glossy; uncommon, in damp woods and marshes.

Pupillidae (chrysalis snails) – one species:

- *Pupilla muscorum* – moss chrysalis snail – H:3–4 mm, W:1.7 mm; shell brown, quite smooth; dry calcareous habitats.

7 (1) Maximum of 3, quite loosely arranged whorls
Vitrinidae (semi-slugs)

Three genera containing one species each, of which one is rare, while another is very rare and is only found in Ireland. In semi-slugs, the body is large in comparison with the shell and is unable to fully retract into the shell. The most common species is:

- *Vitrina pellucida* – winter semi-slug (or pellucid (glass) snail, or western glass snail) – H:2.5–3.5 mm, W:4.5–6.0 mm; shell translucent and pale greenish with more than two whorls, upper surface of living animal pale; found in woods and grassland. Where the animal's body is dark and has black speckling on the mantle, this may be the rare *Phenacolimax major* (greater semi-slug, or greater pellucid glass snail) – H:4–6 mm, W:5–7 mm) which is mainly found in southern England in ancient woodland.

A recent introduction belonging to the family Oxychilidae (true glass snails – see also couplets 12 and 20) also has a shell that is too small for adults to retract into (here the shell is situated towards the tail of the animal, rather than being more central on the body as in the Vitrinidae): *Daudebardia rufa* – Alsatian semi-slug: L:17–20 mm; body grey above and whitish underneath; recorded from two woodlands in Wales.

– More than 3 whorls, which may be quite tightly arranged 8

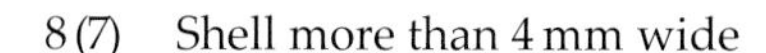

8 (7) Shell more than 4 mm wide 9

– Shell less than 4 mm wide and with at least 3.5 whorls 17

Note that any snails keying out here with fewer than 3.5 whorls are probably juveniles and are not covered by this key.

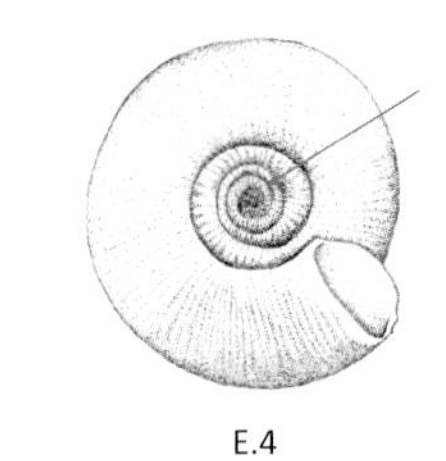

E.4

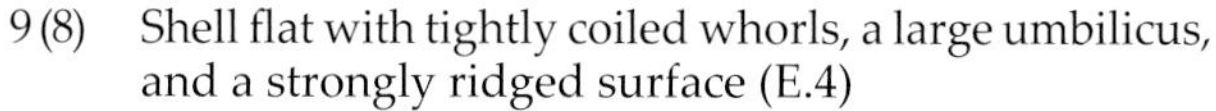

9 (8) Shell flat with tightly coiled whorls, a large umbilicus, and a strongly ridged surface (E.4) Discidae (disc snails)

There is one British species:

- *Discus rotundatus* – rounded snail (or radiated snail, or rotund disc) – H:2.5–3.0 mm, W:5.5–7 mm; shell pale (yellowish) brown (sometimes white) with darker radial stripes; mainly in woods (often coniferous) in leaf litter and rotting logs, also damp grassland and marshes.

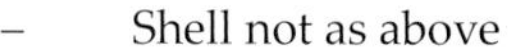

– Shell not as above 10

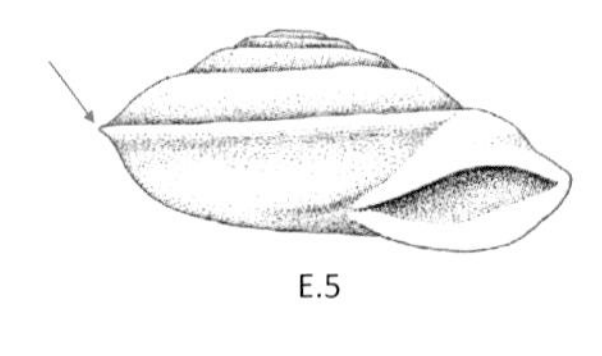

E.5

10 (9) Shell with ridge (keel) around widest part (E.5) two families

Two families key out here (at least in part):

Helicidae (helicid snails – part, see also couplets 13 and 15) – one species keys out here:

- *Helicigona lapicida* – lapidary snail – H:7–9 mm, W:12–20 mm; shell opaque, pale/medium brown, flattened with a large umbilicus; calcareous soils and rocks, also woodlands.

Hygromiidae (leaf snails – part, see also couplets 11, 12, 15 and 16) – two genera with one species each key out here, including one very rare species with a very pointed spire to the shell (*Trochoidea elegans* – top snail – H:5–8 mm, W:7–10 mm). The more common species is:

- *Hygromia cinctella* – girdled snail – H:6–7 mm, W:10–12 mm; shell yellowish white to pale brown, slightly translucent with a conical spire and small umbilicus; found in woods, an introduced species that is expanding its range.

– Shell without keel around widest part 11

11 (10) Shell flattened (not globular) two families

Two families key out here (at least in part):

Helicodontidae (cheese snails) – one species:

- *Helicodonta obvoluta* – cheese snail – H:5–7 mm, W:11–15 mm; shell brown with a sunken spire producing a slight dip in the middle; only in southern England, in woodlands associated with large rotting logs.

Hygromiidae (leaf snails – part, see also couplets 10, 12, 15 and 16) – one species keys out here:

- *Helicella itala* – heath snail (or common heath snail) – H:5–12 mm, W:9–25 mm; shell white or pale ginger, with low spire and a very large umbilicus; exposed habitats such as short calcareous grasslands and dunes.

– Shell globular (not flattened) 12

12 (11) Shell glossy, or waxy and transparent, or translucent three families

Snails from three families key out here:

Gastrodontidae (gloss snails) – three species from the genus *Zonitoides* key out here, which have shells with a slightly raised spire and a large umbilicus. One recent introduction is found in glasshouses and gardens (*Z. arboreus* – quick gloss snail – H:1.7–3.0 mm, W:4.5–6.0 mm; shell pale yellowish brown and translucent). The other two species are larger and more common:

- *Z. excavatus* – hollowed glass snail – H:2.8–3.4 mm, W:5.3–6.0 mm; shell pale brown and slightly translucent, umbilicus large and deep; under refuges in woodland and grassland, especially in areas with acidic soils.
- *Z. nitidus* – black gloss snail (or shiny glass snail) – H:3.5–4.0 mm, W:6–7 mm; shell pale brown, glossy and translucent with the very dark body showing through creating an almost black appearance, umbilicus large but shallow; common, but restricted to very wet areas (sometimes submerged).

Hygromiidae (leaf snails – part, see also couplets 10, 11, 15 and 16) – two species (one of which is rare) from two genera key out here, both having globular shells with convex whorls and moderate indentations between them. The more common species is:

- *Zenobiellina subrufescens* – brown snail (or dusky snail) – H:4–6 mm, W:6–10 mm; shell pale brown, thin and transparent; woodland and well vegetated riverbanks, especially in upland areas.

Oxychilidae (true glass snails – part, see also couplets 7 and 20) – seven species from three genera key out here. There are two *Aegopinella* species which have waxy shells with relatively loose whorls; one *Nesovitrea* species which has distinct radial grooving; and four *Oxychilus* species which have glossy shells with relatively tight whorls and can be difficult to separate. All species are common and widespread:

- *Aegopinella nitidula* – smooth glass snail (or waxy, or dull glass snail) – H:4 mm, W:7–10 mm; shell brown and waxy with 4.5 whorls and a large umbilicus; leaf litter in woods and hedges, often in disturbed habitats.
- *A. pura* – clear glass snail (or delicate glass snail) – H:2 mm, W:3.5–4.5 mm; shell white or colourless (can be brown) and translucent with 3.5 whorls and a large and deep umbilicus, under 20× magnification the shell shows a cross-hatched microsculpture with fine radial lines crossed by spiral lines; damp habitats such as leaf litter in deciduous woods.
- *Nesovitrea hammonis* – rayed glass snail – H:2 mm, W:3.5–4.2 mm; shell pale brown, glossy and transparent with a large umbilicus; wide range of habitats including marshes and both coniferous and deciduous woods including quite acidic sites.
- *Oxychilus alliarius* – garlic glass snail – H:2.5–3.5 mm, W:6–9 mm; shell pale yellowish brown or greenish (may have whitish streaks or be whitish underneath), glossy and translucent; wide range of habitats including quite acidic sites; usually smells strongly of garlic if disturbed.
- *O. cellarius* – cellar glass snail – H:5 mm, W:9–12 mm (occasionally up to 12 mm); shell very pale brown, glossy and transparent, very flattened; wide range of habitats including woods and gardens
- *O. draparnaudi* – Draparnaud's glass snail (or dark-bodied glass snail) – H:6–7 mm, W:11–16 mm; shell yellowish brown, opaque, with relatively large and deep umbilicus, very flattened; common in gardens, parks and woods, introduced.
- *O. navarricus* – glossy glass snail – H:4–5 mm, W:8–10 mm; shell brown and very glossy, edge of the mantle is jet-black; woods and hedgerows; may have faint smell of garlic if disturbed.

– Shell not glossy, or waxy and transparent, or translucent 13

13 (12) Shell more than 25 mm wide Helicidae (helicid snails) (part)

Three species of Helicidae from two genera key out here (see also couplets 10 and 15). One of these species is rare (and legally protected) and only found on chalk: *Helix pomatia* – apple (or Roman, or Burgundy, or edible) snail – H:30–50 mm, W:32–50 mm; shell creamy white often with pale brown bands, mouth with a thickened slightly pigmented rim. A relatively recent introduction from south-eastern Europe is becoming more common: *Helix lucorum* – Turkish snail – H:25–45 mm, W:30–60 mm; shell with reddish brown bands which often meld together obscuring much of the whitish background, mouth thickened with a reddish or brownish rim. The most common species is:

- *Cornu aspersum* – common garden snail (or brown garden snail) – H:25–35 mm (occasionally can be as small as 20 mm and as large as 40 mm), W:25–40 mm (occasionally can be as wide as 45 mm); shell pale brown or yellow, often with dark bands or blotches, mouth with a thickened white rim; often in parks and gardens as well as woods, rocks and dunes.

– Shell less than 25 mm wide 14

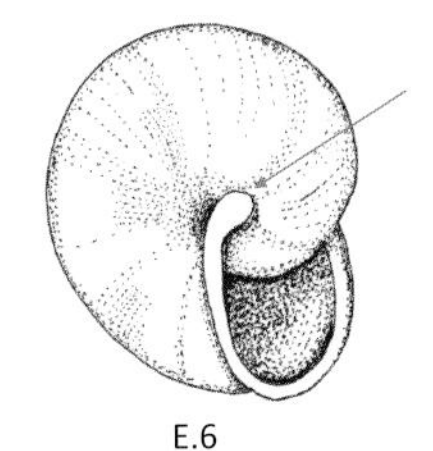

E.6

14 (13) Small umbilicus at least partially obscured by extended lip of mouth (E.6) 15

– Large umbilicus not obscured by lip (E.7), if umbilicus is narrow then it may be quite deep 16

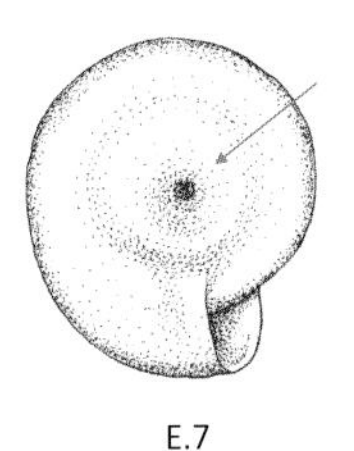

E.7

15 (14) Shell usually over 15 mm wide Helicidae (helicid snails) (part)

Four species from three genera of Helicidae key out here (see also couplets 10 and 13):

- *Arianta arbustorum* – copse snail (or orchard snail) – H:10–22 mm, W:14–28 mm; shell brown or yellow usually with lighter flecks (may have dark brown band at edge of whorls), mouth with well-defined white lip; widespread in damp places (grassland and woods).
- *Cepaea hortensis* – white-lipped snail (or garden banded snail) – H:10–17 mm, W:14–20 (occasionally up to 22 mm wide); shell brightly coloured (can be quite variable), may have dark bands, lip of mouth usually white (can be brown); widespread being found in woods and grassland, usually in damper areas than *C. nemoralis*. This species can usually be distinguished from *C. nemoralis* by being generally

smaller and by the lip colour; where brown-lipped individuals of *C. hortensis* occur, these are usually a minority among the population and of the same size (usually under 20 mm wide) as those with the more typical lip colour. When in doubt, dissection may be required to confirm identification.

- *Cepaea nemoralis* – brown-lipped snail (or grove snail) – H:12–22 mm (occasionally up to 28 mm), W:18–25 mm (occasionally up to 32 mm); shell brightly coloured (can be quite variable), may have dark bands, lip of mouth usually brown (can be white); widespread being found in woods and grassland. This species can usually be distinguished from *C. hortensis* by being generally larger and by the lip colour; where white-lipped individuals of *C. nemoralis* occur, these are usually a minority among the population and of the same size (usually over 20 mm wide) as those with the more typical lip colour. When in doubt, dissection may be required to confirm identification.
- *Theba pisana* – white snail (white garden snail) – H:9–20 mm, W:12–25 mm; shell white or ginger (occasionally pink) often with darker banding (note juveniles have a keel); exposed coastal sites, often on posts and plant stems in dry, hot weather.

– Shell usually less than 15 mm wide

Hygromiidae (leaf snails) (part)

Three species from three genera key out here including two introduced species, one of which is rare and declining (see also couplets 10, 11, 12 and 16). The more common species are:

- *Ashfordia granulata* – silky snail – H:5–7 mm, W:7–9 mm; shell whitish/pale brown, slightly glossy, thin and translucent, hairy (with long straight hairs), umbilicus very small (partially obscured by lip); damp woods, hedges and marshes.
- *Hygromia limbata* – hedge snail – H:9–11 mm, W:12–17 mm; shell creamy-white/yellow to dark brown (often with contrasting bands), slightly glossy and translucent, umbilicus very small and partially obscured by lip; woods, hedges, gardens, and roadsides, introduced to SW England, but expanding its range.

16 (14) Shell white or pale brown (often patterned)

Hygromiidae (leaf snails) (part)

Three genera of calcicolous species (found on neutral or calcareous soils) containing four species key out here; one of which is a very rare introduction (see also couplets 10, 11, 12 and 15). The more common species are:

- *Backeljaia gigaxii* – eccentric snail – H:4–8 mm, W:6–15 mm; shell quite flattened, white or pale brown (often with brown incomplete bands or blotches), opaque, with a fairly large umbilicus; dry open grassy habitats
- *Cernuella virgata* – striped snail – H:6–19 mm, W:8–25 mm; shell globular, white or ginger (often with dark bands or blotches), opaque, with narrow but deep umbilicus; fairly dry open habitats including coastal grassland, dunes and roadsides.
- *Xeroplexa intersecta* – wrinkled snail – H:5–8 mm, W:7–13 mm; shell quite flattened (although not as much as in *B. gigaxii*, white to ginger (often with dark bands or blotches), opaque, with fairly large umbilicus; dry and open habitats

– Shell not white or pale brown
Hygromiidae (leaf snails) (part)

Four species from three genera key out here, of which one is very rare (see also couplets 10, 11, 12 and 15). The more common species are:

- *Monacha cantiana* – Kentish snail – H:10–14 mm, W:15–20 mm; shell creamy white (may be darker near mouth), translucent, with very small umbilicus; disturbed habitats including road verges and open waste ground.
- *Trochulus hispidus* – hairy snail – H:5–6 mm, W:5–12 mm; shell cream to brown, slightly glossy and translucent, hairy when fresh (if hairs have rubbed off, some may persist in the umbilicus) with fairly large umbilicus; widespread in grasslands, woods and wetlands.
- *T. striolatus* – strawberry snail – H:6.5–10 mm, W:11–15 mm; shell creamy yellow to dark reddish brown, opaque or slightly translucent, with medium umbilicus; damp habitats in woods, hedgerows, gardens, verges and waste ground.

17 (8) Shell flattened 18

– Shell not flattened two families

Two families key out here (at least in part):

Euconulidae (hive snails) – two species from the genus *Euconulus*, which have translucent shells:

- *E. alderi* – shiny hive snail – H:2.0–2.5 mm, W:2.3–2.8 mm; shell brown, glossy and translucent, with very small umbilicus; in woods and especially widespread in wetlands. This species is subject to some debate regarding its status and a possible split into two separate species
- *E. fulvus* – tawny glass snail (or brown hive snail) – H:2.0–2.5 mm, W:3.0–3.5 mm; shell pale yellowish

brown, translucent, with very small umbilicus; widespread, usually in damp areas in woodlands, grasslands and marshes.

Valloniidae (grass snails – part, see also couplet 19): two species from two genera key out here:

- *Acanthinula aculeata* – prickly snail – H:1.8–2.0 mm; W:2.0–2.3 mm; shell brown with thin white lip, spiny on prominent ribs on sides of whorls, with small umbilicus; in leaf litter and under fallen wood in woodlands, hedgerows and scrub.
- *Spermodea lamellata* – plated snail – H:2 mm, W:2.0–2.3 mm; shell pale brown, with ribs (without spines) on sides of whorls, with small umbilicus; in leaf litter in woodlands, uncommon mainly in northern Britain.

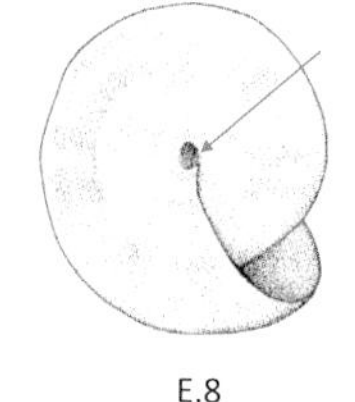

E.8

18 (17) Umbilicus wide (E.7) 19

– Umbilicus narrow (E.8) with lip extending to its edge Pristilomatidae (crystal snails)

Three species of the genus *Vitrea* (crystal snails) key out here:

- *V. contracta* – milky crystal snail – H:1.2–1.3 mm, W:2.0–2.5 mm; shell whitish, thin and translucent, with small, deep umbilicus; widespread especially in dry and calcareous habitats.
- *V. crystallina* – crystal snail – H:1.5–1.7 mm, W:3–4 mm (usually less than 3.5 mm); shell thin, glassy and transparent (sometimes slightly greenish), small umbilicus (narrower than in *V. contracta*); widespread in many habitats.
- *V. subrimata* – H:1.2 mm, W:2.3–3.0 mm (occasionally up to 3.5 mm); shell thin, glassy and transparent, with very small umbilicus (almost closed); calcareous habitats in the north.

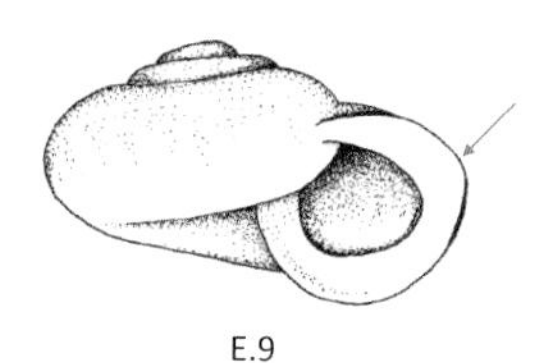

E.9

19 (18) Lip of mouth thickened (E.9) and white Valloniidae (grass snails) (part)

Three species of Valloniidae from the genus *Vallonia* key out here (see also couplet 17):

- *V. costata* – ribbed grass snail – H:1.2–1.3 mm, W:2.2–2.7 mm; shell white to grey, large umbilicus, and (in fresh shells) prominent ribbing on the whorls; calcareous grassland and sometimes in open woodland and scrub.
- *V.* cf. *excentrica* – eccentric grass snail – H:1.1 mm, W:2.0–2.3 mm; shell pale, glossy and translucent, large umbilicus; dry calcareous habitats including grassland and dunes.
- *V. pulchella* – smooth grass snail – H:1.2 mm, W:2.0–2.5 mm; shell white and translucent, large umbilicus; damp open areas. This species is very

similar to *V. excentrica* but is usually larger with more than 3 whorls, compared to *V. excentrica* which usually has up to 3 whorls.

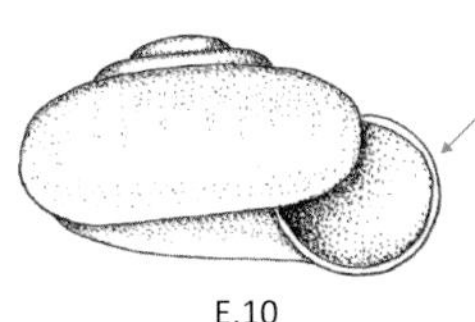

E.10

– Lip of mouth not thickened (E.10) 20

20 (19) Shell tightly coiled three families

Four species from four genera in three families key out here:

Helicodiscidae (mould snails) – one species which is subterranean and a rare introduction:

- *Lucilla singleyana* - smooth coil - H:1 mm, W:1.8–2.5 mm; shell glossy and translucent, yellowish brown, very flattened with large umbilicus; deeply subterranean, may be found in flood debris.

Punctidae (dot snails) – two species from two genera:

- *Paralaoma servilis* – pinhead spot snail – H:1 mm, W:1.5–2.0 mm; shell golden brown with regular ribbing on whorls and wide umbilicus; uncommon introduced species but expanding its range.
- *Punctum pygmaeum* – dwarf snail – H:0.6–0.9 mm, W:1.2–1.7 mm; shell light brown with large and deep umbilicus; widespread and common in leaf litter.

Pyramidulidae (pyramid snails) – one rare species:

- *Pyramidula umbilicata* – rock snail – H:1.5–2.0 mm, W:2.5–3.0 mm; shell dark reddish brown (can be paler – purple or grey – in older animals) with very wide and deep umbilicus; dry limestone rocks.

– Shell loosely coiled
Oxychilidae (true glass snails) (part)

Smaller individuals of two species from two genera may key out here – *A. pura* and *N. hammonis* – see couplet 12 for species descriptions.

## Key F Mites

With at least 2,100 species of mites recorded in the UK (Baker, 2002), and many more undoubtedly yet to be discovered, this key cannot be comprehensive. Mites are also generally very small and their features are difficult to see without a microscope and sometimes need the use of special techniques – see Shepherd & Crotty (2020) for further details. Because of their small size (often under 1 mm in length), lengths are not given here unless the species is particularly large. Where lengths are given (L) they are from the front of the head to the end of the abdomen, ignoring any appendages (e.g. palps and legs). The taxonomy is also quite complex and, because there are so many species, there are additional taxonomic levels that do not occur in some of the other groups covered by this book (see Table 4.1). The nymphs of mites (juvenile forms) may look different from the adults (often having only six legs, compared to adults with eight), which adds to the challenges of identification. This key, therefore, will help you classify a mite found under a log or stone to a taxonomic level that is variable depending on the ease of identification. It focuses on groups that can be relatively readily distinguished, generally on mature mites, and is intended to help you start to get experience of mite identification; the couplets may not include rare exceptions. See F.1 for the generalised morphology of adult mites: F.1a – underside of whole animal; F.1b – mouthparts; F.1c – leg. In general, the darker and harder (i.e. more sclerotised) an individual is, the more likely it is to be an adult. Nymphs are generally less well sclerotised, paler in colour and with underdeveloped genital shields. In some groups the nymphs look completely different from the adults (and may even be larger). For instance, one nymphal stage in the Astigmatina

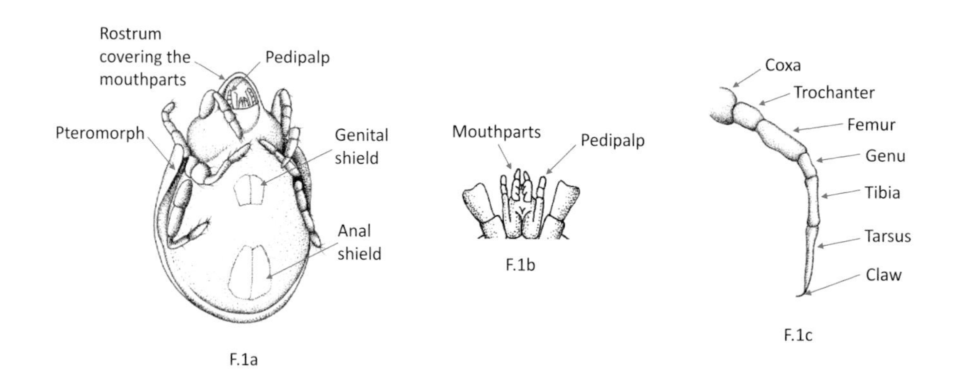


F.1a

F.1b

F.1c

(a dispersal stage called a hypopus – plural hypopi), have no mouthparts and have a plate of suckers at the rear which they use to attach themselves to other animals in order to hitch a ride. Some examples of common genera or species are given; these are indicative of the species that might be keyed out at each point but are not the only species possible, so should not be taken as a definite identification. For further detailed identification of soil mites see Shepherd & Crotty (2020) and a variety of web-based resources (see Chapter 6). See Section 3.5 for further information on mites.

1 Mouthparts a barbed prong projecting forwards from the head — ticks

Ticks (not covered by this key) are not commonly found under logs and stones, they are more likely to be found on vegetation or attached to mammals. For further information and identification see the Bristol University Tick ID webpages (see Chapter 6).

– Mouthparts not as above (see F.1b) — 2

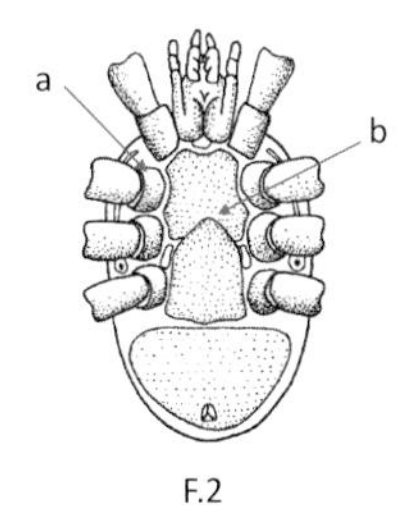

F.2

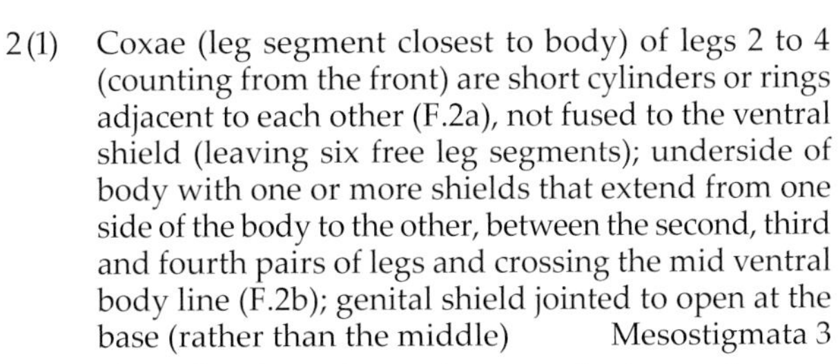

2 (1) Coxae (leg segment closest to body) of legs 2 to 4 (counting from the front) are short cylinders or rings adjacent to each other (F.2a), not fused to the ventral shield (leaving six free leg segments); underside of body with one or more shields that extend from one side of the body to the other, between the second, third and fourth pairs of legs and crossing the mid ventral body line (F.2b); genital shield jointed to open at the base (rather than the middle) — Mesostigmata 3

The order Mesostigmata has nymphs that can be very abundant in soil samples. Although these will key out here, it is not usually possible to identify them further as they can look different from the adults. They tend to be pale, with smaller shields that have extensive areas of soft integument between them. They also lack genital shields.

– Coxae (leg segment closest to body) not like short cylinders or rings (they can be longer tubes, bead-like, triangular or like other leg segments) and are fused to the ventral shield (leaving five free leg segments); underside either with no distinct ventral shield or with shields that have a break down the middle line (F.3); body may fold up to conceal legs — 4

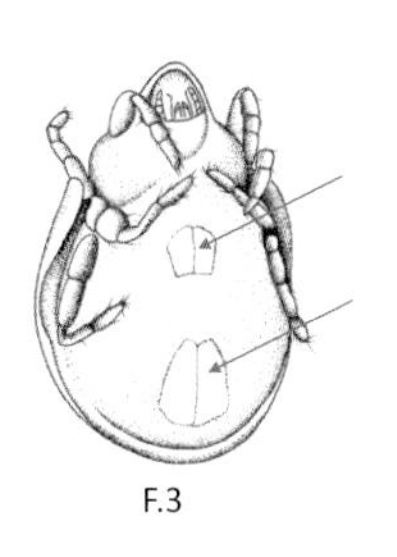
F.3

3 (2) Short legs that retract into pockets (pedofossae) under the body; mostly dorso-ventrally flattened (i.e. flattened top to bottom) and often round in shape — Uropodina (part)

Members of the cohort Uropodina include the tortoise (or turtle, or flying saucer) mites. Common examples include:

- *Cilliba cassidea* – tortoise mite – disc-shaped, with all the legs hidden under a continuous round dorsal shield that extends on both sides and round the posterior end; common but very hard to distinguish from others in the genus. Note that the Mesostigmata includes other Uropodina-like mites, although these are generally less perfectly round than *Cilliba*.

– Legs unable to retract into pockets; can be any shape (often tear-drop-shaped, but including a few that are dorso-ventrally flattened and with a round body shape) other Mesostigmata

A few more common or distinctive examples include:

- *Epicriopsis horrida* – yellow body with orange tubercles and long white hairs (like a hairy tangerine!); feeds on fungi.
- *Pergamasus* species – large (L: up to 1.5 mm), robust, chestnut brown and tear-drop-shaped with a triangular genital shield (males have swollen second pair of legs with spurs underneath to grip females during mating); very common in leaf litter. Common species are *P. crassipes* and *P. longicornis* but note that there are many similar species in the genus.
- *Sejus togatus* – brown oval-bodied mite with a pointed anterior end and blunt posterior end which has two pointed horn-like protuberances (one at each side) bearing setae, the pattern of the dorsal shields appears as an X, upper surface covered in bumps each of which has a seta arising from it.
- *Veigaia cerva* – body with a single yellow dorsal shield (with a curved incision in each side), mouthparts consist of very large pair of dark brown pincers; very common in leaf litter.

4(2) Whole animal like a hard seed or box, with the legs arranged as a bunch, and a hinge in the body allowing the legs to be folded away under the head so that they are completely enclosed and not visible Phthiracaroidea

The superfamily Phthiracaroidea (box mites) belongs to the suborder Oribatida (and are included in the 'lower' macropyline oribatids). Common examples include:

- *Steganacarus magnus* – with a knobbly body is one of at least 23 species that can fold their legs completely so that they are not visible; common in soil. Some species may be shiny and smooth, while others have a more matt surface and/or long setae.

– Not as above, without a hinge between the head and the body, unable to enclose legs under the head, although it may be able to fold the legs into the body or behind flaps 5

5 (4) Body without mouthparts; head reduced to a stump with four bristles; body flattened especially on the underside which bears a plate of suckers at the rear Astigmatina hypopi

These are a nymphal stage of the Astigmatina (suborder Oribatida), which do not feed and are adapted for dispersal by attaching themselves to insects and other arthropods.

– Body with mouthparts; without a rear ventral plate of suckers 6

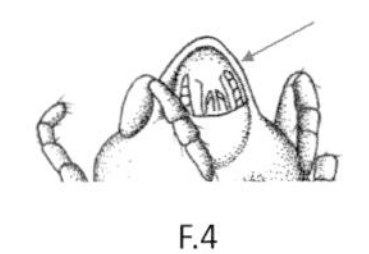

F.4

6 (5) Mouthparts (when seen from above) capable of being completely covered by a hood-like structure wrapping over the top and sides (F.4 – viewed from beneath); never with eyes; generally well sclerotised and yellowish brown to black/brown in colour (although some adults – see first part of couplet 7 – and many nymphs may not be) 7

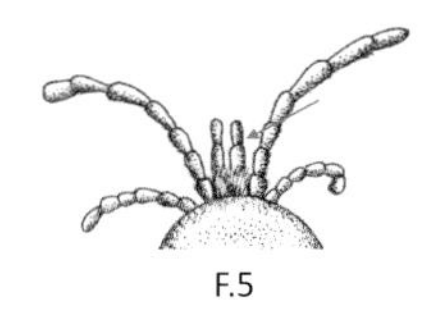

F.5

– Mouthparts not covered by hood-like structure and visible from above (F.5); may have eyes 10

7 (6) Disk shaped, with projecting broom-like hind legs covered with setae Prostigmata (part)

These are members of the family Scutacaridae, which contains many tiny (often whitish) species common under logs and stones.

– Body not disk shaped and without broom-like back legs Oribatida (part) 8

8 (7) Legs with distinct 'knees' – the genu (the third segment counting from the end of the leg – not including the claw – see F.1c) is shorter than the other leg segments; anal and genital shields (which are each made up of a single pair of plates) are separated by a large expanse of ventral shield (as in F.1a) Brachypylina 9

Members of this subcohort are sometimes termed 'higher' brachypyline oribatids.

– Without distinct 'knees', most leg segments of similar length; anal and genital shields often with extra plates (i.e. more than two each), and abut each other, either with no gap between them or at most only a small gap (i.e. not separated by an expanse of ventral shield) Macropylina

Members of this cohort are sometimes termed 'lower' macropyline oribatids. An example is:

- *Platynothrus peltifer* – leathery looking, chestnut brown body with two strong ridges running longitudinally down its back flanked by white hairs, appears quite angular, and often has fragments of soil on its back; very common.

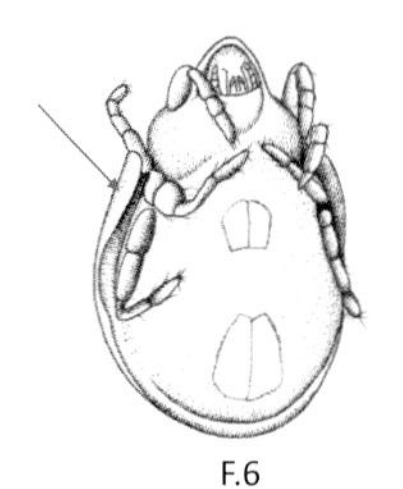

F.6

9 (8) With pteromorphs (wing-like or flap-like structures at the side the body – F.6) behind which the legs may sometimes be hidden — Poronoticae

Despite the apparently distinctive pteromorphs in this subcohort of the Oribatida, many species are difficult to separate. Examples include:

- *Euzetes globulus* – very dark brown/black, round bodied and domed with a tiny head and a glossy appearance; common in soil and leaf litter. This is perhaps the most photographed of any UK species. There are other similar species, such as *Achipteria nitens* which is black, with long hairs, bigger pteromorphs with spikes, and is a little less globular. Members of the family Galumnidae are similar to the previous two species, but have hinged ear-like pteromorphs that project forwards.

– Without pteromorphs — Pycnonoticae

One example of this family is:

- *Damaeus onustus* – L:1.3 mm; hard round, black body with long knobbly legs, cast skins from earlier moults may be stuck to its back; common in leaf litter. There are several similar species.

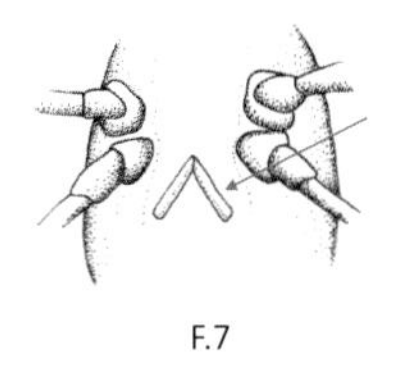

F.7

10 (6) Body often narrowed at the front; generally with single claws on legs, often white with legs usually bunched together and in the shape of tapering tubes, V-shaped genital shield (F.7), rarely have eyes — Astigmatina

Members of the cohort Astigmatina belong to the order Sarcoptiformes, and are generally small and white. One common example is:

- *Tyrophagus putrescentiae* – mould mite – pear-shaped and translucent with numerous short, thick setae on the upper surface and 2–4 pairs of longer, whip-like setae at the rear end; found in a wide range of habitats associated with fungi, including often in soil.

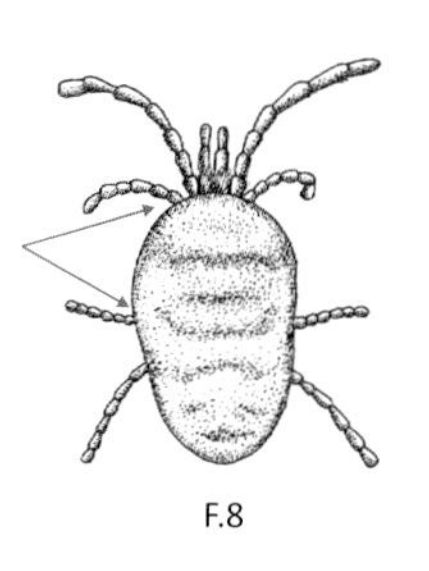

F.8

– Body with broad 'shoulders'; generally with a pair of claws on the legs, sometimes with a brush between them; often colours other than white; usually with a wide gap between legs 2 and 3 (F.8) and often with constrictions at the joints; genital shields usually a divided oval shape; often have eyes — Prostigmata (part) 11

This is a suborder of the order Trombidiformes. The suborder Endeostigmata (order Sarcoptiformes) will also key out here. These are generally tiny, white or greenish (occasionally brownish red), parallel sided or globular and may have a single median eye.

11 (10) Body furry all over Parasitengonina (velvet mites) 12

– May have hairs but not furry all over other Parasitengonina

Common examples include:

- *Anystis baccarum* – the money spinner (or whirligig) mite – body red, head narrow with wider body posteriorly, all legs bunched together (i.e. coxae adjacent to each other), runs very fast on flat surfaces such as stones, paving slabs or tree bark; common under stones. One of five similar species.
- *Labidostoma luteum* – bright yellow body, with ventral shields anteriorly forming a neat regular block of eight and a large pustule on each side of the body above leg pair three and behind a lateral eye.
- *Linopodes motatorius* – brown with paler legs (the first pair of which are extremely long, and which it waves about looking like antennae); common on rotting wood and under stones.
- *Penthaleus major* – dark body with red legs, dorsal red spot (the anus) on abdomen; common under rocks and in moss.
- *Riccardoella oudemansi* – the slug mite – white fast-moving mites found on slugs where they feed on the secretions of the slug and sometimes its blood; common.
- Members of the family Bdellidae (snout mites) have obvious erect, slender and elbowed pedipalps on each side of the head.

12 (11) At least one pair of legs longer than the body; most have one pair of eyes but some may have two pairs Erythraeoidea

One example of this superfamily of velvet mites is:

- *Leptus trimaculatus* – black body with three white spots, red legs.

– All legs shorter than the body; always with two pairs of eyes, some of which may be on stalks Trombidioidea

One example of this superfamily of velvet mites is:

- *Trombidium holosericeum* – very large for a mite (L: up to 4 mm), bright red and furry velvet mite, strawberry-shaped with wide shoulders and eyes on stalks; common under stones. There are several other similar species that are difficult to separate.

## Key G Harvestmen

There have been several new species of harvestmen added to the British list in recent years, which are not included in the latest synopsis of the group (Hillyard, 2005), although some are covered by Richards (2010). A number of species of harvestmen are arboreal and hence unlikely to be found under logs and stones, while others may mostly or occasionally frequent the ground surface and shelter beneath refuges by day. The current key should allow the identification of adult specimens to family and in many cases species. In juveniles, body parts are still growing and so can cause confusion in identification. Separating adults from juveniles can be tricky, involving close examination of whether the front edge of the genital operculum (the genital cover on the underside between the bases of the legs) is free (in adults) or still fused (in juveniles). See G.1 for the general morphology of harvestmen. The sizes given are body lengths (L), not including legs and palps. Where a specimen does not match the description given, consult Hillyard (2005), Wijnhoven (2009), Richards (2010), and the interactive key by Richards & Burkmar (2017). Further descriptions and images can be found on the harvestmen pages of the British Arachnological Society website (see Chapter 6). A checklist and further information on many of the species (especially recent introductions) can be found in Davidson (2019). Distribution maps can be found in Hillyard (2005) and on the Spider and Harvestman Recording Scheme website (see Chapter 6). See Section 3.5 for further information on harvestmen.

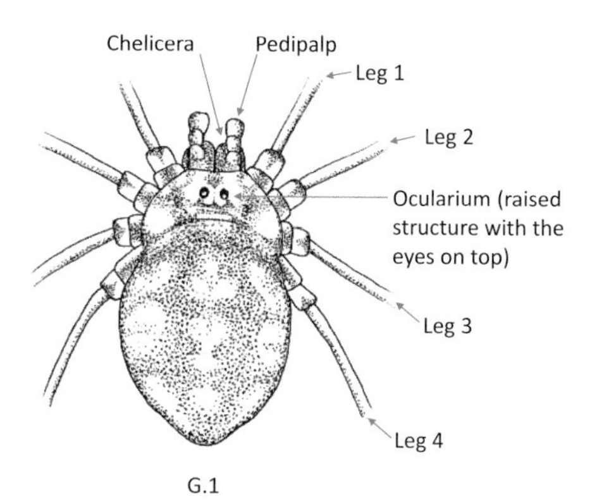


G.1

1 Front of cephalothorax (combined head and thorax) extended forming a hood covering the palps and chelicerae which are not visible from above (this may be difficult to see – look from above); no ocularium; body very flattened; whole animal often covered in soil particles Trogulidae

There are two genera each with one species; both are specialist snail predators:

- *Anelasmocephalus cambridgei* – L:2.8–4.0 mm; body somewhat flattened, grey-brown (juveniles purplish and often not covered in soil); across England (up to southern Cumbria) and Wales, in leaf litter and under logs and stones in calcareous habitats.
- *Trogulus tricarinatus* – L:5–10 mm; body highly flattened, sandy-brown (juveniles purplish and often not covered in soil); especially in the south of England and Wales (but has been recorded from Cumbria), in leaf litter and under logs and stones in calcareous habitats. These two species can be separated on size and on the number of tarsal segments on the first two pairs of legs (three in *A. cambridgei* and two in *T. tricarinatus*).

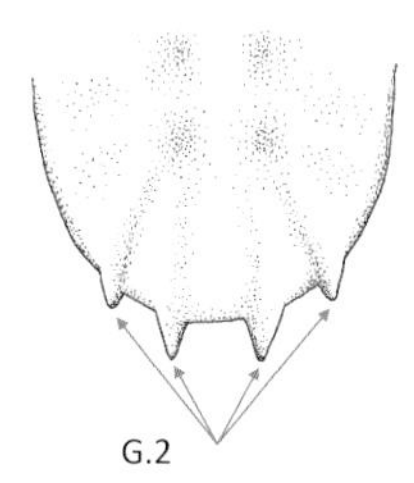
G.2

– Front of cephalothorax not extended into a hood; eyes on a raised ocularium; body may be flattened; whole animal not usually covered in particles of dirt (see couplet 2 for an exception) 2

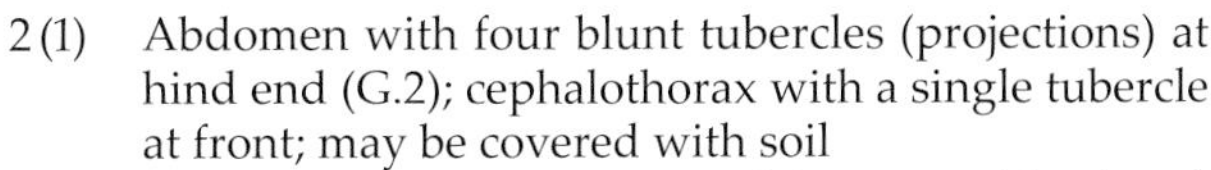

2 (1) Abdomen with four blunt tubercles (projections) at hind end (G.2); cephalothorax with a single tubercle at front; may be covered with soil Sclerosomatidae (part)

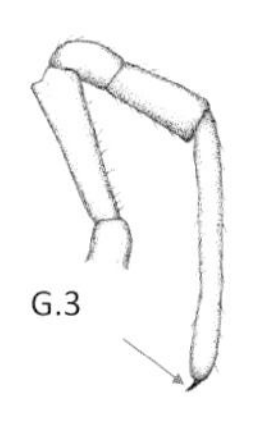
G.3

A single species keys out here (see also couplets 4 and 5):

- *Homalenotus quadridentatus* – L:3.5–5 mm; body sandy brown with many spiky tubercles on the top of the legs and around the front of the body, with rows of pale blunt tubercles surrounded by squarish dark marks running down the abdomen; in leaf litter and under logs and stones in open woods and calcareous grassland, south from Lincolnshire and into Wales.

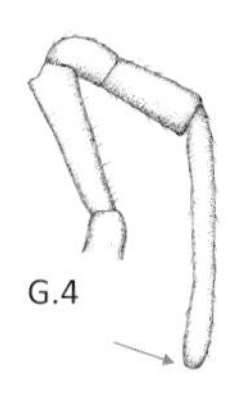
G.4

– Abdomen without such projections at the hind end 3

3 (2) Tarsus of pedipalp with a claw at the end (G.3) 4

– Tarsus of pedipalp without a claw at the end (G.4) 9

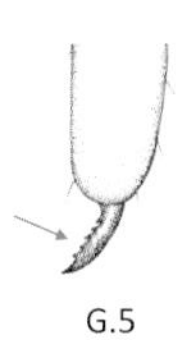
G.5

4 (3) Claw at end of the pedipalp toothed (G.5); body smooth, lacking tubercles Sclerosomatidae (part) 5

Nine species of Sclerosomatidae key out here (see also couplet 2). These species were previously included within the family Leiobunidae but have been moved following a recent revision of the group. They have particularly long legs and are more often found climbing on structures above ground such as on large logs, rather than under logs and stones, although juveniles may be found in the ground layer, and adults hide in cavities during the day.

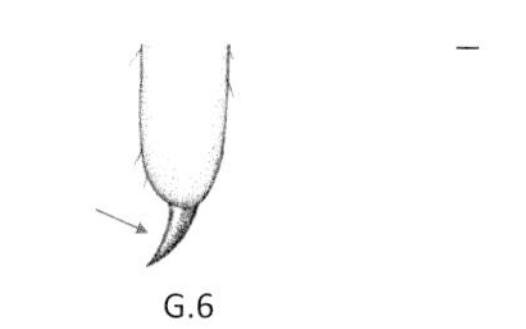
G.6

– Claw at end of pedipalp not toothed (G.6); body not smooth, tubercles present Phalangiidae 6

Fifteen species of Phalangiidae key out here.

A rare member of another family (Phalangodidae) will also key out here. This species is very small and mite-like and highly restricted in distribution:

- *Scotolemon doriae* – L:1.5–1.7 mm; body brownish orange-yellow, pale tarsi, very small eyes, rather mite-like in appearance; currently restricted to Devon and Guernsey, subterranean and found in soil under stones.

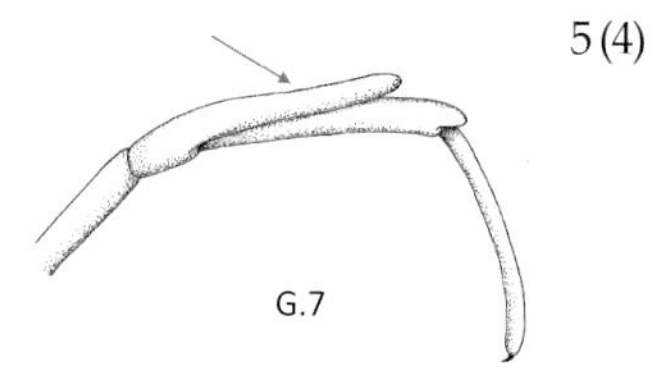
G.7

5 (4) Pedipalp with a long projection at patella (joint between the femur and tibia) (G.7) which is at least half the length of the tibia (best seen from above as the palps are held out to the front) *Dicranopalpus* species

There are three species, all of which are introductions to Britain. The two common species were once thought to be a single species and are difficult to separate:

- *D. ramosus* – L:3–6 mm; bodies differ between sexes – the smaller males (L:3–4 mm) are orange-brown, grey, yellow or silver with a black saddle, and may have a black stripe across the eyes, while females (L:6 mm) are silver-grey or brown with brown or orange patterning; an established introduction, which has extended its range across Britain, found in vegetation including bushes and trees in parks and gardens. The similar species *D. caudatus* is smaller (L:2–3 mm) and males usually lack the dark eye stripe. *D. larvatus*, the most recent introduction, is also smaller than *D. ramosus* (L:2.6–3 mm) and the males have a cream band in the middle of the body between the darker rear end and the brown area around the eyes; the species is restricted to the south of England, the Isle of Wight, Guernsey and the Scilly Isles.

– Pedipalp without a long projection at the patella two genera

There are five *Leiobunum* species, of which three are relatively recent introductions and have rather local distributions. These species often aggregate under over-hanging structures including under rocks. They can exhibit bobbing behaviour when disturbed to distract predators. There is also one *Nelima* species which is a well-established introduction. The three most common species are:

- *L. blackwalli* – L:2.6–6 mm; bodies differ between the sexes – the smaller males (L:2.6–3.6 mm) are pale orange-brown, while the females (L:4–6 mm) are pale silvery-brown with a darker saddle, in both sexes the eyes are surrounded by pale rings with a dark area between them; widespread, often in low vegetation and on tree trunks, but can be surface active.
- *L. rotundum* – L:3–6.5 mm; bodies differ between sexes – the smaller males (L:3–4 mm) are orange or reddish brown with contrasting dark legs, while the females (L:4.5–6.5 mm) are brown with gold or silver marks and often a dark saddle, in both sexes the eyes are surrounded by dark rings with a white area between them; widespread, often in vegetation or on walls, but can be surface active.
- *N. gothica* – L:2.5–4.5 mm; body pale brown with silvery patches and dark spots; widespread but uncommon, ground active in low vegetation and under logs and stones (often in cavities in walls during the day), including in grassland, dunes, and parks and gardens.

The three rarer introductions can be distinguished as follows:

- *L. gracile* – L:6.5 mm; bodies differ between the sexes – males are silvery-white towards the front then dark brown with a dark metallic blue saddle, while females are mottled cream and brown with a dark saddle with metallic bands; currently restricted to northern Scotland, in human-disturbed habitats, usually on walls but has been found under logs and stones.
- *L. limbatum* – L:7.4 mm; bodies differ between the sexes – males are orange-brown with a darker edge and slightly darker indistinct saddle, females are paler and have dark spots on the cephalothorax and grey bands across the abdomen; currently restricted to Lancashire and is not yet confirmed as an established British species.
- *L.* species A (as yet unnamed/unidentified) – L:9 mm; bodies differ between the sexes – males are metallic green (pale yellowish to light brown underneath), while females are paler with metallic gold and whitish marks and a dark metallic green saddle, both sexes have very long legs; the species appears to be expanding its range, currently east Midlands and south Yorkshire with some records from elsewhere in southern England, in urban brownfield sites (sometimes forms very large aggregations in porches and roof spaces).

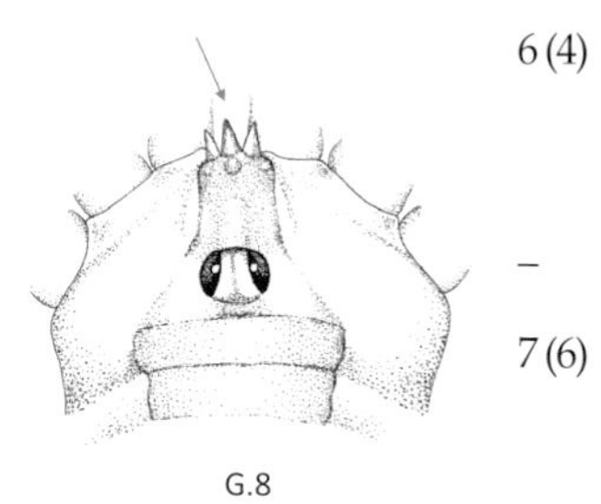
G.8

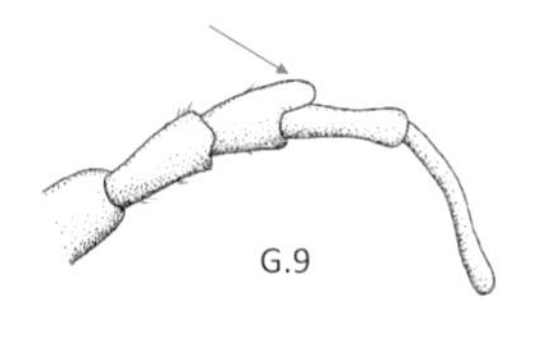
G.9

6 (4) Front edge of cephalothorax with a 'trident' formed from an array of (usually) three (sometimes more) distinct, conical, pointed tubercles (projections) in the centre line (G.8) 7

– Front edge of cephalothorax without such a trident 8

7 (6) Patella of pedipalp with a short extension along the tibia (G.9) *Lophopilio palpinalis*

- *Lophopilio palpinalis* – L:2.8–5.0 mm; body yellowish or reddish brown usually with a distinct saddle and pairs of dark patches towards the rear of the abdomen, pedipalps with very obvious downwardly and inwardly pointing spine-tipped tubercles and the patella with a short extension along the tibia; widespread, in the ground layer of woods, hedgerows and gardens.

– Patella of pedipalp without a short extension along the tibia four genera

There are four genera, containing six species:

- *Lacinius ephippiatus* – L:3.5–5.5 mm; body pale yellowish brown or grey with a darker saddle, femur of pedipalp with large tubercles, the three trident tubercles are of equal length, with extra tubercles behind them; widespread and common, in the ground layer of woods, marshes and meadows.
- *Odiellus spinosus* – L:6.5–11.0 mm; body yellowish grey to greyish brown with a distinctive saddle outlined in black, the trident has three large forward facing tubercles (view from the side), femur of pedipalp with large ventral tubercle; an introduced species that is spreading northwards, found in Wales and England mainly from Yorkshire southwards, on the ground and in vegetation in human-disturbed habitats such as parks, gardens and brownfield sites.
- *Oligolophus hanseni* – L:3.3–5.0 mm; body dark, greyish brown with an indistinct saddle, area around the eyes dark, 'trident' usually appears to have five conspicuous tubercles; common, found in woods, usually up in trees.
- *O. tridens* – L:3.5–5.5 mm; body yellowish brown with lighter spots and a darker saddle, area around the eyes pale, trident fairly vertical; very common and widespread, surface active in woods, hedgerows and gardens.
- *Paroligolophus agrestis* – L:3–5 mm; body silvery-grey to pinkish brown with a pale central stripe, may have a brown saddle, genital operculum (covering) underneath body between the leg bases with a distinctive notch at the front end (the two sides forming an almost circular hole in females – G.10a – and a very slight notch in males – G.10b); very

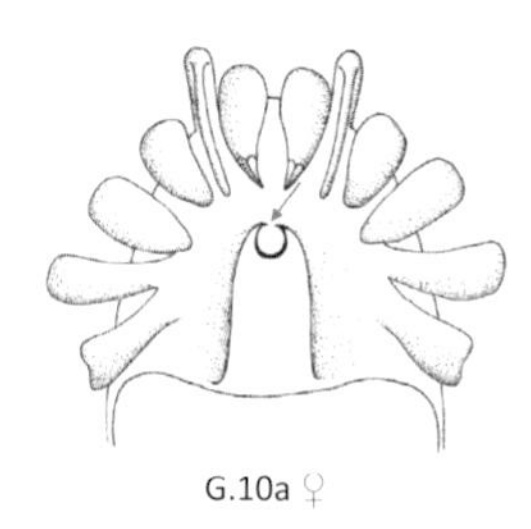
G.10a ♀

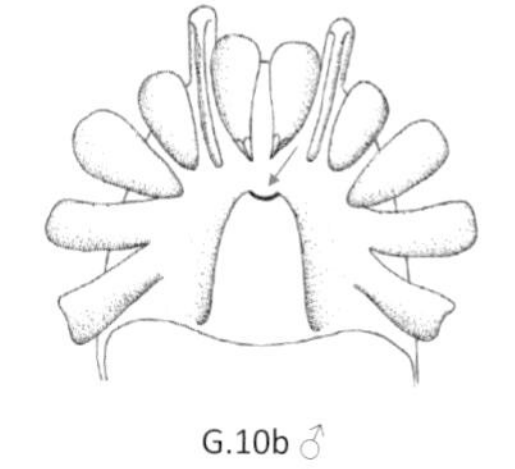
G.10b ♂

common and widespread in vegetation including trees and on the ground, wide range of habitats including woods, grassland, hedgerows, heathland, dunes, parks and gardens.

- *P. meadii* – L:2.3–3.8 mm; body straw coloured or greyish with rows of white-tipped, stout, pointed tubercles running across the abdomen, trident with the central tubercle longer (×2) than the outside pair; not common, found in the ground layer of dry habitats including heaths, downs and dunes.

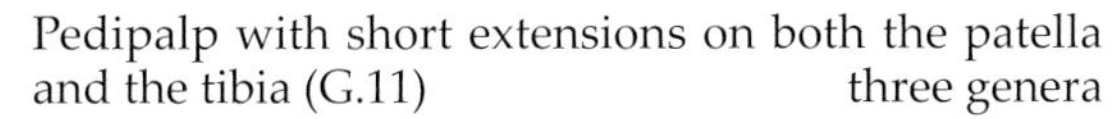

8 (6) Pedipalp with short extensions on both the patella and the tibia (G.11) three genera

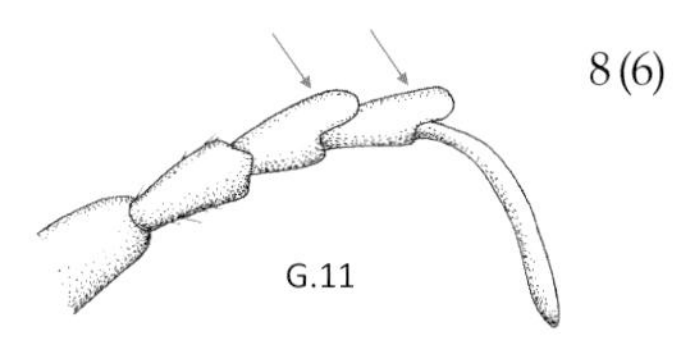

G.11

There are three genera, each containing one species:

- *Megabunus diadema* – L:2.6–4.8 mm; body silvery-white with darker mottling, very distinctive ocularium with a 'crown' consisting of two rows of 5–6 long spines (G.12 – note that *diadema* is Latin for crown), pedipalp with short extensions on both the patella and tibia; widespread especially in wetter upland areas, on the ground, under logs and stones, and in vegetation, in woods, heaths and moors.

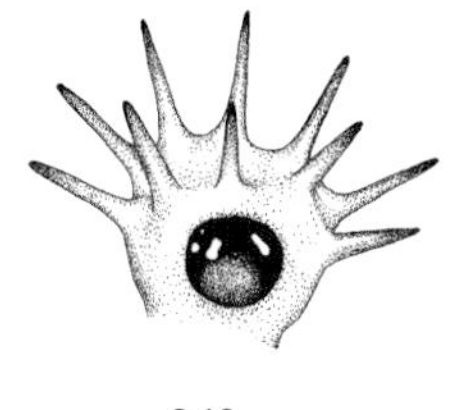

G.12

- *Platybunus pinetorum* – L:5–9 mm; bodies differ between the sexes with smaller males (L:5–6 mm) being brown or black, while females (L:7–9 mm) are pale grey to dark brown with a white-edged dark saddle, the ocularium is noticeably wide with a deep groove between the eyes, pedipalp with short extensions on both the patella and tibia; a recent introduction which is spreading rapidly, found in England (Northamptonshire, Derbyshire and Yorkshire) and more widely in Scotland, usually above ground level in woods and grassland, also human-disturbed sites such as quarries and gardens.
- *Rilaena triangularis* – L:3.7–7.0 mm; body pale orange-brown with whitish spots, saddle is pale (darker towards the rear) or almost absent, pedipalp with short extensions on both the patella and tibia, the male has a distinctive horn on the chelicerae just above the pincers; widespread and common, on the ground and in vegetation, in woods and grasslands.

– Patella of pedipalp without short extensions on the patella or tibia three genera

There are three genera, containing five species:

- *Mitopus morio* – L:4.0–8.5 mm; body pale brown-grey with rows of dark spots across the abdomen, often with a dark saddle (in females the saddle may be edged with black and white); widespread and common, in woods, grasslands, heaths, moors and hedgerows, in all levels including low vegetation. This is a very variable species in both size and colour.

- *Opilio canestrinii* – L:6–9 mm; body orange-red or orange-brown with orange bases to the legs, females with two rows of black and white bars along the abdomen, a relatively recently introduced species which has spread rapidly across Britain, on walls in urban habitats such as brownfield sites and parks and gardens.
- *O. parietinus* – L:5–9 mm; body grey or brown with an indistinct saddle (may be missing in males) and a pale central stripe along the abdomen; widespread especially in England and Wales, commonly associated with urban areas and human-disturbed habitats, often above ground level on walls and fences.
- *O. saxatilis* – L:3.2–6.0 mm; body yellowish grey with dark brown mottling, with a pale central stripe along the abdomen; widespread, often in ground and low vegetation layers of open woods, grasslands, heaths, dunes and gardens. This species may be separated from *O. parietinus* by being smaller and having shorter legs, and by the number of tubercles in each row on the ocularium: 3–4 per row in *O. saxatilis* and 5–7 in *O. parietinus*.
- *Phalangium opilio* – L:4–9 mm; body pale grey or yellowish brown with a saddle which may be indistinct (especially in males), there is a pair of tubercles above the chelicerae and just below where a trident would be, adult males have very distinctive upwardly-pointing horns on the base of the chelicerae (away from the pincers) which are very variable in size; widespread and common across Britain, in low vegetation as well as higher on trees and walls, in open sites such as woods, grasslands, hedgerows, dunes, brownfield sites and gardens.

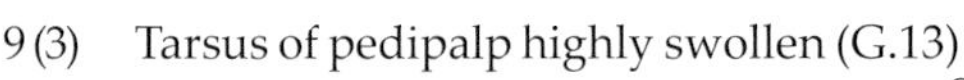

9 (3) Tarsus of pedipalp highly swollen (G.13) Sabaconidae

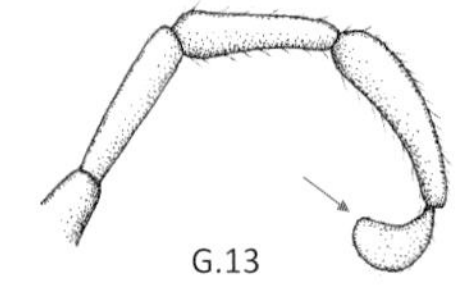

G.13

There is one established introduced species:

- *Sabacon viscayanum ramblaianus* – L:2.5–4.6 mm; body grey-brown with some paler areas, quite spiny, the pedipalps are longer than the body; in leaf litter and under decaying logs in damp woods, originally restricted to a few sites in south Wales, now expanding through Wales and south-west England.

– Tarsus of pedipalp slender (not swollen) (G.14) Nemastomatidae

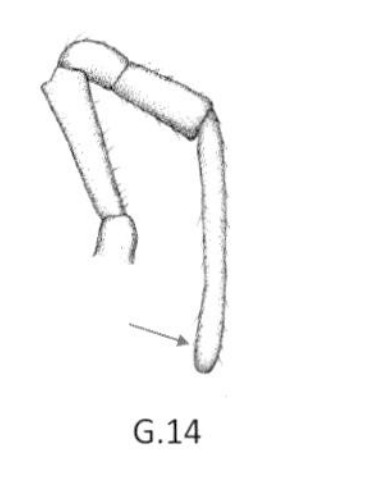

G.14

There are four genera, each with one species (two of which are relatively recent introductions with very restricted distributions). The common species are:

- *Mitostoma chrysomelas* – L:1.5–2.5 mm; body yellowish brown, legs brown or pale brown, the long pedipalps are held high up and forwards; common and widespread, in leaf litter and under logs and stones across Britain. One recent introduction (*Nemastomella bacillifera*) is also brownish (although darker in background colour) with paler markings (although these include both golden and whitish marks), with four pairs of long, robust spines along the back and up to 10 shorter spines across the end of the abdomen, the legs have swollen patellae (joints between the femur and tibia); restricted to Devon.

- *Nemastoma bimaculatum* – two-spotted harvestman – L:2.0–2.8 mm; body black (or dark brown) with two squarish, silvery white patches towards the back of the cephalothorax (entirely black specimens have also been recorded), legs dark with paler joints; very common and widespread (possibly the most common species found under logs and stones), in leaf litter and under logs and stones in a range of different habitats, including upland areas. One recent introduction (*Histricostoma argenteolunulatum*) is also dark (although with a bronze tint) with two rows of five spines along the abdomen, and pale spots but has two pairs of marks (the second, smaller pair, towards the end of the abdomen); restricted to Kent.

## Key H Pseudoscorpions

This key enables the identification to family and, for common pseudoscorpions found under logs and stones, to genus and species. There are two species that are particularly common, with the number of records an order of magnitude larger than any other British species. These are *Chthonius ischnocheles* (couplet 3) and *Neobisium carcinoides* (couplet 4). Many pseudoscorpions can be found in leaf litter and on the underside of turned-over logs and stones. Although some species are more usually encountered under the bark of standing trees, their phoretic behaviour may mean that following an association with the flies upon which they hitch a ride, even some of these may be found at ground level. Pseudoscorpions are small (under 5 mm in length) and so are often overlooked; lengths (L) given are for adult specimens. This key identifies adult specimens that can be distinguished by the presence of reproductive structures on the underside of the animal just behind the last pair of legs (for an example, see H.1a). Colour is not usually a useful characteristic in separating species of pseudoscorpions since most are brownish or sometimes greenish yellow. Where there are useful differences, this is mentioned in the key. See H.1b for the general morphology of pseudoscorpions. Where descriptions do not fully cover individuals, consult Legg & Jones (1988), or Legg & Farr-Cox (2016). The former has more detailed species descriptions but is somewhat out of date, while the latter includes recently discovered species, as well as images of almost all of the British species; a new rare species, which was not included, was found in 2017. The

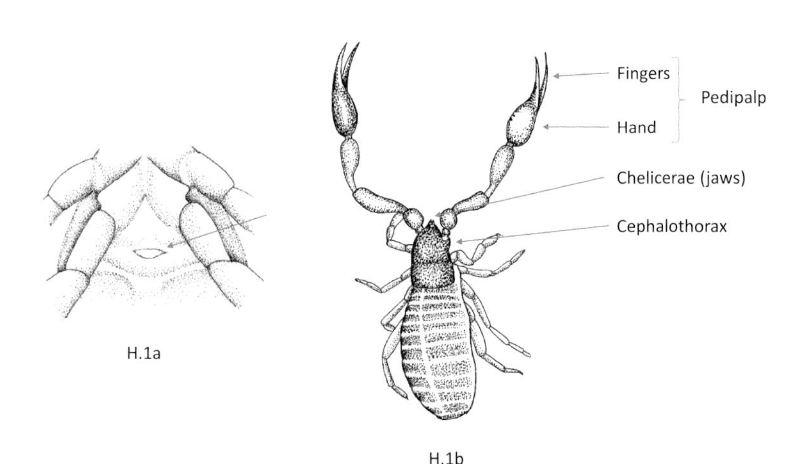


H.1a

H.1b

checklist is maintained on the Pseudoscorpion Recorders Group section of the British Arachnological Society website (see Chapter 6). Descriptions and distribution maps of each species are also given at this site; the latest published atlas is Jones (1980). Legg (2019) provides an annotated checklist and gives further consideration to the more recent introductions (his website also provides futher details on the British species – see Chapter 6). See Section 3.5 for further information on pseudoscorpions.

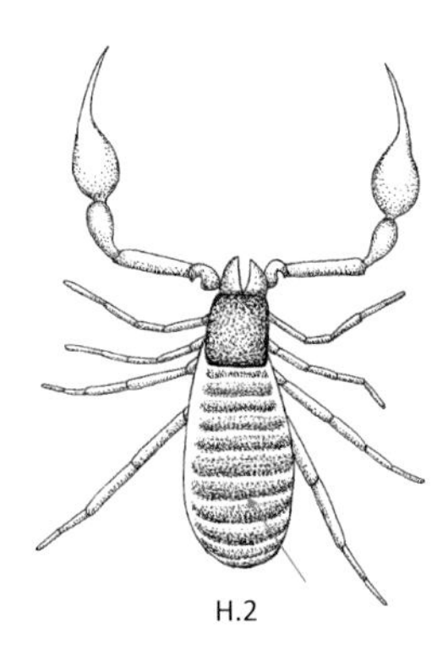

H.2

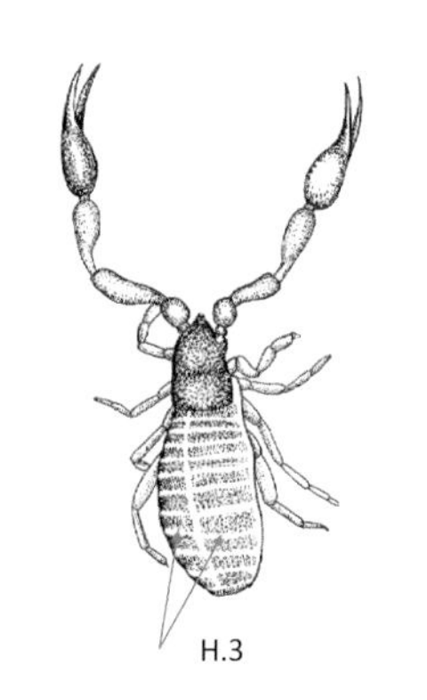

H.3

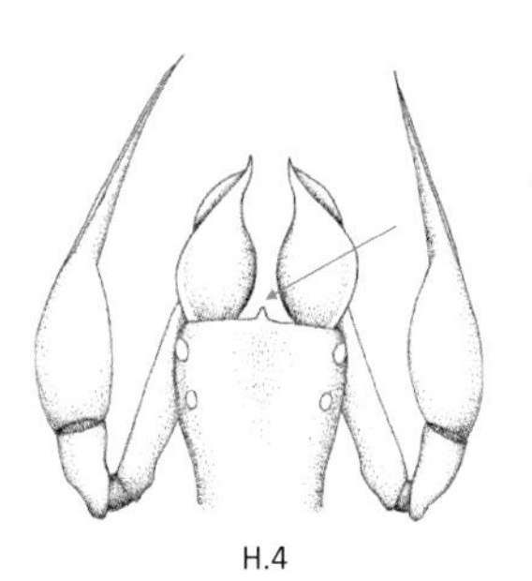

H.4

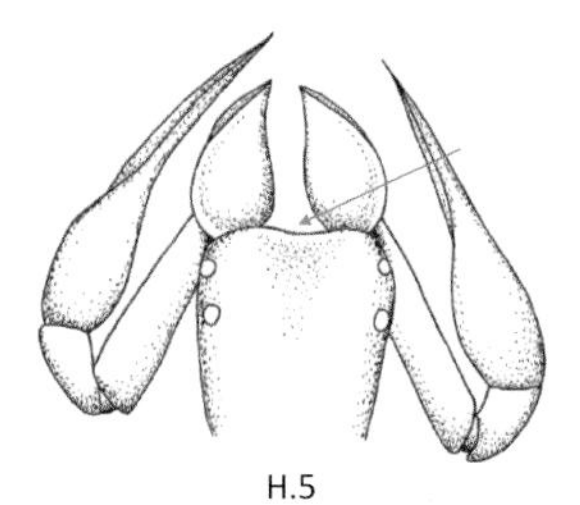

H.5

| | | |
|---|---|---|
| 1 | Single plates on back of abdomen (H.2) | 2 |
| – | Double plates on back of abdomen (H.3) | 5 |

2 (1) Large jaws, chelicerae (jaws – first pair of mouthparts) more than two-thirds the length of the cephalothorax (combined head and thorax) — Chthoniidae 3

Two genera containing six species of which two are very uncommon.

– Small jaws, chelicerae less than two-thirds the length of the cephalothorax — Neobisiidae 4

3 (2) Epistome (mid-point of front of cephalothorax) pointed (H.4) — two species

Of these two species, one (*Chthonius halberti* – Halbert's chthoniid) is very rare. The (very) common species is:

- *Chthonius ischnocheles* – common chthoniid – L:1.6 mm (males), L:2.5 mm (females); two pairs of large eyes, fingers of pedipalp nearly twice as long as hand; widespread and very common under stones and debris.

– Epistome absent (mid-point of cephalothorax not pointed) (H.5) — two genera

These two genera contain four species, of which one (*Ephippiochthonius kewi* – Kew's chthoniid) is uncommon and coastal (found under debris and stones on the upper shore). The more common species are:

- *Chthonius orthodactylus* – straight-clawed chthoniid – L:1.9 mm (males), L:3.0 mm (females); two pairs of large eyes, and fingers of pedipalp just over one and a half times as long as the hand; locally abundant in south-east England and south Wales, often in tussocky grassland rather than under logs and stones. This species can be difficult to separate from *Ephippiochthonius tetrachelatus*, although its size and habitat can be helpful.
- *C. tenuis* – dark-clawed chthoniid – L:1.3 mm (males), L:2.3 mm (females); hand of the pedipalps dark olive-green contrasting with the pale pink legs, two pairs of large eyes, and fingers of pedipalp just

over 1.5 times as long as the hand; widespread in southern England and Wales, often under stones on well drained habitats.

- *Ephippiochthonius tetrachelatus* – dimpled-clawed chthoniid – L:1.3–1.9 mm; hand of pedipalp reddish, two pairs of large eyes, and fingers of pedipalp only slightly longer than the hand (which has a distinct depression on the upper surface); common (more so than *Chthonius orthodactylus*), found under stones near the coast and in human-disturbed habitats including gardens and waste ground.

4 (2) Two pairs of eyes — two genera

These two genera contain five species, of which two (*Neobisium carpenteri* – Carpenter's neobisiid – and the recently discovered *N. simile* – quarry neobisiid), are rare and may have been confused in the past, while a third (*Microbisium brevifemoratum* – bog neobisiid) is currently known from only two localities (both sphagnum bogs). The more common species are:

- *Neobisium carcinoides* – moss neobisiid – L:2.2–3.0 mm; two pairs of eyes, epistome (mid-point of front of cephalothorax) indistinct (e.g. H.5), and fingers of pedipalps longer than hand; widespread and very common, found under stones and in moss in woods, grasslands, moorland, heathland, dunes, saltmarshes and the seashore.
- *N. maritimum* – shore neobisiid – L:3.2 mm; two pairs of large eyes, epistome prominent (e.g. H.4), and fingers of pedipalp nearly as long as hand; seashore under stones and in rock crevices often in the splash zone.

– One pair of eyes — two genera

These two genera contain one species each, which are difficult to separate on appearance, although they are very different in size when adult and one has a local distribution:

- *Roncocreagris cambridgei* – Cambridge's two-eyed neobisiid (chelifer) – L:1.2–2.0 mm; locally distributed in the south-west of England, in deciduous litter, and under stones at the coast.
- *Roncus lubricus* – reddish two-eyed neobisiid (chelifer) – L:2.5–3.0 mm; in moss and under stones and leaf litter in woodlands.

5 (1) Eyes distinct — four families

These four families contain five species, two of which are found mainly in buildings, another is a rare introduction associated with stored products, while yet another is a very rare species recorded only once from a nest in an old tree. The most common species under refuges belongs to the family Cheliferidae, and is:

- *Dactylochelifer latreillii* – marram grass chelifer – L:2.3–3.1 mm; cephalothorax and pedipalps chocolate-brown, abdomen brown; east and south-east coast of England, found especially among dune grasses and under seashore debris.

– Eyes absent (or indistinct) — Chernetidae

There are seven genera containing eleven species, of which one is an unestablished introduction. Several of these species are difficult to identify and separate, and only a few are likely to occur under logs and stones, these are:

- *Chernes cimicoides* – common tree chernes – L:2.2–2.7 mm; cephalothorax and abdomen dull and brown; widespread in England and southern Wales, the commonest species under the bark of deciduous trees and so might be found under the bark of fallen wood.
- *Lamprochernes* species – shining claw pseudoscorpions (two very similar species) – L:1.5–2.2 mm; cephalothorax and abdomen glossy; in human-disturbed sites and rotten wood.
- *Pselaphochernes dubius* – small chernes – L:1.5–1.7 mm; cephalothorax and abdomen dull and green-brown; widespread, under stones and woodland litter often in calcareous sites.

## Key I Spiders

It is often impossible to identify spiders to species without examining the reproductive organs under a microscope. Therefore, the current key enables identification of spider families, together with some genera and species commonly found under logs and stones. Adult spiders are easier to identify, even to family, than are juveniles, where key characters may often be soft and pale and difficult to distinguish. Mature male spiders can be recognised by the expanded terminal segments to the palps which show complex structures projecting from them (in contrast to immature males where the segment is smooth). Mature females have an obvious genital structure (the epigyne) on the underside at the front of the abdomen; this is absent in immature females. See I.1 for the general morphology of spiders (I.1a – full body; I.1b – leg). Lengths (L) are given from the front of the head to the tip of the abdomen (not including any legs, palps or mouthparts). Genera and species are given here as examples only and should not be taken as confirmed identification without consulting specialist works such as Locket & Millidge (1951, 1953), Locket *et al.* (1974), and Roberts (1993). Distribution data are given in Harvey *et al.* (2002a, 2002b) and Bee *et al.* (2017). The latter also provides information on the majority of British species as well as images of most species. Further images can be found in Jones (1983), Roberts (1995) and the British Arachnological Society website (see Chapter 6). A checklist (Lavery, 2019) is also available on this website. See Section 3.5 for further information on spiders.

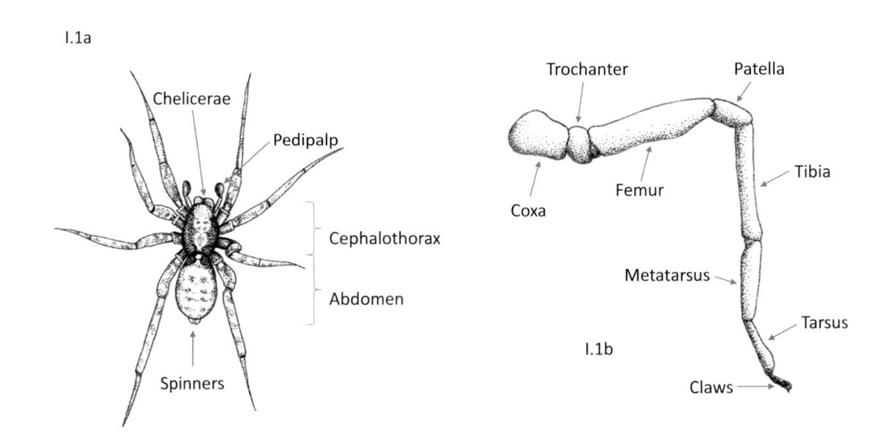

1 With 6 eyes 2

– With 8 eyes (most spiders have 8 eyes but some may be small or hard to see) 5

2 (1) Less than 3 mm long; pink or peach coloured all over Oonopidae

Oonopidae – goblin spiders – the single genus is *Oonops* containing two species, one of which is found indoors. The more common species is:

- *O. pulcher* – L:1–2 mm; pale to dark pink, widespread under logs and stones.

– More than 3 mm long and not pink 3

3 (2) Eyes arranged in a tight circle Dysderidae

Dysderidae – woodlouse spiders – there are two British genera and four species in this family, including one that is very rare (recorded from a single site in Essex):

- *Dysdera crocata* – woodlouse spider – L:9–15 mm; red-brown cephalothorax and legs, creamy abdomen, large prominent chelicerae (prey on woodlice); under refuges in human-disturbed sites such as gardens.
- *D. erythrina* – lesser woodlouse spider – L:7–10 mm; red-brown cephalothorax and legs, creamy abdomen, large prominent chelicerae (prey on woodlice), unlike *D. crocata* does not have spines on the upper surface of the femora; widespread although somewhat local in distribution, more common than *D. crocata* at the coast and in moorland.
- *Harpactea hombergi* – L:5–7 mm; dark brown cephalothorax, grey abdomen, chelicerae less prominent than *Dysdera* species, body tubular in shape; widespread and common, under detritus at the base of hedges and trees.

– Eyes not arranged in a circle (arranged in three pairs) 4

4 (3) Abdomen with dark band along the middle made up of diamonds (may be indistinct in one species which has iridescent green chelicerae) Segestriidae

Segestriidae – tubeweb spiders – there are three species, one of which is a coastal species (from south-west England and Wales), while another is found in the south of England. The most common species is:

- *Segestria senoculata* – snake-back spider – L:6–10 mm; brown cephalothorax with darker head end, pale brownish grey abdomen with darker central band of diamonds; widespread and common, found on brick walls more usually than under logs or stones, although can also be found in holes in dead wood, usually within a tubular web made in a hole, with radiating strands at the entrance.

– Abdomen not patterned as above — Scytodidae

Spitting spiders include a single UK species, found in buildings (*Scytodes thoracica* – spitting spider – L:3–6 mm; cephalothorax and abdomen pale yellow with black markings; uncommon, found in the south of Britain).

5 (1) Eight eyes in a cluster — 6

– Eight eyes in two or three rows (if any eyes are in a cluster then these are arranged on the top of a raised part of the cephalothorax and separated from the other eyes) — 7

6 (5) Huge chelicerae projecting forward (about as long as the cephalothorax) — Atypidae

Atypidae – purseweb spiders – one species:

- *Atypus affinis* – purseweb spider – L:7–15 mm; shiny, greenish brown cephalothorax, brown abdomen; scattered distribution from the south of England to southern Scotland (mainly southern England), usually within its sock-like web which can be covered in debris, often found on sunny, short-grazed grassland.

– Small chelicerae with short fangs projecting inwards; the middle pair of eyes towards the front are larger than the others — Zodariidae

Zodariidae – ant-hunting spiders (or ant spiders) – four species, of which three are very rare:

- *Zodarion italicum* – L:1.6–4.3 mm; brown cephalothorax with black striations, shiny black abdomen with pale yellow sides; restricted to south-east England, found with ants in grassland and waste ground, often in a retreat made of debris.

Three other families key out here: one (Pholcidae – cellar spiders – two species) lives in buildings, the other two are rare (Oecobiidae – discweb spiders – one introduced species restricted to buildings; Mysmenidae – dwarf cobweb spiders – one recently introduced species found under rocks only from two sites in Wales).

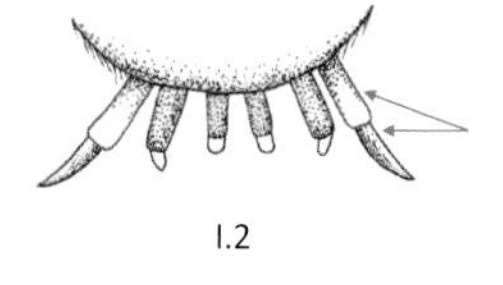

I.2

7 (5) Uppermost or outermost pair of spinners longest and clearly of at least two segments – seen most clearly from below (I.2) — 8

– All spinners about the same length, with only one segment (I.3) — 9

I.3

8 (7) Spinners arranged in a row, the end ones made up of two segments (I.2) Hahniidae (part)

Hahniidae – lesser cobweb spiders – five genera with ten species (see also couplet 16), one of which is found in wetlands and pond margins, and three of which are rare. Many of these species are difficult to separate. Their webs form small tangled sheets, with no retreats, across depressions and between stones. The more common species are:

- *Hahnia helveola* – L:2.2–3.0 mm; pale yellow cephalothorax (sometimes with a darker central patch with lines radiating from it), yellowish abdomen with broad dark chevrons; widespread and common, in woods and sometimes grassland, heathland and moorland.
- *H. nava* – L:1.5–2.0 mm; dark brownish black cephalothorax, black hairy abdomen; widespread, especially in the south of England, in grassland and under stones in open habitats.
- *Iberina montana* – L:1.5–2.0 mm; brownish grey cephalothorax with darker central patch with lines radiating from it, dark brown to black abdomen (which may have faint pale chevrons); widespread and common, damp woods, occasionally in grassland, heathland, wetlands and sand dunes.

– Spinners not in a row Agelenidae

Agelenidae – funnelweb spiders – five genera and 14 species, of which, two are rare and five are usually restricted to buildings. Species that are more common under logs and stones are:

- *Agelena labyrinthica* – labyrinth spider – L:8–12 mm; pale margins to the cephalothorax with two dark brown bands either side of a pale central band, dark grey abdomen with a paler central band and white chevron markings; widespread, especially in southern England and Wales, often found at the entrance to a funnel web with a large tubular retreat low down in the vegetation.
- *Coelotes* species make tubular burrows with a collar of silk around the entrance. Females may guard white disc-shaped egg-sacs within the burrow. There are two species, the most widespread of which is *Coelotes atropos* – L:7–13 mm; dark reddish brown shiny cephalothorax with a darker head end, greyish brown abdomen with a darker central stripe which is broader towards the front and darker mottling on either side; widespread and common in Wales and western England, under logs in woodland, moorland and heathland (often in a web retreat). *C. terrestris* (L:7–13 mm) is found in similar habitats (although mainly in south-east England) and has

a darker abdomen than *C. atropus* and without as clear a central stripe.

- *Eratigena* and *Tegenaria* species – house spiders – L:5–20 mm; brown centre and margins to the cephalothorax with two darker wavy bands between them, pale brown abdomen with darker mottling (creating pale chevrons); widespread, may build a web on vegetation with the retreat hidden in crevices or under stones (the spider often sits in a retreat at the back of a sheet of web shaped like a funnel), found in woods, on waste ground and brownfield sites. The two commonest species found under refuges may be separated on size and habitat, with *E. agrestis* (hobo spider – L:7–15 mm) being found in scrubby grasslands in human-disturbed areas such as wasteland, while the smaller *T. silvestris* (wood house spider – L:5–7 mm) is more often encountered in woods. *Eratigena* and *Tegenaria* species can be found under logs and stones, often with silk or egg cases nearby. The egg-sacs are spherical with a rough surface; in *E. agrestis* this is due to the inclusion of debris under the outer layer of silk, while other species have prey items attached to the outside of the egg-sac.
- *Textrix denticulata* – toothed weaver – L:6–7 mm; dark brown to nearly black cephalothorax with a pale central band, dark grey to black abdomen with a reddish central band bounded by a pale semicircular mark towards the front and two lines of pale spots towards the rear; widespread if scattered distribution, with webs containing a tubular retreat among stones, walls and low vegetation.

I.4

9 (7) Middle pair of eyes in front row extremely large (often described as being like headlights – I.4), can be seen from the front or (as here) when viewed from above **Salticidae**

Salticidae – jumping spiders – 21 genera and 41 species, of which several are quite rare and others are more often associated with cracks in walls than being under refuges on the ground. Some, however, are found under and among debris, including on shingle shorelines. More common species found under logs and stones are:

- *Euophrys frontalis* – L:2–5 mm; brown cephalothorax with darker head area (males have orange hairs around the front eyes and white hairs around the palps), cream abdomen with a pattern of black spots; widespread, grassland and heathland.
- *Talavera aequipes* – L:2–3 mm; body dark brown to black covered with pale yellow hairs, legs pale yellow with black rings; scattered throughout England

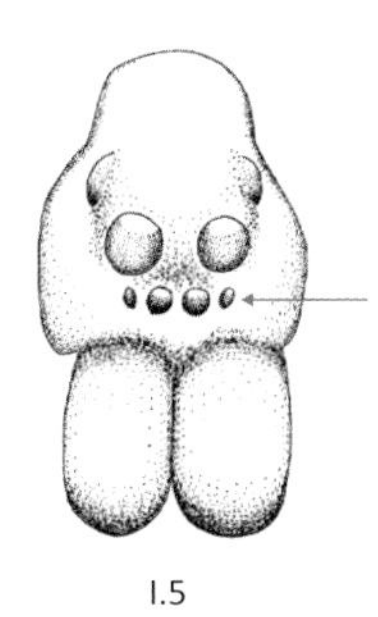

I.5

especially in the south, in open stony sites such as quarries.

– All eyes in front row small and more or less the same size (I.5), easiest to see from the front 10

10 (9) Eyes in three distinct rows (e.g. I.6 – as seen from above, or I.7 – as seen from the front) six families

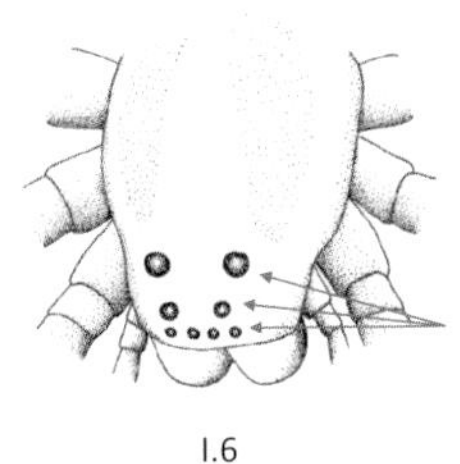

I.6

Of these six families, two (each containing one species) are rare (Eresidae – velvet spiders, which includes the ladybird spider *Eresus sandaliatus* – and Oxyopidae – lynx spiders) and one (also with a single species) is restricted to buildings (Zoropsidae – false wolf spiders). The more common families are:

Lycosidae – wolf spiders – eight genera and 38 species, of which 24 are rare, scarce or uncommon. All Lycosidae have a very obvious eye pattern with the rear (larger) four eyes being positioned in a square (I.6). Note that the front row of eyes is not easily seen from above (in contrast to *Zora* species). Lycosidae are not often found under refuges, being ground-running species that are more often found on or near to logs and stones. Females may be seen carrying an egg-sac attached to the spinners. Two genera with widespread common species are more often found in coastal habitats (*Arctosa*) or wet areas (*Pirata* – pirate wolf spiders). Three common genera found under logs and stones in drier habitats are:

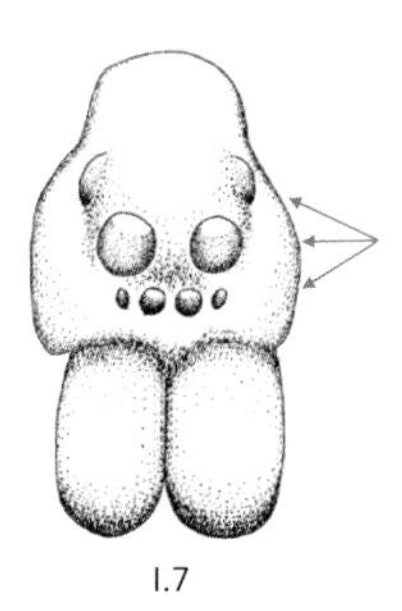

I.7

*Alopecosa* with four species, only one of which is common:

- *A. pulverulenta* – L:5–10 mm; dark brown cephalothorax with a pale central band which extends onto the reddish brown abdomen; widespread, in open areas such as grassland, heathland and gardens.

*Pardosa* with 15 species, of which eight are rare, scarce or uncommon – L:3.5–9 mm; stripe down centre of body (often with stronger colour contrast than *Trochosa*), hind legs longest; in a wide range of habitats. *P. amentata*, *P. monticola* and *P. palustris* are widespread and especially common on stony ground with sparse vegetation. *Pardosa* species can be difficult to separate.

*Trochosa* with four species of which two are rare. The two more common species are:

- *T. ruricola* – rustic wolf spider – L:7–14 mm; dark brown cephalothorax with pale central band containing darker marks, grey-brown abdomen with paler central stripe near the front, male darker than the female; widespread and common, under refuges in damp habitats including grassland and marshland.

- *T. terricola* – ground wolf spider – L:7–14 mm; dark brown cephalothorax with reddish brown bands in the centre and along the sides, reddish brown abdomen with some (indistinct) marking; widespread and common, under logs and stones in woodland, grassland, heathland and urban waste ground.

Pisauridae – nurseryweb spiders – two genera and three species, of which two are semi-aquatic. The more terrestrial species is:

- *Pisaura mirabilis* – nurseryweb spider – L:12–15 mm; greyish brown with striking cream band, bordered with brown, pale yellowish grey abdomen with a wavy-edged central band; widespread and common, in open woodland, grassland and heathland, more often seen on the ground or on low vegetation rather than under refuges. This species is very obvious and frequently noticed, especially if carrying a large egg-sac held below the body or when a female is guarding her spiderlings within a pyramidal-shaped web in grass or similar vegetation.

Miturgidae – ghost spiders – one genus and four species, of which two are rare and one has a very local distribution (see also couplet 17). The most common is:

- *Zora spinimana* – L:4.5–6.5 mm; pale brown cephalothorax with a pale central band bordered with darker brown, pale brown abdomen with scattered darker markings and a paler central band; widespread and common, mainly grassland but also hedges and open woods. The eye pattern of *Zora* species resembles that in I.7, but the front row of eyes can be seen clearly when viewed from above (in contrast to Lycosidae).

– Eyes in two rows or with a very different arrangement 11

Note that two families that key out in the first part of this couplet may be confusing here, with eyes in two or three rows depending on the angle from which it is viewed – one (Zoropsidae – false wolf spiders) is only found in buildings, the other Miturgidae (ghost spiders) is described in the first half of this couplet (view from above to more easily see the three rows of eyes and to distinguish these from Lycosidae).

11 (10) Eyes approximately equal in size 12

– Eyes not approximately equal in size (the back middle two being small in comparison with the others) Nesticidae

Nesticidae – comb-footed cellar spiders – two species, one of which is a very rare, recently discovered species from a mine in the south-west of England. The common species is:

- *Nesticus cellulanus* – cavity spider – L:3–6 mm; pale yellow cephalothorax with dark brown central stripe (which is pinched in the middle), pale yellow abdomen with darker rings either side of a rough central stripe; widespread if a somewhat local distribution, often in caves, cellars, etc., but can be found under refuges in damp habitats, including debris such as sheet metal and wood. The web is a loose framework of silk threads. Females may be seen carrying large egg-sacs attached to their spinners.

An aquatic species from another family also keys out here (Dictynidae – mesh web spiders – see also couplet 16).

12 (11) Small (2–3 mm) and ant-like in appearance — Phrurolithidae

Phrurolithidae – ant-like sac spiders – two species from one genus:

- *Phrurolithus festivus* – L:2.5–3.0 mm; dark brown cephalothorax with black head end, dark grey to black abdomen with six patches of white hairs (two near the cephalothorax, three centrally, which may be joined, and one near the spinners); widespread and fairly common, under stones in grassland and gardens, sometimes associated with ants.
- *P. minimus* – L:2.0–3.0 mm; reddish brown cephalothorax, similar abdomen to *P. festivus* but with (sometimes) fewer and less distinct white marks; mainly concentrated in south-east England, under stones in chalk grassland and sometimes open woodland.

– Not ant-like in appearance — 13

13 (12) Legs I and II with rows of curved spines on the upper surfaces of the tarsi and metatarsi — Mimetidae

Mimetidae – pirate spiders – there are four species from one genus, which can be difficult to separate:

- *Ero* species – L:2–4 mm; pale brown cephalothorax with a darker brown head end and sides, globular creamy-brown to orange abdomen with black mottling; often found on other spiders' webs preying on the host.

– Legs I and II without such curved spines — 14

14 (13) Pale brown with pair of darker arrowhead markings on abdomen — Anyphaenidae

Anyphaenidae – buzzing spiders – one genus with three species, which are found in vegetation above the ground, two of which are relatively new additions to the British fauna.

– Not patterned as above — 15

15 (14) Abdomen silver and spherical — Theridiosomatidae

Theridiosomatidae – ray spiders – one species: *Theridiosoma gemmosum* – ray spider – L:1.5–3.0 mm; widespread but local, in southern Britain, usually found in its web attached to low vegetation above the ground.

– Abdomen not silver — 16

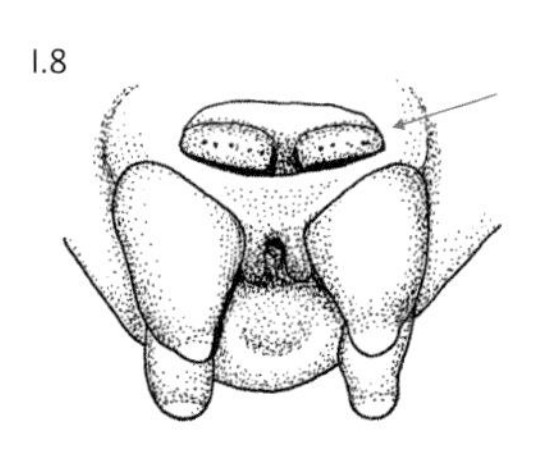
I.8

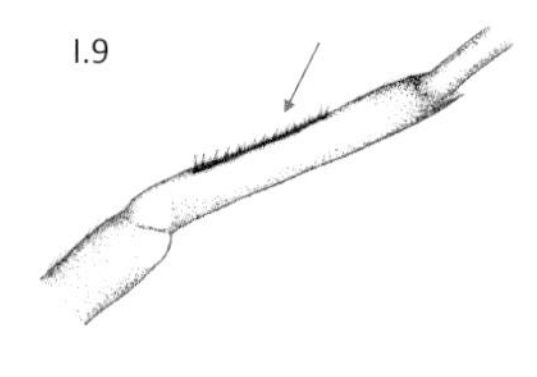
I.9

16 (15) Spider with cribellum (plate used for carding silk) on lower side of abdomen just in front of the spinners (I.8) and a calamistrum (I.9) (row of hooks on the hind leg used for carding silk) – note that these features are easiest to see in females — five families

Of these five families, one (Uloboridae – cribellate orbweb spiders – three species) is found either in hot-houses or off the ground in its webs in vegetation. The more common families under refuges are:

Amaurobiidae – laceweb spiders – contains three species, two of which are more often found on the walls and window frames of buildings, and are associated with rather unstructured and lacey webs, which are bluish when fresh. The most common species under logs and stones is:

- *Amaurobius fenestralis* – L:4–9 mm; dark reddish brown cephalothorax with darker head end, dark brown patch with a pale yellowish border on top of abdomen and paler chevron pattern towards the rear; widespread and common, under stones and tree bark.

Cybaeidae – soft spiders – two genera, each with a single species, one of which is rare. The more common species is:

- *Cryphoeca silvicola* – L:2.5–3.0 mm; brown cephalothorax with darker head end, dark grey to black abdomen with a central line of paler chevrons; widespread and common, in woods and under stones on moorland.

Dictynidae – mesh web spiders – seven genera with 15 species, many of which are found off the ground in webs within vegetation, under water (see couplet 11), or are rare. The webs are small and tangled frameworks of threads, which may be bluish when fresh but which gather dust and debris over time and so may begin to look quite dirty. A common genus (with two species) that can be found under refuges is:

- *Argenna* species – L:1.5–3.0 mm; pale to dark brown cephalothorax with darker streaks, dark grey to black abdomen with three lines of tufts of white hairs; widely scattered across England and Wales, under stones in grassland, dunes and waste ground (more often near the coast). *A. subnigra* (L:1.5–2.5 mm) is more widespread and generally smaller than *A. patula* (L:2.5–3.0 mm).

Hahniidae (lesser cobweb spiders) – two genera containing three species from this family key out here (see also couplet 8), two of which are rare. The more widespread species is:

- *Cicurina cicur* – L:5–7 mm; glossy yellow-brown cephalothorax, pinkish brown abdomen; widespread but local, under logs and stones in damp areas of woodland.

– Spider without cribellum or calamistrum — 17

17 (16) Eyes dark and beady, often with a paler ring round them — 18

– Eyes pale or pearly, often with a darker ring round them — 20

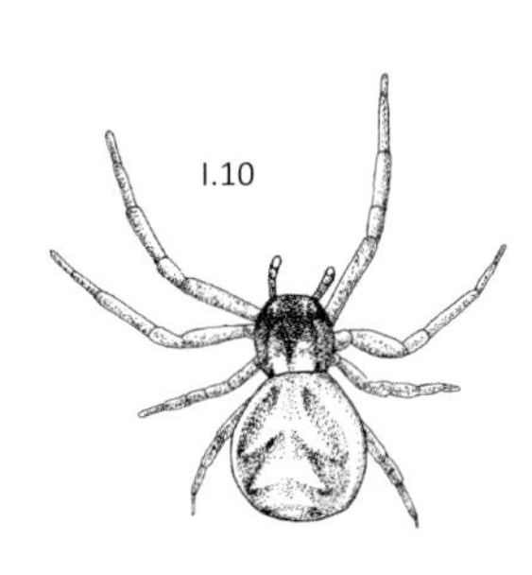


18 (17) Crab-like in overall appearance (I.10); the first and second pairs of legs are longer and often stouter than the third and fourth pairs — 19

– Not as above — Sparassidae

Sparassidae – huntsman spiders – one species which is mainly found in vegetation (especially oak) above the ground.

19 (18) First two pairs of legs longer and distinctly stouter than the rear two pairs; can move sideways; slow moving — Thomisidae

Thomisidae – crab spiders – 27 species from eight genera (six of which contain a single species which are either rare or tend to sit on flowers and within foliage in wait for prey). The other two (more speciose) genera contain species that can be found under logs and stones:

*Ozyptila* (nine species, which can be difficult to separate, including six species that are uncommon, local or rare) and *Xysticus* (11 species, which can be difficult to separate, including eight species that are either uncommon, scarce or rare, or that are usually found in vegetation above the ground). The more common species are:

- *O. atomaria* – L:3–6 mm; yellowish brown cephalothorax with two dark brown lines (which may be indistinct), brownish grey abdomen with darker markings towards the rear and a flat front edge; widespread, among debris and under stones in grassland and heathland. *O. trux* (L:3–5 mm) is very similar although the two marks on the cephalothorax may be better defined and the pale brown abdomen has black and white mottling; widespread and the most common species in the genus, in woods, grassland and heathland.
- *O. praticola* – L:2.5–4.0 mm; dark brown to black cephalothorax with a reddish central band (male darker than the female), greyish brown abdomen mottled with black; widespread in England and Wales, among debris in woods, grassland, and human-disturbed sites such as gardens.
- *Xysticus* species – L:3–8 mm; cephalothorax with a pale central band flanked by darker latteral bands, abdomen pale and creamy with darker brown markings; widespread and common, in open habitats. Three common species are *X. cristatus* (L:3–8 mm; widespread across Britain), *X. erraticus* (L:4–8 mm; widespread but with a scattered distribution) and *X. kochi* (L:4–8 mm; widespread but mainly in southern England and the Welsh coast).

– First two pairs of legs longer but not much stouter than the rear two pairs; fast moving Philodromidae

Philodromidae – running crab spiders – four genera and 18 species of which 16 are mainly found in vegetation off the ground (often on lower tree branches, or in shrubs or tall plants). The species active at ground level are:

- *Rhysodromus fallax* – sand running spider – L:4–6 mm; sandy cephalothorax with brown pattern, sandy abdomen with darker central mark towards the front and two darker bands containing white marks towards the rear; rare, found on the ground in sand dunes.
- *Thanatus striatus* – L:3–5 mm; yellowish brown cephalothorax with darker central band towards the rear and two darker bands at the sides, pale brown abdomen with darker central band and two dark lines towards the rear; uncommon, found on the ground in grassland, heathland, dunes and saltmarsh habitats.

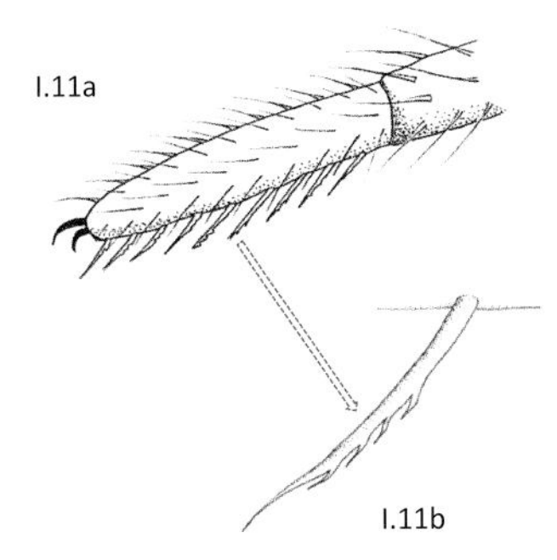


20 (17) Last segment of the fourth leg with a row of thickened bristles underneath forming a comb (I.11a and I.11b – this can be difficult to see – use a good light and high magnification) Theridiidae

Theridiidae – comb-footed spiders – 25 genera (13 of which contain only one species each) and 58 species, of which 55 species are found in buildings, on vegetation above the ground, or are rather rare, scarce, uncommon or local. Their webs are a loose framework of threads, which may be three dimensional with threads hanging down. The more common ground active species are:

- *Enoplognatha thoracica* – L:2.5–4.0 mm; brown to dark brown cephalothorax, black abdomen (sometimes with small white spots); widespread and locally common, under stones in a range of habitats.
- *Pholcomma gibbum* – L:1.25–1.5 mm; orange-brown cephalothorax, globular greyish brown abdomen with four reddish spots (male has a reddish brown plate on the upper and lower surfaces of the abdomen towards the front end); widespread and common, on the ground in a range of habitats. The webs are attached to the ground with sticky threads.
- *Robertus lividus* – L:2.5–4.0 mm; shiny dark brown cephalothorax, greyish black abdomen with four reddish spots; widespread and common, under stones in woodland and moorland. At first sight, this species can be confused with a linyphiid (see couplet 23), however *R. lividus* has many more spines on the legs, darker tarsi than the femora, and a hairier abdomen.

– Without a comb on the fourth leg 21

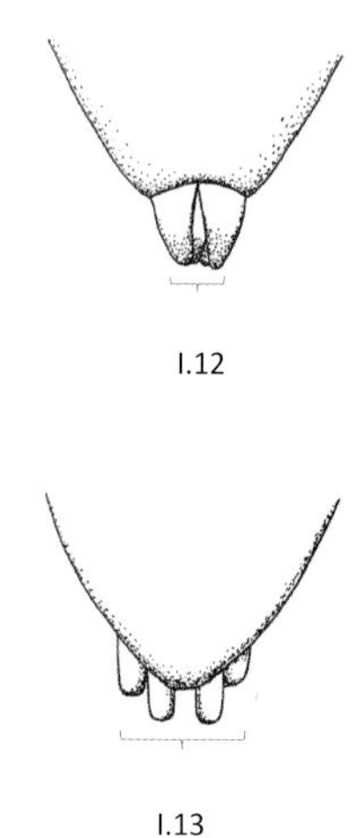
I.12

I.13

21 (20) Abdomen narrow and pointed behind with spinners protruding (I.12 or I.13); usually fairly even in colour or with a characteristic pattern of chevrons 22

– Abdomen not narrow 23

22 (21) Spinners very close together, barely separated at base (I.12); abdomen often with a chevron-like pattern (may be pale and indistinct) three families

Three families key out here:

Cheiracanthiidae – eutichurid spiders – one genus containing three species, one of which is rare, and another occurs in low vegetation. The species occurring most frequently under stones is:

- *Cheiracanthium virescens* – L:5–9 mm; cephalothorax yellowish brown, abdomen grey-green with a darker stripe towards the front; uncommon, found under stones in dry open areas including heathland, dunes and human-disturbed sites.

Clubionidae – sac or (foliage) spiders – two genera containing 23 species, of which 11 are rare, scarce or uncommon, and two are relatively common species which are normally found associated with vegetation in wetlands. Several species are difficult to separate. Spiders under logs and stones are often enclosed in a tight tubular web, and females may be found associated with their egg-sacs within silk cells. The most common genus found under refuges on the ground is:

- *Clubiona* (10 common species) – L:3–11 mm; brownish cephalothorax (can be greyish brown, yellowish brown or reddish brown) sometimes with a darker head end, *C. corticalis* (bark sac spider – L:6–10 mm) and *C. comta* (L:3–6 mm) have a dark brown central stripe down the abdomen which ends in four or five chevrons – other common species may have the darker band but the chevrons are faint or absent; widespread and common, under refuges in a wide range of habitats. Females of *C. diversa* can often be found under logs and stones, guarding their egg-sacs situated within cells made of white silk.

Liocranidae – running foliage spiders – five genera containing 12 species, of which eight are rare, scarce or rather local. Several of these species are difficult to separate. The more common (although still relatively uncommon) ground-active species are:

- *Agroeca* species (six species of which three are rare) – L:3–8 mm; pale brown cephalothorax with darker streaks and margins, reddish brown abdomen with a darker central mark near the front and chevrons towards the rear; widespread, in a range of habitats.

- *Liocranum rupicola* – L:5.5–8.5 mm; pale brown cephalothorax with two darker bands along the sides, dark brown abdomen with a paler diamond shape towards the front (within which is a darker central line) and paler chevrons towards the rear; under stones in dry habitats.

– Spinners more widely separated (I.13); spiders usually dark with no clear pattern on abdomen; eyes often oval rather than round Gnaphosidae

Gnaphosidae – ground spiders – 11 genera with 34 species, of which 28 are rare, scarce or uncommon, or are found in buildings. The more common species are:

- *Drassodes* species (three species of which one is scarce) – L:9–18 mm; reddish brown cephalothorax, reddish brown (slightly greyish) abdomen; widespread and fairly common, in open habitats. Of the two common species, *D. cupreus* can look rather coppery and is more widespread across Britain (under stones in heathland and grassland) than *D. lapidosus* which is widespread in southern Britain (on scree and human-disturbed sites such as gardens and waste ground). Females may be found together with their egg-sac within a silken chamber.
- *Drassyllus pusillus* – L:3–5 mm; dark brown cephalothorax, dark brown to black abdomen; widespread and common, open heaths and downland.
- *Haplodrassus signifer* – L:6–9 mm; pale to dark brown cephalothorax, brown to nearly black abdomen often slightly metallic with lighter chevron pattern at the rear; widespread and common, under stones in grassland and heathland.
- *Micaria pulicaria* – L:2.7–4.5 mm; black cephalothorax with lines of white hairs, black (iridescent) abdomen with a white line across the centre of the back (may also have white spots down the centre); widespread and common, grassland, heathland, dunes and brownfield sites, as well as saltmarshes and woods.
- *Zelotes latreillei* – L:4.5–8 mm; black cephalothorax and abdomen; widespread and common, under debris including stones in open habitats such as grassland and heathland. The round, flattened egg-sacs are attached to the underside of logs or stones.

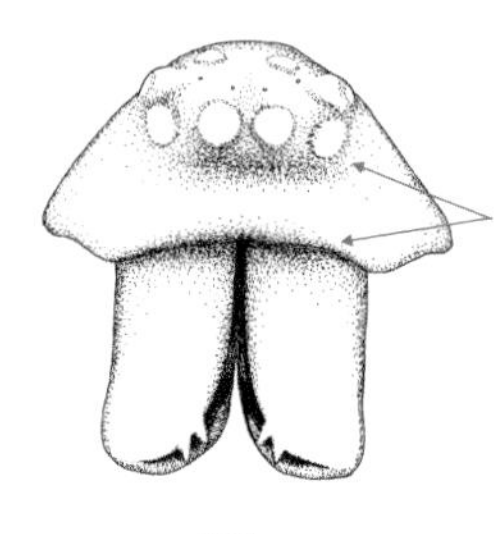

I.14

23 (21) Variable in shape but spinners not clearly protruding; head sometimes elevated at front (males of some species have protuberances on their heads); may have regular grooves on the side of the chelicerae; when viewed from the front, the gap above the base of the chelicerae up to the front row of eyes is at least as large as (and often larger than) the gap between the front and second row of eyes (I.14); spiders usually, but not always, tiny (<4 mm) Linyphiidae

Linyphiidae – money spiders – 123 genera containing around 280 species, of which the genera *Agyneta* (16 species), *Erigone* (10 species), *Lepthyphantes* (2 species), *Monocephalus* (2 species), *Palliduphantes* (4 species), *Porrhomma* (11 species), *Tenuiphantes* (7 species), and *Walckenaeria* (21 species) are very common around logs and stones. Linyphiidae can be very challenging to identify to species, requiring high-power microscopy and more detailed keys. Many species are very small and difficult to separate. Money spiders often produce horizontal sheet webs attached to low-lying vegetation under which they hang waiting for prey (but as these webs can be quite small and may be up against logs, they often get damaged when turning refuges over). Some of the more obvious species that may be found under refuges on the ground are:

- *Oedothorax* (five species) – L:2.0–3.2 mm; pale to chestnut brown, usually paler legs and darker abdomen, male cephalothorax modified to a greater or lesser extent (generally domed behind the eyes – males of the commonest species in wetter habitats – *O. gibbosus* – occur in two forms, both have a raised cephalothorax but in one form it is greatly enlarged and has a deep notch containing long black hairs); widespread and common.
- *Poeciloneta variegata* – L:1.8–2.6 mm; yellow-brown cephalothorax with black margins and lines, brown to brownish grey abdomen patterned with white spots and black stripes, widespread and fairly common especially in Wales and northern England, under stones in grasslands and heathland.
- *Stemonyphantes lineatus* – L:4.0–6.8 mm; yellowish brown cephalothorax with black stripes in the centre and along the sides, greyish white abdomen (may be tinted pink, yellow or green) sometimes with three dark stripes (which may be in a series of spots); widespread and common, under stones in grassland and gardens.
- *Tenuiphantes* (seven species, which require high magnification to separate) – L:1.6–2.8 mm; dark brown cephalothorax, lighter brown or grey abdomen with black spots joined by dark wavy lines (the strength of the pattern depends on the

species); widespread and common, in a wide range of habitats including woods, grassland, heathland, wetland and gardens.

– Without the features above two families

These two families are:

Araneidae – orbweb spiders – 16 genera with 32 species, all of which build webs on vegetation (and other structures) above the ground. They are rarely found under logs and stones although one (*Nuctenea umbratica* – walnut orbweb spider – L:8–14 mm; flattened body which is mottled brown with black and white marking) is frequently encountered under the bark of standing trees.

Tetragnathidae – long-jawed orbweb spiders – four genera with 14 species, of which eight build circular orb webs in vegetation above the ground, two build webs in buildings or caves, and one is uncommon. The three more common species found at ground level are *Pachygnatha* species, which do not spin webs as adults – although immatures will:

- *Pachygnatha clercki* – L:5–7 mm; glossy brown cephalothorax with a darker central band, olive-brown abdomen with a creamy central band bordered by dark wavy bands; widespread and common, often in wetter habitats such as marshes, bogs and the margins of freshwater courses. A similar but smaller and less common species is *P. listeri* (L:3–5 mm), which is found in deciduous and mixed woodland.
- *P. degeeri* – L:2.5–3.8 mm; very dark brown pitted cephalothorax, whitish abdomen with brown central band with wavy margins (sometimes with a central line of white markings; widespread and common, in a range of habitats including woodland and human-disturbed sites such as waste ground.

## Key J Woodlice

This key will enable the identification of common British woodlice found outdoors; note that there are several introduced species found only in hot-houses that are not (generally) covered here. It is worth noting that there are five widespread and very common British species. These are *Armadillidium vulgare* (common pill woodlouse), *Oniscus asellus* (common shiny woodlouse), *Philoscia muscorum* (common striped woodlouse), *Porcellio scaber* (rough woodlouse) and *Trichoniscus pusillus* (common pygmy woodlouse). The latter species was thought to exist in two forms and records for both of these are often listed under this species name. However, it is now clear that these are actually two distinct species (*T. pusillus* and *T. provisorius*) which are not easy to separate (especially in the field) and are now usually recorded as *T. pusillus* agg. Sizes given are the maximum lengths (L) for each species from the front of the head to the end of the abdomen ignoring the antennae and legs. See J.1 for the general morphology of woodlice. Where any specimens do not match the descriptions, consult Hopkin (1991), Oliver & Meechan (1993) and Gregory (2009). All three texts provide species descriptions, and while the former two provide detailed identification keys, the latter gives both distribution data and images of all British species. The British Myriapod and Isopod Group website hosts a checklist and provides further species descriptions and distribution maps (see Chapter 6). See Section 3.5 for further information on woodlice.

**sexing woodlice**
Male woodlice usually have modified pleopods (visible on the underside) bearing long thin genital projections, which are absent in females. Females have a brood pouch (marsupium) between the first few pairs of legs in which they carry the fertilised eggs. This pouch may be difficult to see unless the female is carrying eggs or young.

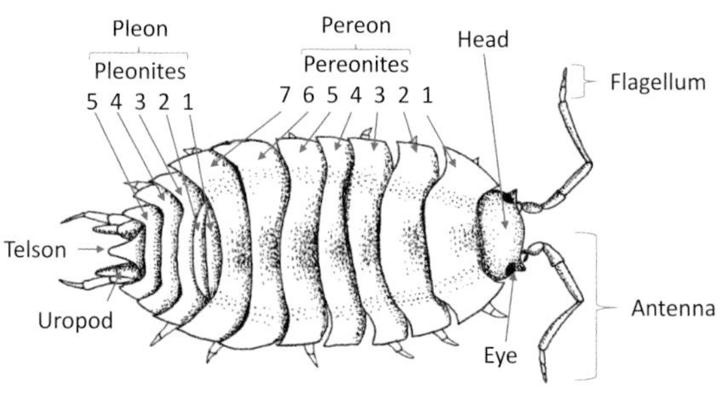


J.1

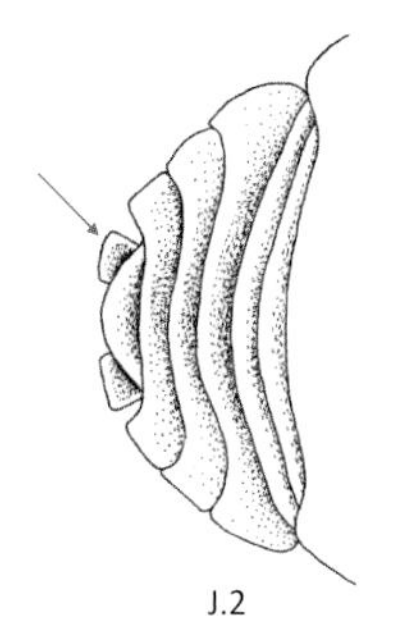
J.2

1 Hind end rounded; uropods (arrowed in J.2) do not project beyond end of body); can roll into an almost perfect ball — Armadillidiidae

There are seven species of Armadillidiidae (pill woodlice) from two genera plus an introduced hot-house species. The common species are:

- *Armadillidium album* – beach pill woodlouse – L:6 mm; body sandy coloured; coastal sand dunes.
- *A. depressum* – southern pill woodlouse – L:20 mm; body slightly flattened and grey (sometimes with yellow flecks); associated with synanthropic areas including gardens under paving and under limestone rubble in disused quarries and railway cuttings.
- *A. nasatum* – striped pill woodlouse – L:12 mm; body dark grey with paler stripes, head with a narrow projection between the antennae raised above the surface of the head; relatively dry human-disturbed areas under debris, often found indoors.
- *A. pictum* – decorated pill woodlouse – L:9 mm; body brown or blackish with yellow or greenish mottling (the edge of the 7th segment having a contrasting darker patch); found under debris in upland grassland, screes and ancient woodland, in northern England, Wales and on the Welsh borders.
- *A. pulchellum* – beautiful pill woodlouse – L:5 mm; body similar to *A. pictum* in terms of colouration, mottling and the darker patch on edge of the 7th segment; natural grasslands, especially at the coast.
- *A. vulgare* – common pill woodlouse (or pill bug) – L:18 mm; body typically uniformly grey but can be variable in colour and even mottled but lacks the darker patch on the edge of the 7th segment seen in *A. pictum* and *A. pulchellum*; common in most habitats in south-east England, becoming more restricted to coastal and human-disturbed sites in its northern range. Be careful not to confuse with millipedes, which have two pairs of legs per segment (Key K). In rolled up pill millipedes (*Glomeris marginata*) the telson overlaps the head and the 'ball' has one flat side. In *A. vulgare* the head and telson do not overlap and the ball is almost perfectly round (Fig. 3.49).
- *Eluma caelata* – L:15 mm; body usually purplish brown, very similar in appearance to *A. vulgare* but with eyes consisting of a single ommatidium (compared to multiple ommatidia in *A. vulgare*); natural coastal habitats and human-disturbed sites such as gardens and railway embankments under debris and burrowed into soil, mainly in the extreme south-east of England.

**ommatidia**

Isopodologists (people who study woodlice) use the term ommatidia (singular ommatidium: from the Greek *ommatidion*, a diminutive term from *omma* meaning eye) for the units that make up the eyes of woodlice. Where present these may be single units or clustered together to form compound eyes.

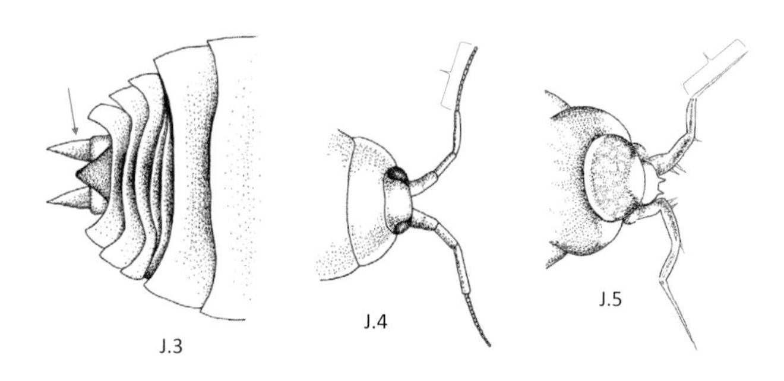


The family Armadillidae (tropical pill woodlice) comprising three introduced hot-house species will also key out here.

– Uropods elongated and project beyond end of body (J.3); do not usually roll into an almost perfect ball 2

Two rare and tiny (L:<3 mm) species (one of which is found only in hot-houses) can roll into almost perfect balls – see couplet 4.

2 (1) Flagellum of antenna (end segments at end of antenna) with 10 or more bead-like segments (J.4) Ligiidae

There are two genera of Ligiidae (rock slaters), each with one species (there is some debate about the taxonomic position of these species, which may result in them being placed in different families):

- *Ligia oceanica* – (common) sea slater – L:30 mm; body greenish grey to grey-brown with large eyes comprising over 40 ommatidia; rocky seashores.
- *Ligidium hypnorum* – carr slater (or moss slater) – L:9 mm; body dark brown and mottled with a shiny surface (occasionally purple); damp habitats, including in woodlands, grasslands and wetlands, mainly in the south and east.

– Flagellum of antenna with six or fewer segments (which may be indistinct) 3

3 (2) Flagellum of antenna comprising a tapered cone (under high magnification may be seen to be up to six indistinct segments with terminal bristles) (J.5) Trichoniscidae 4

Sixteen species from eight genera (plus one hot-house species).

Three hot-house species of the family Styloniscidae will also key out here. These are similar to *Trichoniscus* species (see couplet 5) in having three ommatidia, however their upper surfaces are not smooth, but have rows of small bumps.

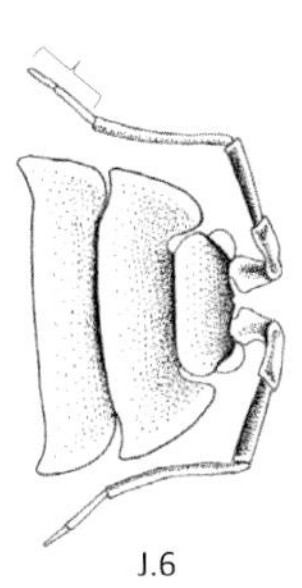

J.6

– Flagellum of antenna composed of two or three distinct segments (not comprising a tapered cone with terminal bristles) (J.6 or J.7) 8

4 (3) Upper surface of body with obvious ridges running down the back; white or dirty white (L:4 mm), body outline smooth (e.g. J.8) *Haplophthalmus* species

There are three British species (which each have eyes consisting of a single ommatidium), two of which are difficult to separate:

- *H. danicus* – spurred ridgeback – L:4 mm; body white or pale creamy yellow, no projection on the 3rd pleonite; often associated with dead wood in damp habitats, including woodland and flood plains as well as human-disturbed sites such as waste ground and gardens.
- *H. mengii* – common ridgeback – and *H. montivagus* – southern ridgeback – are not possible to separate without very close examination (and then only for males) – L:4 mm; body white or pale creamy yellow, with a pair of prominent projections on the 3rd pleonite; under rotting wood and stones – *H. mengii* is more wide ranging (from woodland to coasts to farmyards) than *H. montivagus* which typically inhabits calcareous sites, typically woodlands.

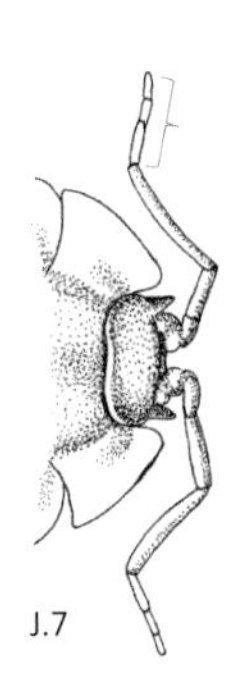

J.7

Another species (*Buddelundiella cataractae* – pygmy pill woodlouse) will also key out here. This rare species is 3 mm long and has three ommatidia in contrast to the single ommatidium in *Haplophthalmus* species (difficult to see except under high magnification). In addition, *B cataractae* can roll into a ball. One hot-house member of the Armadillidae (see couplet 1) – *Reductoniscus costulatus* – looks very similar to *B. cataractae* but has eyes comprising clusters of more than three ommatidia.

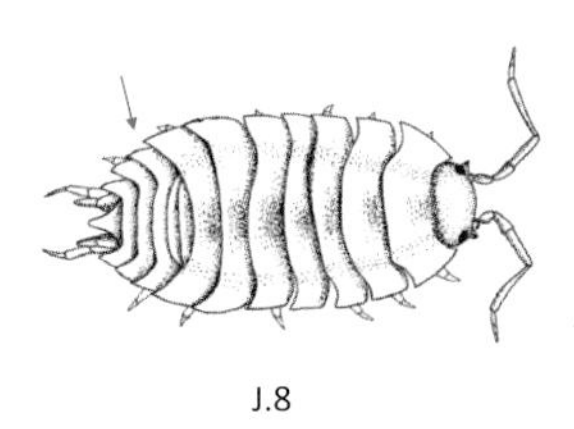

J.8

– Upper surface smooth or with rows of small bumps across it; body outline stepped (e.g. J.9) 5

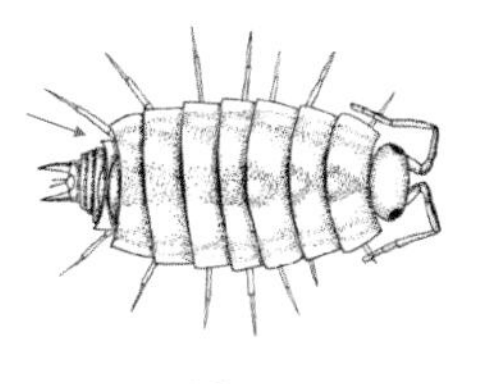

J.9

5 (4) Upper surface smooth; eyes small, each composed of three ommatidia (simple eyes); L:5 mm or less *Trichoniscus* species

There are three British species, two of which are difficult to separate:

- *T. pusillus* – common pygmy woodlouse – L:5 mm; body reddish or purplish brown. There is a very similar species (*T. provisorius* – common pygmy woodlouse), which is smaller (L:3.5 mm), and can only be separated by close examination of the males. However, since *T. pusillus* is parthenogenetic, and *T. provisorius* is sexually reproducing, discrete populations with very few (or no) males are likely to be the former species, while discrete populations with approximately 50% males and 50% females will probably be the latter species. Both species are widespread under dead wood and stones with *T. pusillus* preferring somewhat damper habitats. Since these species require detailed examination to separate them, they are usually collectively referred to as *T. pusillus* agg.

- *T. pygmaeus* – least pygmy woodlouse – L:3 mm; body white or creamy, occasionally pale pink, head paler than the body; soil dwelling in woodlands, grasslands and human-disturbed habitats, found under logs and stones.

A rare species – *Oritoniscus flavus* – may also key out here on the basis of its smooth upper surface, although this species has only a single ommatidium per eye – this is only found in southern Ireland (plus one site in Wales and a second near Edinburgh) – L:9 mm; body purple-red; under logs and stones in waterlogged sites such as those close to watercourses.

– Upper surface with rows of small bumps across it 6

6 (5) No eyes *Metatrichoniscoides* species

There are two species, one of which (*M. leydigii*) is rare and restricted to four sites (two coastal sites, a walled garden and a garden centre – now destroyed – in Oxford), the other is:

- *M. celticus* – L:2.5 mm; body white; in Wales under stones above rocky shores and in a limestone quarry.

*Platyarthrus hoffmannseggii* may also key out here if the flagellum of the antenna is difficult to discern (see couplet 11 for a description of this species).

– Eyes comprise a single ommatidium 7

7 (6) Eyes each consist of a single large black ommatidium three species

Three species key out here:

- *Androniscus dentiger* – rosy woodlouse – L:6 mm; body salmon pink or orange with a double yellow stripe along the back (which does not wash out in alcohol), upper surface with rows of small bumps across it; widespread at the coast and in human-disturbed habitats including gardens and waste ground.
- *Miktoniscus patiencei* – coastal black-eye – L:4 mm; body white with a dark line showing along the back; a coastal species, being found above the strand line and on salt marshes and sea cliffs.
- *Trichoniscoides albidus* – rough pygmy woodlouse – L:4 mm; body reddish purple, eyes a pair of large brownish black ommatidia, dorsal surface of pereon

rough, pleon smooth; in damp soils in woodlands, grasslands and human-disturbed habitats such as gardens.

– Eyes red or orange (colour washes out in alcohol) *Trichoniscoides* species

Three of the four *Trichoniscoides* species key out here (see the previous half of this couplet for the fourth), all of which are difficult to separate:

- *T. helveticus* – Swiss red-eye – L:4 mm; body creamy-white with a pink tinge; found in grassland and deciduous woodland. Two other species are difficult to separate from *T. helveticus*. These are *T. saeroeensis* – coastal red-eye, and *T. sarsi* – Sars' red-eye. Close examination of males is required to separate these three species. *T. helveticus* is most often found in grassland and deciduous woodland, while *T. saeroeensis* is mainly a coastal species, and *T. sarsi* is more urban, being found in human-disturbed habitats.

8 (3) Flagellum of antenna with 3 distinct segments (J.7) 9

– Flagellum of antenna with 2 distinct segments (J.6) 11

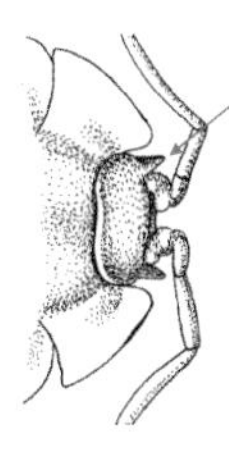

J.10

9 (8) Head with projecting horn-like lobes (J.10); body outline smooth and continuous (J.8) Oniscidae

One species (with two subspecies):

- *Oniscus asellus asellus* – common shiny woodlouse – L:16 mm; body shiny grey-brown, with two rows of yellow patches on back; widespread and common including under rotten logs in woodland and debris in gardens. The subspecies *O. asellus occidentalis* (western shiny woodlouse) is smaller, more brightly coloured and found in south-west England. The two subspecies can interbreed producing intermediate hybrids. Reliable separation of the subspecies requires close examination of males.

– Head without horn-like lobes; body outline stepped (J.9) 10

10 (9) With a dark central stripe down the back Philosciidae

Two species which can be hard to separate (plus a number of introductions – one to the Channel Islands, one to the Isles of Scilly, and five to hot-houses):

- *Philoscia muscorum* – common striped woodlouse (or fast woodlouse) – L:11 mm; body typically dark mottled brown but may be yellow, reddish or greenish, with the head uniformly coloured and distinctively darker than the body, and typically with an oval or flattened yellow spot at the rear of the head; grasslands, hedgerows, woodlands and gardens. The species *Philoscia affinis* is very similar to *P. muscorum* in size and appearance, although typically the head is brown with paler mottling. *P. affinis* is found in similar habitats (and sometimes the same site) to those of *P. muscorum* but favours open woodland and coastal grassland in southern and western Britain.

– Without dark central stripe — Halophilosciidae

Two genera each with one relatively rare species:

- *Halophiloscia couchii* – L:10 mm; body pale pinkish brown; upper seashore under rocks at the base of cliffs, in debris and sometimes among shingle.
- *Stenophiloscia glarearum* – L:6 mm; body white with brown mottling; in shingle on upper shores in the south and south-east of England.

11 (8) Blind (eyeless); body pure white, broad in shape with broad antennae; associated with ants and their nests — Platyarthridae (part)

One species keys out here (a second, introduced species keys out in couplet 13: *Trichorhina tomentosa* – dwarf white woodlouse):

- *Platyarthrus hoffmannseggii* – ant woodlouse – L:5 mm; body white; in ants' nests (especially associated with *Lasius* and *Myrmica* species).

– With eyes; body usually darker — 12

12 (11) Back strongly convex, can roll into an imperfect ball with the antennae and uropods protruding — Cylisticidae

One species:

- *Cylisticus convexus* – curly woodlouse (or false pill woodlouse) – L:15 mm; body pale greyish brown with paler (grey or orange) uropods; rocky coastal sites and stony human-disturbed sites such as waste ground.

– Back not strongly convex, will not roll into a ball — 13

13 (12) Five pairs of pleopodal lungs (J.11) — Trachelipodidae

**pleopodal lungs**
Where present, these appear as white patches on the underside of the pleon (see J.1) and are part of the respiratory system. Note that these may be difficult to see in preserved specimens.

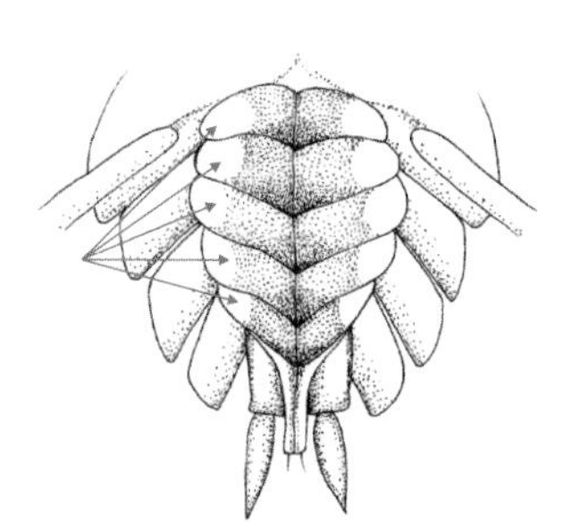

J.11

One species (plus two hot-house introductions):

- *Trachelipus rathkii* – Rathke's woodlouse – L:15 mm; body dark grey or grey-brown with pale orange and grey mottling; damp grassland, scrub, woodland and waste grounds.

– Two pairs of pleopodal lungs (J.12) Porcellionidae 14

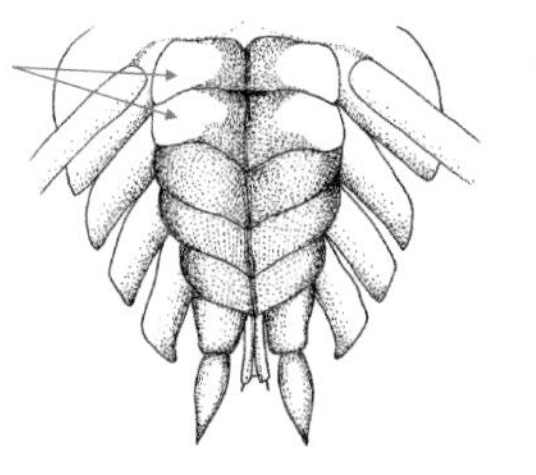

J.12

Seven species (plus three hot-house introductions).

An introduced hot-house species of Platyarthridae (*Trichorhina tomentosa* – dwarf white woodlouse) may key out here – it does not have any pleopodal lungs.

14 (13) Outline of body smooth (J.8) *Porcellio* species

Four species (plus two hot-house introductions):

- *P. dilatatus* – L:15 mm; body grey-brown with paler broad mottled stripes at the sides; calcareous grassland and human-disturbed sites including gardens.
- *P. laevis* – dooryard sowbug – L:20 mm; body shiny, light grey or brownish with some mottling at the sides; human-disturbed habitats such as farmyards and gardens.
- *P. scaber* – (common) rough woodlouse – L:17 mm; body grey, often with paler mottling of white, yellow or grey; common under logs and stones in woodland, grassland, hedgerows and human-disturbed sites.
- *P. spinicornis* – painted woodlouse – L:12 mm; body brownish with a darker central stripe and mottled sides, head black; limestone walls and stones in human-disturbed sites

– Outline of body stepped (J.9) two genera

Four species from two genera (of which one is restricted to a sea cliff in Ireland and another is a hot-house introduction):

- *Porcellionides cingendus* – L:9 mm; body yellowish, reddish or brown, heavily mottled; in grassland, scrub and open woodland (south and west of England, Ireland and lowland Wales).
- *P. pruinosus* – plum woodlouse – L:12 mm; body purple-grey with a greyish bloom; in human-disturbed habitats such as in garden compost heaps, dung heaps in farmyards and under bark in pasture trees.

## Key K Millipedes

This key will enable the identification of families, genera, and for many common animals, species of adult millipedes. In most millipedes, the males can be distinguished because leg pairs 8 and 9 appear to be missing, leaving a gap. When mature, these legs are modified into gonopods, which are structures that transfer sperm into the females during mating; they often also have other features to help grip the female. The gonopods can be external and visible (as in polydesmids) or internal and not visible without dissection (as in julids). Snake millipedes (julids and blaniulids) have glands on most body rings that secrete a defensive secretion when the animal is threatened. The colour of the glands can sometime help with identification but they do vary with preservation and are generally more obvious in paler individuals. New species of millipede are being found fairly frequently at present, many of which are introductions found in hot-houses and botanic gardens and are not covered by this key. The lengths (L) given are for adult specimens. Where an individual does not fully match the description given, then consult Blower (1985) or Lee & Harding (2006). The latter also provides distribution maps. Helen Read and Paul Lee are currently working on an updated synopsis to replace Blower (1985). A checklist and further information are given on the British Myriapod and Isopod Group website (see Chapter 6). See Section 3.5 for more information on millipedes.

**segments or rings**
In millipedes the body rings are actually two segments fused together and hence have two pairs of legs on each of the major body rings.

1 Small (2–3 mm); with tufts of bristles along each side (K.1) Polyxenida

Polyxenida – bristly millipedes – there is a single species:

- *Polyxenus lagurus* – bristly millipede – L:2–3 mm; body pale brown with rows of tufts and bristles; widespread especially in the south and midlands, under the bark of dead (especially coniferous) trees, as well as under stones, in leaf litter and on sand dunes.

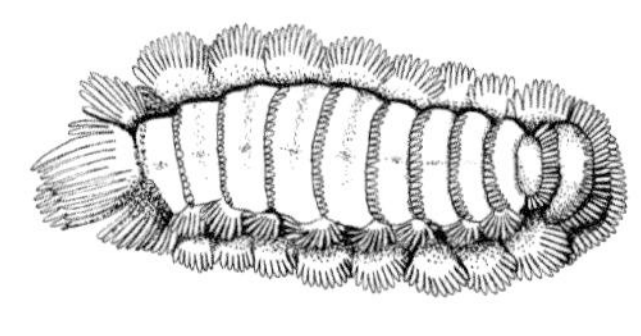
K.1

– Larger; no tufts of bristles along the sides 2

2 (1) Able to roll into a ball; black and shiny (7–20 mm) Glomerida

Glomerida – pill millipedes – there are four species that can roll into a ball when disturbed, three of which are small (L:<7 mm), pale and scarce or rare. The larger, darker, and much more common species is:

- *Glomeris marginata* – pill millipede – L:7–20 mm; body shiny brown or brown-black (edges of the body rings may be light brown, brown or even reddish); widespread and common, especially in

K.2

deciduous woodlands, as well as sand dunes and human-disturbed sites such as waste ground. Be careful not to confuse with woodlice, which have only one pair of legs per segment (Key J). In rolled up pill woodlice (*Armadillidium* species) the head and telson do not overlap and the ball is almost perfectly round. In *Glomeris*, the telson overlaps the head and the 'ball' has one flat side (Fig. 3.54).

– Unable to roll into a ball — 3

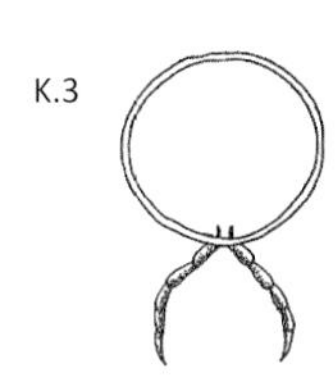

K.3

3 (2) Hemispherical in cross-section (K.2) — Polyzoniida

There is a single species:

- *Polyzonium germanicum* – Kentish pin-head – L:5–18 mm; body yellowish to orange-brown, with a small triangular head, and each eye with three ommatidia; rare, mainly restricted to woodlands in Kent.

– Not hemispherical in cross-section – round (K.3) or flat-backed or arch-backed with the upper surface expanded to the sides to form lobes (K.4 or K.5) — 4

**ommatidia**
Diplopodologists (people who study millipedes) use the term ommatidia (singular ommatidium: from the Greek ommatidion, a diminutive term from omma meaning eye) for the units that make up the eyes of millipedes. Where present these may be single units or clustered together to form compound eyes.

4 (3) Round in cross-section (K.3) — 5

– Not round in cross-section, with a flat-backed or arch-backed appearance with the upper surface expanded to the sides to form lobes (K.4 or K.5) — 13

5 (4) Body heavily calcified and hard especially when preserved; a complete cylinder in cross-section with head more or less continuous with body (K.6) — 6

– Body softer; not necessarily a perfect cylinder; with a distinct notch between the head and body (K.7) — Chordeumatida (part)

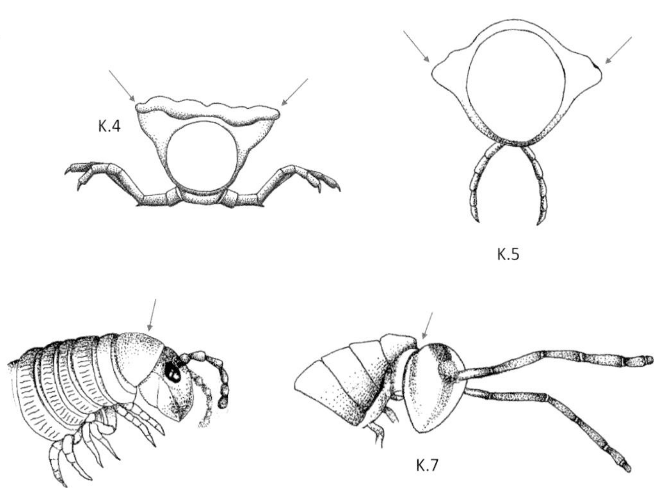

K.4 K.5 K.6 K.7

There are two genera containing five species (see also couplet 13):

- *Chordeuma* species – L:10.8–13.4 mm; body brown to dark brown, eyes equilateral triangles of 24–29 ommatidia; common especially in the south-west of Britain, woodland and waste ground.
- *Melogona* species – L:5.6–10 mm; body creamy white to pale brown, eyes made up of acute triangles of 8–18 ommatidia; common, found in woodlands and human-disturbed sites including under garden debris.

6 (5) Striae (fine grooves) covering the whole surface of each ring (K.8) — Julida 7

There are 12 genera containing 27 species.

– Striae (fine grooves) only on bottom half of each ring (K.9) — 11

7 (6) Telson ('tail') with a projection on the top (K.10, K.11 or K.12) — 8

– Telson ('tail') without a projection on the top (K.13) — four genera

Of these four genera, one (containing a single species) has only been recorded from a single garden locality, and another (also with a single species) is rare. The two common genera are:

*Brachyiulus* – two species, of which one is a recently introduced species recorded from a hot-house. The common species is:

- *B. pusillus* – L:7.2–13 mm; body brown with two pale yellow stripes; common and widespread, in

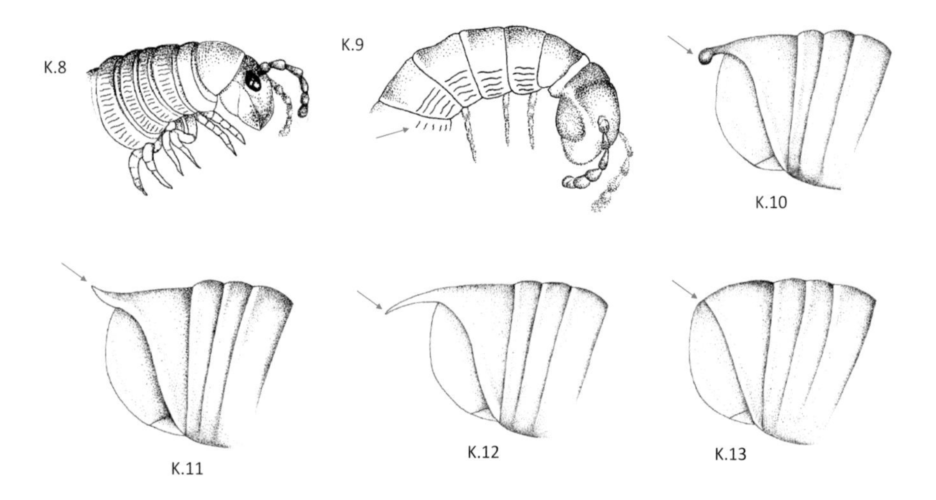

grassland and farmland, as well as coastal sites further north.

*Cylindroiulus* (part – see also couplets 8 and 10) – five species key out here, of which two are scarce. The more common species are:

- *C. britannicus* – L:9.5–15.9 mm; body light to mid-brown; common and widespread, under the bark of, or in the soil near to, dead wood. This species can be difficult to separate from *C. latestriatus* (L:8.6–16.0 mm; body light brown) which is common mainly at coastal sites such as dunes, although it may be found inland in deciduous woods.
- *C. caeruleocinctus* – L:20–30 mm; body brown-black with a blueish bronze tint; common, especially in the south and east of England, from human-disturbed sites such as farmland and parks.

8 (7) Projection on telson rounded and club-shaped (K.10) two species

Two species of *Cylindroiulus* key out here (see also couplets 7 and 10):

- *C. londinensis* – L:20–48 mm (broader than other millipedes at 3.5–4.0 mm diameter); body brown-black to black; widespread (but much less common than *C. punctatus*), mainly in woodlands.
- *C. punctatus* – blunt-tailed snake millipede – L:14–27 mm; body pinkish brown to light brown (stripy in appearance with the stripes going round the body); very common, in or under decaying wood.

– Projection pointed (K.11 or K.12) 9

9 (8) Projection on telson with tip turned up (K.11) two genera

These two genera contain three species, one of which has been recorded from a single woodland in south Wales. The two common species are:

- *Ommatoiulus sabulosus* – L:14–33 mm; body dark brown with two paler orange lines along the body; widespread and common, usually on sandy soils, especially in coastal sites.
- *Tachypodoiulus niger* – white-legged snake millipede – L:15–39 mm; body black with contrasting white legs (immatures may be paler with 2 dark lines along the body); widespread and common, including in woodland.

– Projection with tip turned down (K.12) 10

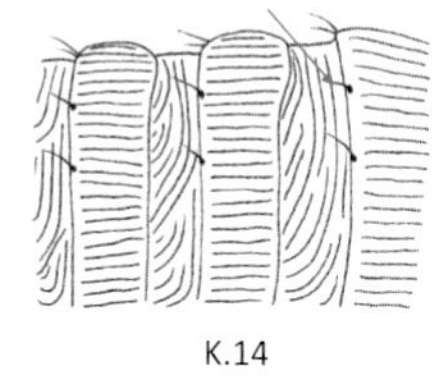
K.14

10 (9) Setae on the hind margin of the body rings (K.14) run the whole length of the body — four genera

These four genera contain six species, of which three are scarce and another is an introduced species recorded from hot-houses and botanic gardens. The common species are:

- *Julus scandinavius* – L:13.1–30.5 mm; brown-black to black (in mature males first pair of legs is very small and the second pair have long flattened projections); widespread and common, in a range of habitats including grassland, wetland, sand dunes and especially woodlands.
- *Ophyiulus pilosus* – L:13.8–30 mm; brown-black to black (mature males with sickle-shaped first pair of legs); widespread and common, in deciduous woodland as well as human-disturbed sites such as farmland and gardens.

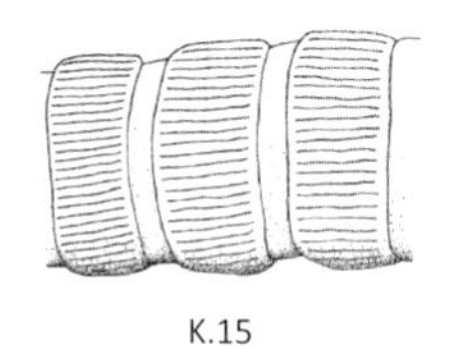
K.15

– No setae on the hind margin of the body rings (e.g. K.15), except (in two species) those just behind the telson – i.e. the rings with no legs (apodous rings) — three genera

These three genera contain six species, of which four are rare. The more common species are:

- *Allajulus nitidus* – L:11.0–24.8; body tan to light brown, setae present on the hind margin of the apodous body rings; widespread but not common, in human-disturbed sites.
- *Cylindroiulus vulnerarius* – L:12.4–21.8 mm; body pale with orange spots along the sides, no ommatidia (unlike the other species keying out here); scarce, in soil, mainly urban sites such as parks and gardens. See couplets 7 and 8 for other more common *Cylindroiulus* species.

11 (6) Eyes formed of ommatidia in a rough equilateral triangle — Nemasomatidae

Two genera, each with one species:

- *Nemasoma varicorne* – L:4–14 mm; body greyish white; widespread, under the bark of coniferous and deciduous trees.
- *Thalassisobates littoralis* – L:9–17 mm (can be as long as 21 mm); body greyish white; only found under stones on shingle beaches.

– No ommatidia or ommatidia in a line or double line (which may form a narrow triangle but not a rough equilateral triangle) — Blaniulidae 12

12 (11) With ommatidia present — three genera

Three genera, each with one species, which are very thin for their lengths and often have prominent spots along their sides. Two of these species are rather scare and the most common species is:

- *Proteroiulus fuscus* – L:6.5–13.4 mm (may be up to 15 mm); body brown to dark brown, with darker brown spots along the sides, ommatidia forming a very narrow triangle (in adults – single line in juveniles); widespread and common, often associated with decaying wood in woods and human-disturbed sites such as gardens and waste ground.

– No ommatidia three genera

Three genera, each with one species, key out here:

- *Archiboreoiulus pallidus* – L:8.5–15 mm; body cream-white with orange spots along the sides, relatively hairy; widespread but scarcer in the west of Britain, across a range of habitats including woods, grassland and farmland. This species is similar to, but hairier than *Boreoiulus tenuis* (L:7.1–8.7 mm but can be as long as 11 mm; widespread, in woods, farmland and human-disturbed sites such as gardens).
- *Blaniulus guttulatus* – spotted snake millipede – L:8–16 mm; body whitish to cream with very obvious deep red spots along the sides, curls into a Catherine wheel shape when disturbed; widespread and common, from woods and more open habitats including agricultural croplands.

13 (4) Ommatidia present Chordeumatida (part)

Ten genera and 13 species (including three possible new species) key out here, of which 11 are rare or have very restricted distributions (see also couplet 5). The two common species are:

- *Craspedosoma rawlinsii* – L:15–16 mm; body deep reddish brown with a darker central band, slight bumps along the side (although not fully flat-backed); widely distributed but not particularly common, in damp areas under stones in woods and in wetlands.
- *Nanogona polydesmoides* – false flat-back millipede – L:17–21 mm; body light brown to dark brown, with obvious keels making it flat-backed (juveniles have very long setae on the body); common and widespread, under logs and stones in woods, coastal sites, quarries and waste ground.

– No ommatidia present 14

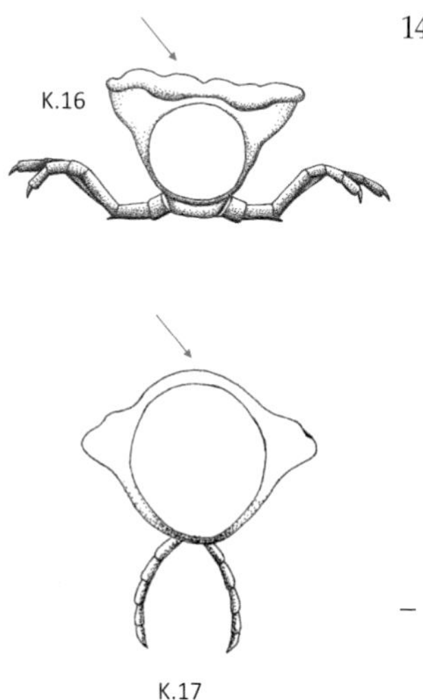


14 (13) Flat-backed (K.16) Polydesmidae

Three genera containing eight species, one of which is rare, while another two are introductions. The more common species are:

- *Brachydesmus superus* – L:8–10 mm; body greyish to light brown (sometimes whitish); widespread in a range of habitats, including human-disturbed sites such as farmland.
- *Polydesmus* species – L:10–21 mm; body brown; widespread and common, in many habitats. There are four widespread species of which *P. angustus* is the most common and largest (L:15–21 mm, can range from 13–28 mm) being found in most habitats especially in woods, also in human-disturbed sites such as farmland, gardens and wasteland.

– Arch-backed (K.17) seven species

Of these seven species, four are introduced hot-house species. The more common species are:

- *Macrosternodesmus palicola* – L:3.5–4.0 mm; tiny with white body covered in small lumps (rougher than *Ophiodesmus albonanus*); widespread, especially in calcareous habitats, and human-disturbed sites such as gardens and parks.
- *Ophiodesmus albonanus* – L:4.5–5.0 mm; body white (smoother than *M. palicola*); widespread, especially in calcareous habitats, and human-disturbed sites such as gardens and parks.
- *Stosatea italica* – L:11–14 mm; body smooth, chestnut brown with paler brown spots on the back; from southern England and southern Ireland in mainly calcareous habitats including human-disturbed sites.

## Key L Centipedes

This key will allow adults to be identified to order, family, and in some cases species. Although, in some instances only, juveniles may also be identified here, care should be taken to check that they are not adults of smaller species. In adult centipedes the sex organs are on the underside of the last segment. These are most conspicuous in female lithobiomorphs (they are indistinct in the males) and may be difficult to see in other groups. The number of pairs of legs should be carefully noted since this is useful in the identification of many centipedes. This is especially important in geophilomorphs (soil centipedes) where it remains constant throughout their life and thus enables even some juveniles to be identified to species. Scolopendromorphs (which include giant centipedes and *Cryptops* species) also have the same number of pairs of legs throughout their lives, but here the juveniles are more difficult to identify to species as features such as the characteristic teeth (combs) on the femur and tibia of the last legs are less fully developed. In lithobiomorphs (stone centipedes) and scutigeromorphs (house centipedes) the number of pairs of legs only reaches 15 in adults. Note that the forcipules (poison jaws) are actually modified legs (comprising a number of segments: basal coxosternite, femoroid, tibia, tarsus, and ending in a poison claw). These forcipules are not counted when recording the number of pairs of legs for this key. Other structures that it is useful to be able to recognise include the plates covering the top of each segment (tergites) and their counterparts underneath each segment (sternites). See L.1 for the general morphology of centipedes. All lengths (L) given are the maximums for adults. Where the descriptions do not fully cover individuals, consult Barber (2008), Barber (2009) and Iorio & Labroche (2015). A checklist may be found on the BMIG website (see Chapter 6) and distribution maps are given by Barber (2022). See Section 3.5 for further information on centipedes.

L.1

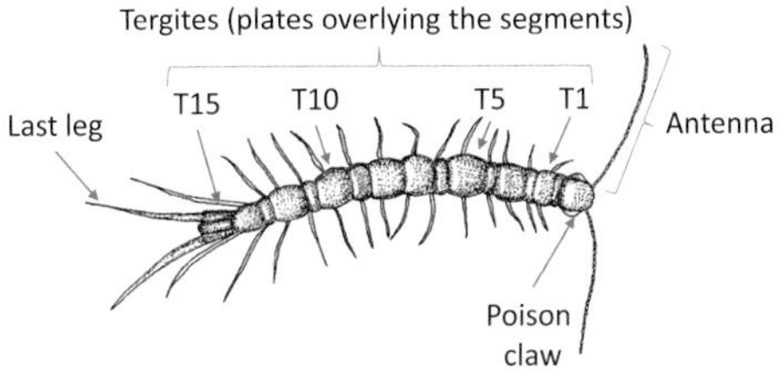

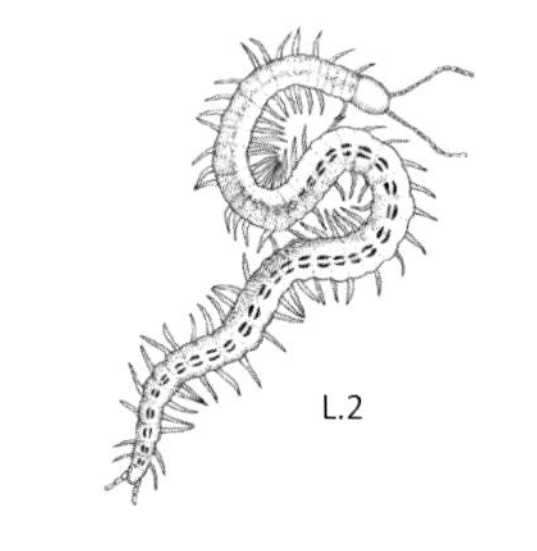
L.2

1 Fewer than 30 pairs of legs; fast moving 2

– More than 30 pairs of legs; slower moving (L.2) Geophilomorpha 9

2 (1) Fifteen pairs of legs (L.3 and L.4) 3

– Twenty-one pairs of legs (L.5) Scolopendromorpha

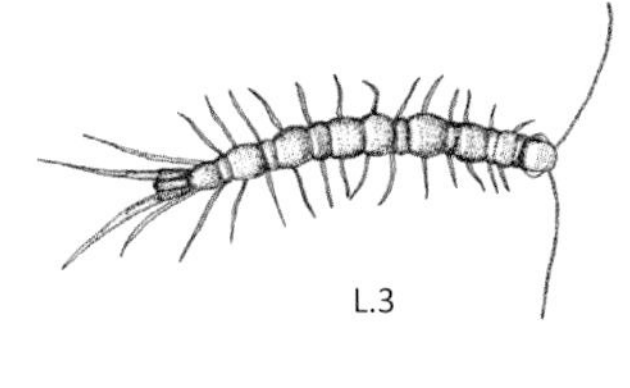
L.3

Two families are found in Britain. The Scolopendridae (giant centipedes – L:150 mm or more; often distinctively coloured and with two pairs of ocelli on each side of the head) includes only occasional introductions to Britain. The Cryptopsidae, which do not have ocelli, includes some hot-house introductions and three native species all in the genus *Cryptops*, which can be difficult to separate (may require close examination of the teeth on the tibiae and tarsi of the last pair of legs):

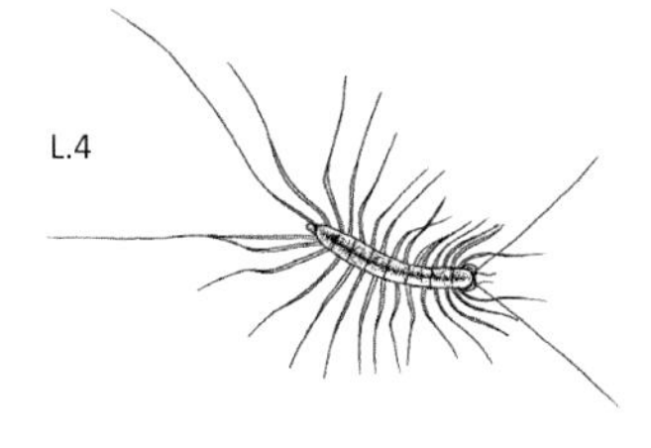
L.4

- *C. anomalans* – greater cryptops – L:50 mm; body ginger, with two parallel sutures running the entire length of the head, and (uniquely among these three species) the suture on the first tergite is X-shaped; mainly human-disturbed sites in southern England and south Wales.
- *C. hortensis* – common cryptops – L:30 mm; body ginger, with no sutures along the length of the head; found in human-disturbed sites up to central Scotland and in woodlands in Southern England.
- *C. parisi* – Paris cryptops – L:>30 mm (does not get as large as *C. anomalans*); body ginger, with two parallel sutures running part way from the base of the head; in human-disturbed sites up to southern Scotland and in woodland in south-west England.

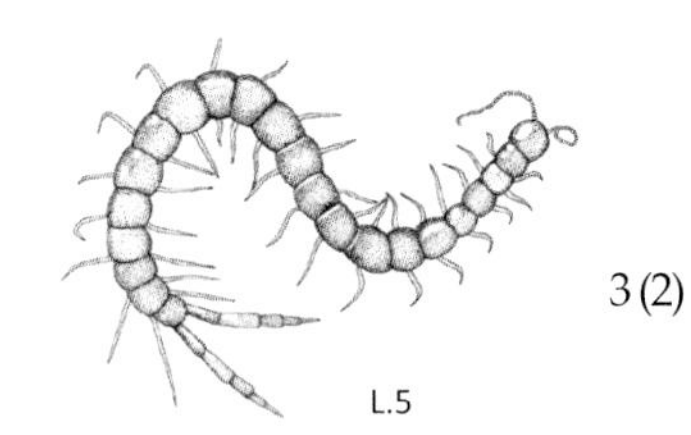
L.5

3 (2) Legs rather delicate (and easily shed) and very long, with many segments (as have the antennae) (L.4); head with compound eyes Scutigeromorpha

There are two introduced species from two genera, which may be found occasionally, usually restricted to buildings (although they may sometimes be found outdoors near buildings). The most commonly seen is:

- *Scutigera coleoptrata* – house centipede – L:30 mm; body dull purplish brown with three purple lines running along the body (so-called 'go faster stripes') and purple bands on the legs; occasional in buildings across England and Wales and up to Scotland, also found outdoors in the Channel Islands and occasionally on outside walls in England. The stripes in this species can distinguish it from the rarer introduction *Thereuonema tuberculata* (Japanese house centipede), which has been recorded once in Britain.

– Legs relatively robust and not very long (L.3) Lithobiomorpha 4

**organ of Tömösváry**
These are paired sensory organs situated on the head of some centipedes and millipedes, as well as some hexapods such as springtails. Their true function is uncertain but it seems most likely that they evolved as chemoreceptive or more strictly olfactory organs.

4 (3) One simple eye (ocellus) on each side of head (take care to distinguish the pigmented ocelli from the smaller and unpigmented organ of Tömösváry, which lies to the front of the ocelli) Henicopidae

There is one genus containing two species, one of which is a hot-house introduction (which does not have ocelli). The other species is:

- *Lamyctes emarginatus* – one-eyed centipede – L:10.5 mm; body chestnut to dark brown, triple claws on last legs (the main claw plus two smaller accessory claws); cultivated and damp habitats, mostly seen during the summer and autumn.

– At least three ocelli on each side of the head Lithobiidae 5

There are 17 species from one genus (*Lithobius*), some of which are difficult to separate.

5 (4) Three (sometimes four) ocelli on each side of head; reddish brown; last two pairs of legs swollen; curls up when disturbed *Lithobius microps*

- *L. microps* – least lithobius – L:9.5 mm; body (light) chestnut brown, antennae with about 25 segments; human-disturbed sites in England and Wales as well as woodland in the south.

– More than four ocelli on each side of head 6

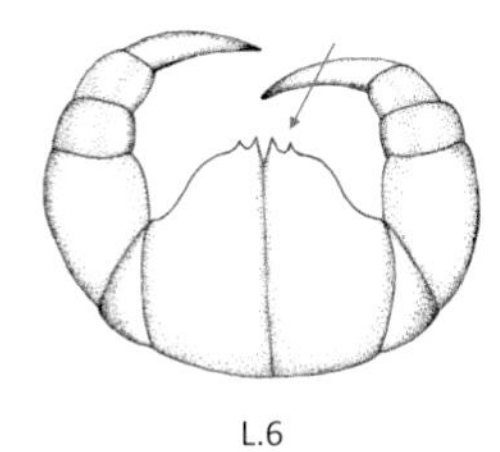

L.6

6 (5) Fewer than four teeth (two or three) on each coxosternite of the forcipule (i.e. on the underside of the first segment of the poison jaws) (L.6); animal under 20 mm long (take care that small animals are not juveniles of larger species) 7

– Four or more teeth on each coxosternite of the forcipule (L.7); animal usually more than 20 mm long 8

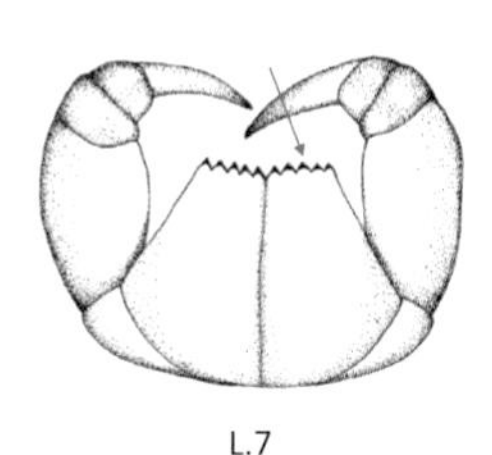

L.7

7 (6) Single claw on fifteenth (last) leg (L.8) *Lithobius* species

Four *Lithobius* species key out here, one of which has only one modern record. The more common species may be difficult to separate:

- *L. crassipes* – thick-legged lithobius – L:13.5 mm; body chestnut brown, with 20 antennal segments and 9–13 ocelli; widespread in much of Britain except the south-west, in woods, grassland, heathland moorland and farmland.
- *L. curtipes* – curling lithobius – L:11 mm; body chestnut brown, with 20 antennal segments and 6–9 ocelli; quite widespread in England and Wales, in woodland and limestone grassland and in the uplands in Wales, absent from much of Scotland.
- *L. muticus* – broad-headed lithobius – L:15 mm; body dark brown, with 34–43 antennal segments and 10–14 ocelli; woodland and scrub.

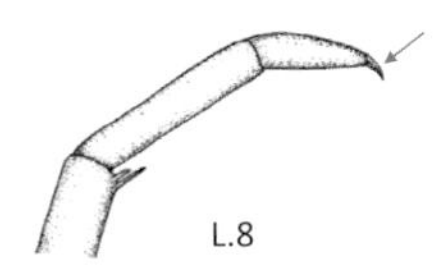

L.8

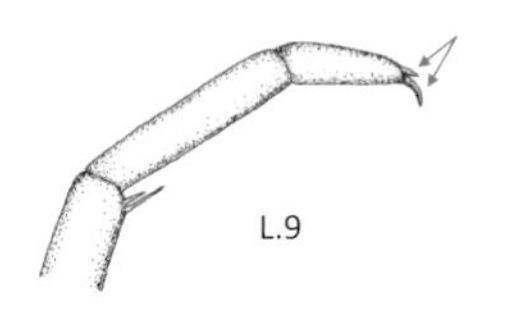
L.9

– Double claw (the main claw plus a smaller accessory claw) on fifteenth (last) leg (L.9) *Lithobius* species

Six *Lithobius* species key out here, some of which can be difficult to separate (see also the tabular key below):

- *L. borealis* – western lithobius – L:12.5 mm; body chestnut to dark brown; in western Britain found in moorland and heathland, also in woods, scrub and farmland.
- *L. calcaratus* – black lithobius – L:15 mm; body dark brown, can be blackish or greyish; dry grassland, moorland and some coastal areas.
- *L. lapidicola* – sandy lithobius – L:8 mm; body chestnut brown; rare, coastal areas in eastern England (East Anglia, Essex and Kent), also found in hot-houses.
- *L. macilentus* – virgin lithobius – L:14 mm; body chestnut brown, only females have been found in Britain; mainly woodland habitats.
- *L. melanops* – garden lithobius – L:17 mm; body chestnut brown (may have a darker central stripe); common in human-disturbed sites, also in coastal areas.
- *L. tricuspis* – three-spined lithobius – L:14 mm; body brown with darker head; rural sites in the south-west of England and south Wales.

These species can also be separated on the basis of the numbers of ocelli and antennal segments, and whether they have projections on the hind angles of the 9th and/or 11th tergites (the plates overlying the top of the segments – L.10):

**locating tergite numbers**
The easiest way to check the tergite numbers in lithobiomorphs is to check which two neighbouring segments appear roughly the same size (i.e. both being squarish rather than one being square and one being much narrower) – these are tergites 7 and 8.

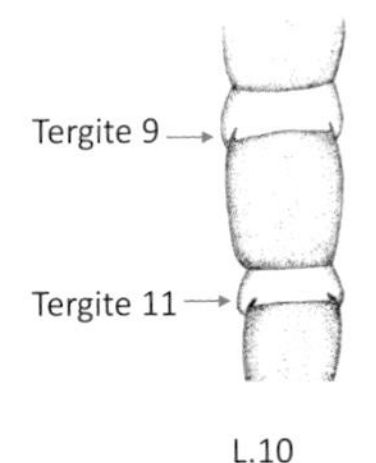

L.10

| Species | Number of ocelli | Number of antennal segments | 9th tergite* | 11th tergite* |
|---|---|---|---|---|
| *Lithobius borealis* | 8–12 | 28–34 | No projections | Projections |
| *L. calcaratus* | 7–9 | 39–50 | No projections | No projections |
| *L. lapidicola* | 10–11 | 26–34 | No projections | No projections |
| *L. macilentus* | 7–9 | 39–45 | Projections | Projections |
| *L. melanops* | 10–13 | 32–42 | Projections | Projections |
| *L. tricuspis* | 10–12 | 40–45 | Projections | Projections |

* See marginal note to ensure correct identification of tergite numbers

8 (6) Single claw on fifteenth (last) leg (L.8) *Lithobius* species

Four *Lithobius* species key out here, including one rare introduced species, which has been found in a Scottish churchyard and a greenhouse. The three common species are:

- *L. forficatus* – common lithobius – L:30 mm; body chestnut brown (pale blue when freshly moulted); widespread and common in many habitats including urban sites and gardens, tends to run quickly when disturbed.
- *L. pilicornis* – greater lithobius – L:35 mm; body chestnut to dark brown; from urban areas across the country and a wide range of habitats in the south-west of England.
- *L. variegatus* – variegated lithobius (or variegated centipede) – L:30 mm; body lightish brown with purple markings and striped legs; very common (less so, or even absent, in the east of Britain), especially in woodland and moorland, tends to be found in rural rather than urban sites, likely to freeze when first disturbed.

These species can also be separated on the basis of the number of ocelli and whether they have projections on the hind angles of the 9th and/or 11th tergites (the plates overlying the top of the segments – L.10):

| Species | Number of ocelli | Number of antennal segments | 9th tergite* | 11th tergite* |
|---|---|---|---|---|
| *Lithobius forficatus* | 20–30 | 35–43 | Projections | Projections |
| *L. pilicornis* | 20–40 | 29–34 | No projections | Projections |
| *L. variegatus* | 13–18 | 35–46 | Projections[†] | Projections |

* See marginal note to ensure the correct identification of tergite numbers
[†] Also has projections on tergite 7

– Double claw (the main claw plus a smaller accessory claw) on fifteenth (last) leg (L.9) *Lithobius* species

Two *Lithobius* species key out here, including one introduced vagrant species that has only been recorded twice and may now be extinct. The more common species is:

- *L. piceus* – long-horned lithobius – L:21 mm; body chestnut brown; mainly woodland in south-east England and south Wales.

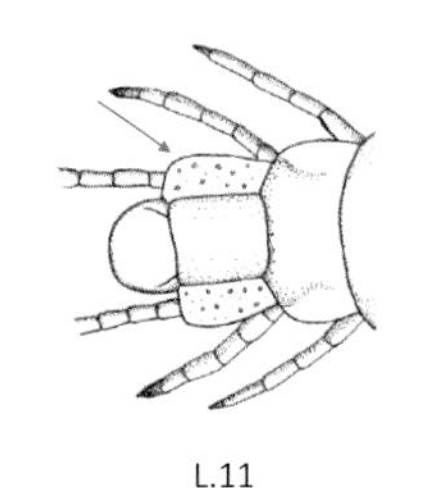
L.11

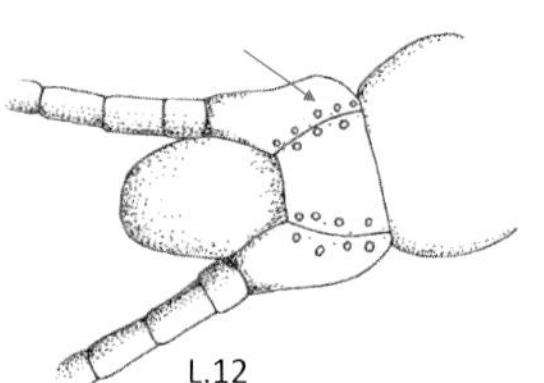
L.12

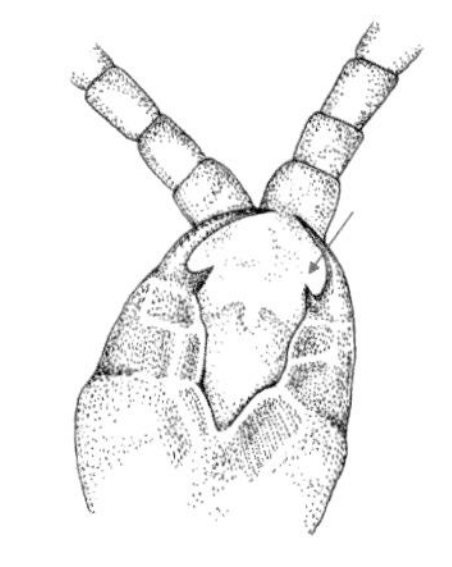
L.13

9 (1) Coxal pores on the last legs obvious, spread over at least most of the underside of the coxae (first segment of the leg) (L.11) 10

– Coxal pores on the last legs confined to the edge of the coxae next to the underside of the last segment (L.12) (note that these can sometimes be difficult to see and there may be an additional single isolated pore away from the main group in some species) 12

10 (9) Numerous coxal pores on the last legs on lower surface only; very large projection at the base of the poison jaw (L.13 – view from below) Linotaeniidae

There are three species in one genus:

- *Strigamia acuminata* – shorter red centipede – L:30 mm, 37–39 body segments in males and 41 in females; body red; woodlands, especially in the south of Britain.
- *S. crassipes* – longer red centipede – L:50 mm, 49–51 body segments in males and 51–53 in females; body red; woods, especially in the south of Britain.
- *S. maritima* – maritime red centipede – L:40 mm, 47–49 body segments in males and 49–51 in females; body red; throughout Britain and Ireland, sometimes very abundant, under stones, and in shingle and crevices on the seashore.

– Coxal pores on the last legs spread over the whole upper and lower surface of the coxae 11

11 (10) Head wider than long Himantariidae

Two species in one genus, one of which is rare (apparently confined to the Falmouth area of Cornwall). The more common species is:

- *Haplophilus subterraneus* – western yellow centipede – L:70 mm (more usually 60 mm), 77–83 pairs of legs; yellow to pale brown, head darker; human-disturbed habitats up to southern Scotland, as well as woodlands in the south-west.

– Head narrower than long Mecistocephalidae

Three genera, each containing one species, all of which are introductions to hot-houses.

One species from the family Geophilidae will also key out here:

- *Pachymerium ferrugineum* – red-headed centipede – L:50 mm; body reddish yellow with darker head end; coastal shingle, rare, on the east and south coasts of England and the Channel Islands.

12 (9) Head wider than long — Dignathodontidae

Two species from one genus:

- *Henia brevis* – southern garden centipede – L:19 mm, 53–57 pairs of legs; body whitish yellow; mainly human-disturbed sites in southern England.
- *H. vesuviana* – white-striped centipede – L:50 mm, 63–75 pairs of legs; body robust, greenish grey with a central white line, and both head and tail ends light reddish brown; coastal (above the high tide line) and elsewhere in southern England, generally in human-disturbed areas inland.

– Head narrower than long — 13

13 (12) Last leg with claws; two or more coxal pores (up to 12 or more) on each of the last legs (these may be difficult to see in one species since the pores open up into pits – use oblique lighting – easier to see on living specimens) — Geophilidae 14

There are four genera containing 13 species.

– Last leg without claws; only two coxal pores on each of the last legs — Schendylidae

There are two genera containing five species, of which one has been recorded only once from a greenhouse. Some of the more common species are difficult to separate:

- *Hydroschendyla submarina* – seashore schendyla – L:40 mm; body reddish brown; seashores at (or below) the high-water mark. Superficially similar to *Strigamia maritima* (see couplet 10) but lacks the large projection at the base of the poison jaw as found in that species (L.13).
- *Schendyla dentata* – toothed schendyla – L:12 mm, 39 pairs of legs; body colourless (translucent), with two teeth on the forcipule (one towards the base – on the femoroid – and one at the base of the terminal segment, the poison claw); in urban and semi-urban sites under stones.
- *S. nemorensis* – common schendyla – L:20 mm; body colourless to pale yellow, darker head; common in gardens and coastal and rural sites.
- *S. peyerimhoffi* – lesser shore schendyla – L:18 mm; body colourless to pale yellow, darker head end; under stones on muddy seashores. Difficult to separate from *S. nemorensis*.

A species from the family Geophilidae will also key out here:

- *Arenophilus peregrinus* – least shore centipede – L:12 mm; body yellowish white, darker head; rare (only found three times – on the Isles of Scilly and at two sites in Cornwall).

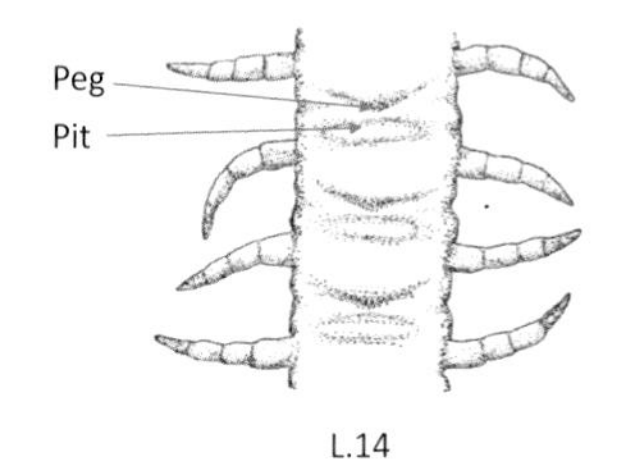

L.14

14 (13) With carpophagus structure (L.14) (transverse pit and peg on underside of front segments – largest on segments 8/9 – this is best observed on dry specimens) several genera

There are two genera containing nine species, of which one has been recorded only once from Shetland. Some of the more common species are difficult to separate (see also the tabular key below):

- *Geophilus alpinus* (previously known as *G. insculptus*) – common geophilus – L:40 mm (usually 30 mm); body pale yellow with darker head end, and with an isolated coxal pore on the last pair of legs away from the main group; widespread across a range of habitats.
- *G. carpophagus* – luminous centipede – L:60 mm; body greenish grey or brownish grey (sometimes with a purple tint), with both head and tail ends reddish brown; found in human-disturbed areas and at the coast.
- *G. easoni* – Eason's centipede – L:40 mm; body chestnut brown; under logs and stones in woods and moorland.
- *G. electricus* – Linnaeus' centipede – L:40 mm (or longer); body pale yellow with darker head end; human-disturbed sites.
- *G. osquidatum* – western geophilus – L:30 mm; body yellow with darker head end; tends to have a distinctly southern and western distribution in Britain, in various habitats including gardens.
- *G. seurati* (previously known as *G. gracilis*) – beach geophilus – L:30 mm; body yellowish with darker head end; under stones on the seashore.
- *G. truncorum* – small geophilus – L:20 mm (usually 12–14 mm); body pale yellow to pale brown with darker head; distinctly rural, often found under dead bark and leaf litter, woodland, grassland and moorland.
- *Eurygeophilus pinguis* – cut-short centipede (or Devonshire paradox) – L:20 mm; body pale yellowish brown, very distinctly short (low number of pairs of legs: 35–37) but fairly broad in comparison to the other species listed here, 6–10 coxal pores plus a single isolated coxal pore, poison claw without tooth at base – smooth and scimitar-like (unlike the *Geophilus* species described here); north Devon and west Cornwall, often associated with deciduous trees.

These *Geophilus* species can also be separated on the basis of the size and distinctiveness of the carpophagus pit, the number (and position) of the coxal pores, and the number of pairs of legs:

| Species | Carpophagus pit | Number of coxal pores | Number of pairs of legs |
|---|---|---|---|
| *Geophilus alpinus* | Indistinct; over three-quarters the width of the sternite | 4–7* | 45–53 |
| *G. carpophagus* | Distinct; nearly half the width of the sternite | 4–8† | 51–57 |
| *G. easoni* | Distinct; nearly half the width of the sternite | 6–12 | 47–51 |
| *G. electricus* | Indistinct; about three-quarters the width of the sternite | 10–18*‡ | 65–73 |
| *G. osquidatum* | Indistinct; about two-thirds the width of the sternite | 3–4 | 53–63 |
| *G. seurati* | Indistinct; about three-quarters the width of the sternite | 4 | 51–57 |
| *G. truncorum* | Indistinct; just over half the width of the sternite | 2 | 37–41 |

* plus an isolated single coxal pore
† difficult to see
‡ distributed on upper (4–6 pores) as well as the lower (6–12 pores) surface (all other species in this table have coxal pores only on the lower surface)

– Without carpophagus structure (pegs may be present but not pits) several genera

There are three genera containing four species, some of which can be difficult to separate (see also table below):

- *Geophilus flavus* – long-horned geophilus – L:45 mm; body bright yellow with darker head end, with long antennae composed of elongated segments, carpophagus pegs present; widespread and common in many areas of Britain, in a range of habitats.
- *G. pusillifrater* – scarce geophilus – L:13 mm; body pale with slightly darker head; only recorded on a few occasions, on coastal shingle in the south of England.
- *Nothogeophilus turki* – Turk's geophilus – L:13 mm; body yellowish white with darker head end; rare, Isles of Scilly and Isle of Wight.
- *Stenotaenia linearis* – larger urban geophilus – L:55 mm; body yellow with darker head end, poison claw without tooth at base – smooth and scimitar-like (unlike *N. turki* and the *Geophilus* species described here); across Britain, most common around the London area, can be abundant in human-disturbed areas.

These species can also be separated on the basis of the number (and position) of coxal pores and the number of pairs of legs:

| Species | Number of coxal pores | Number of pairs of legs |
|---|---|---|
| *Geophilus flavus* | 6–10 | 49–57 |
| *G. pusillifrater* | 2* | 41–43 |
| *Nothogeophilus turki* | 3–5 | 37–39 |
| *Stenotaenia linearis* | In pits† | 63–79 |

* plus an isolated single coxal pore
† difficult to see, easier to find in living specimens

## Key M Hexapoda

The Hexapoda include the insects and insect-like arthropods (such as springtails, diplurans and proturans) that have three pairs of legs. The classification of some of these groups is subject to a certain amount of debate and the three non-insect hexapod groups are sometimes included within a single class (the Entognatha), while other taxonomists give each group class status (Collembola, Diplura and Protura), alongside the Insecta as four classes within the subphylum Hexapoda. We have followed the former convention. This key will enable the identification of hexapods to class or order. Where lengths are given (L), these are body lengths taken from the front of the head to the end of the abdomen (not including any appendages (e.g. antennae, legs or tails – cerci). Note that in some winged insects the forewings are just as membranous as the hindwings (e.g. in many Hymenoptera – bees, wasps and ants), while in others (especially in beetles – Coleoptera – and earwigs – Dermaptera) they are thickened and hardened to form a cover and are called elytra (singular elytron – from the Greek *elutron* meaning a cover or sheath). In many true bugs (Hemiptera) the forewings are only thickened at the bases and are termed hemelytra. In groups such as cockroaches (Dictyoptera) and grasshoppers and crickets (Orthoptera), the leathery thickened forewings are called tegmina (singular tegmen – from the Latin *tegere* meaning to cover). See M.1 for the general morphology of hexapods. Where specimens do not fully meet the descriptions here, consult Tilling (2014). Brock (2019) contains illustrations of many British hexapods, especially of insects. The Amateur Entomologists' Society and the Royal Entomological Society websites provide more information about hexapods (see Chapter 6). See Section 3.5 for further information on hexapods.

M.1

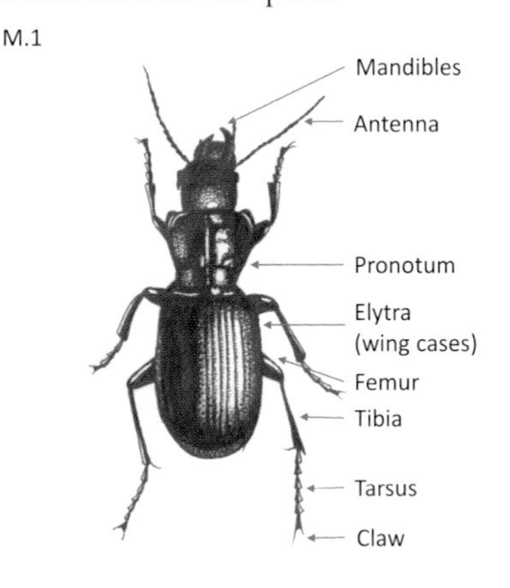

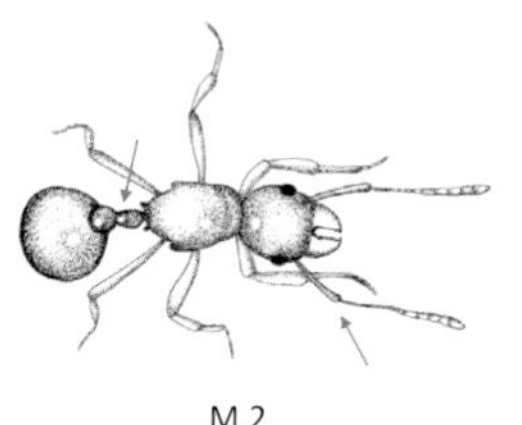
M.2

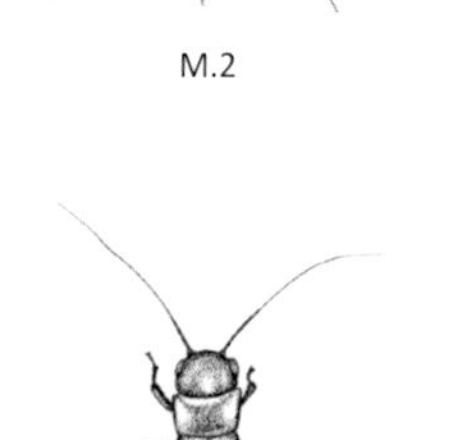

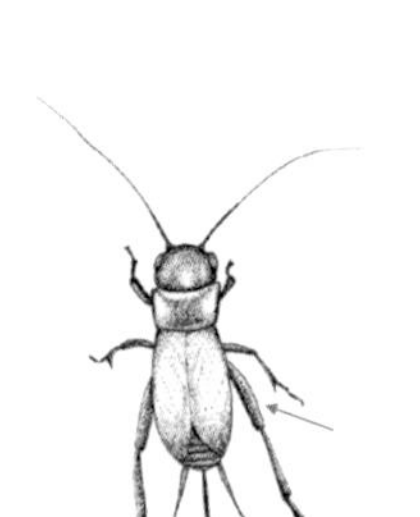

M.3

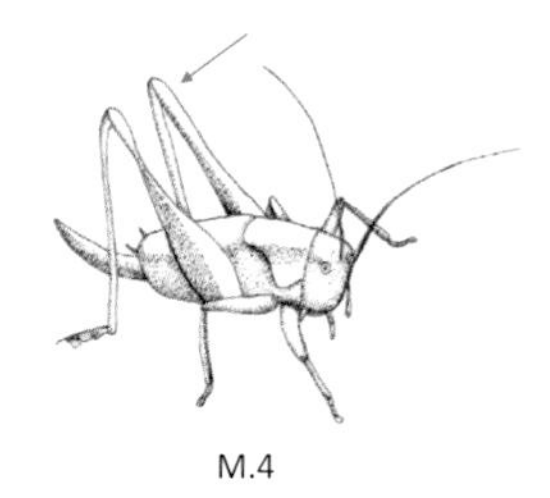
M.4

1 Wingless or wings not obvious (in some species wings may be hidden under hard wing-cases – see couplet 5) 2

– Wings obvious, not hidden under wing-cases 12

2 (1) Distinct waist behind thorax; antennae often elbowed (M.2) Hymenoptera (wingless ants) Key R

– Not as above 3

3 (2) Body grasshopper-like with large hind legs, which often overlap the top of the body (M.3 or M.4) Orthoptera (orthopterans: grasshoppers and crickets) Key Q

– Not as above 4

4 (3) With not more than six abdominal segments and always with a ventral tube (M.5); antennae usually with four segments, never more than six segments; small, often unpigmented Collembola (springtails) Key N

– With more than six abdominal segments; not as in M.5 5

5 (4) Abdomen wholly or partly covered by two hardened wing-cases which do not usually overlap one another at rest; wings usually completely covered (M.6, M.7, M.8, or M.9) 6

– Abdomen not so covered 7

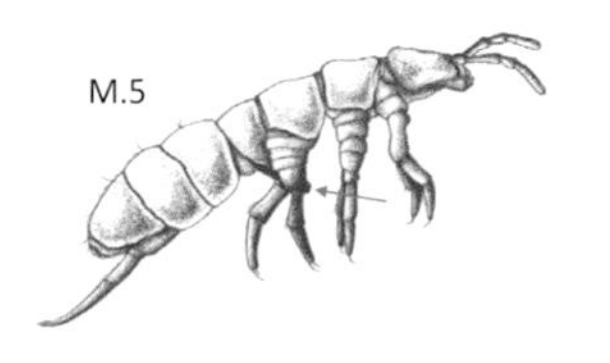
M.5

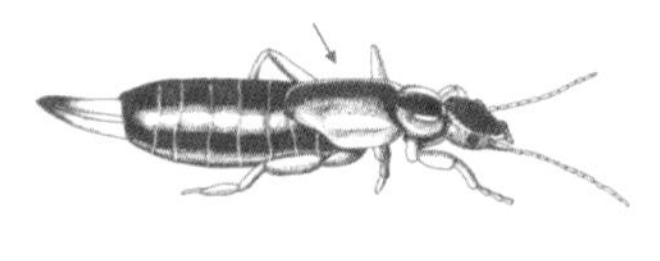

M.6

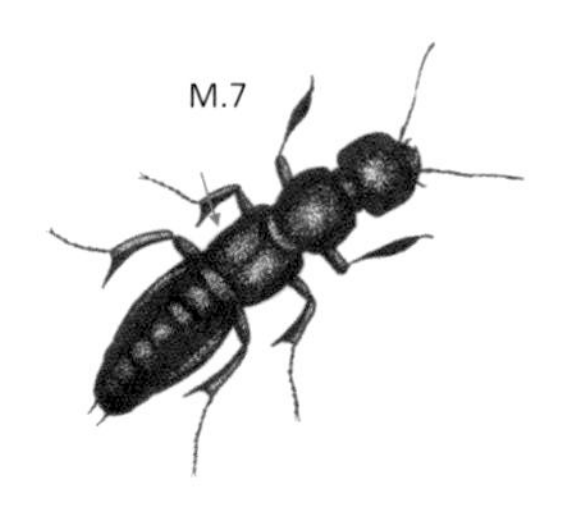
M.7

M.8

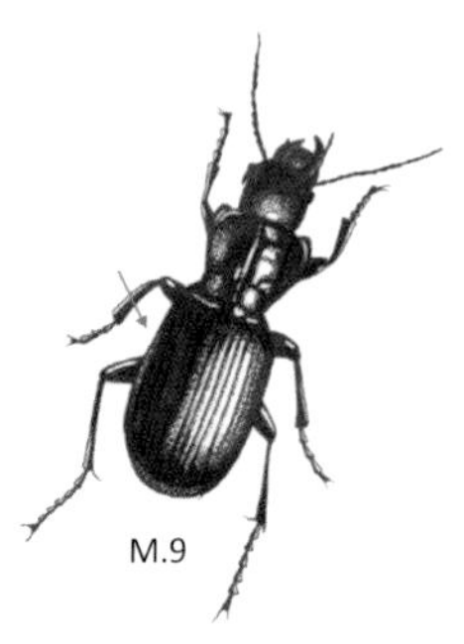
M.9

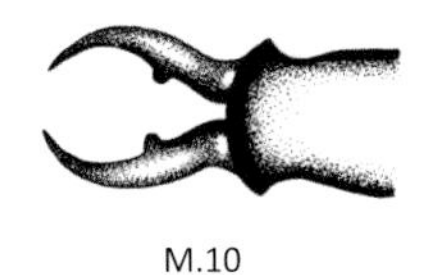
M.10

6 (5) Pair of curved forceps (may only be slightly curved) at hind end of body (M.6 – shape of forceps as in M.10 or M.11) Dermaptera (earwigs) Key O

– Without curved forceps (e.g. M.7, M.8 or M.9) Coleoptera (beetles) Key S

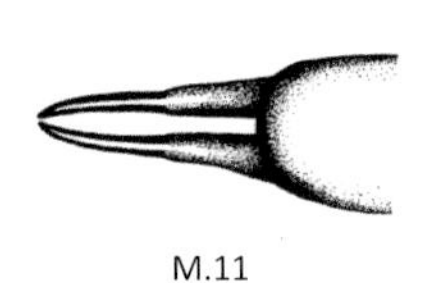
M.11

7 (5) Mouthparts a needle-like proboscis, usually resting between legs and pointing backwards (view from the side – as here – or from below) (M.12) Hemiptera (true bugs)

Hemiptera are not often found under logs and stones but may be associated with vegetation close by. See Southwood & Leston (1959) and Unwin (2001) for further information.

– Not as above 8

M.12

8 (7) Antennae absent or very small (M.13) 9

– Antennae long, usually more than three times the length of the head (e.g. M.14) 10

9 (8) Body less than 2 mm long; head tapers to a point in front; the front pair of legs often held forward in a similar position to antennae (M.15) Protura (proturans or coneheads)

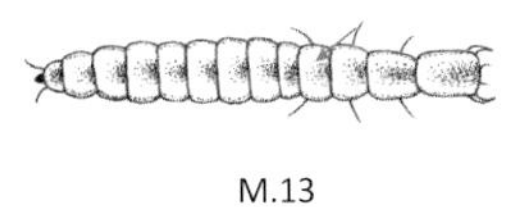
M.13

There are three British families containing six genera and 12 species. They have been relatively poorly studied and many species are difficult to separate. See Section 3.5 for further information. Note that some bush crickets are also called coneheads, providing an example of why using scientific names is usually preferable to using the common name.

– Not as above insect larvae Key V

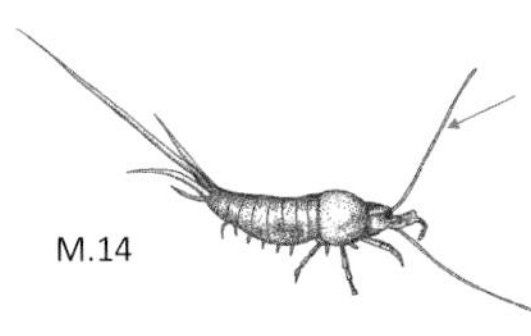
M.14

10 (8) With three long 'tails' on abdomen (M.16) Archaeognatha (jumping or three-tailed bristletails)

There are four genera containing seven species, two of which are very rare. *Dilta* species (L:7–11 mm; antennae much shorter than the body) are the commonest species found outdoors. *Petrobius* species (L:13–14 mm; antennae at least as long as the body) are more often found around coastal rocks. See Delaney (1954) for further details.

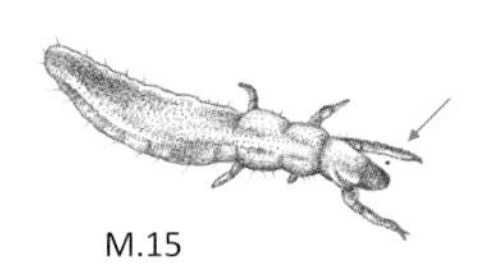
M.15

Species within the Zygentoma may also key out here. These include *Lepisma saccharina* (silverfish), *Ctenolepisma longicaudata* (silverfish or paperfish) and *Thermobia domestica* (firebrat), which are usually found in buildings. See Delaney (1954) for further details.

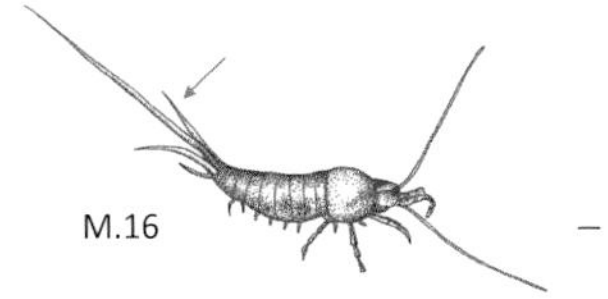
M.16

– Not as above 11

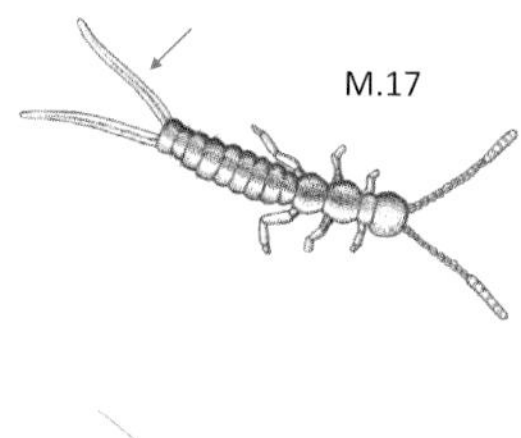
M.17

11 (10) With two long 'tails' on abdomen (M.17)
Diplura (two-tailed or two-pronged bristletails)

There are 12 species from one genus. One species that is common throughout the British Isles is *Campodea staphylinus* (3.9–4.6 mm). See Delaney (1954) and Section 3.5 for more information.

– Not as above — other groups

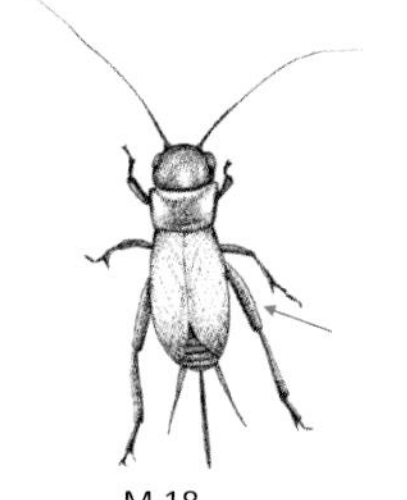
M.18

12 (1) Only one pair of wings — Diptera (true flies)

Diptera adults are rarely found under logs and stones, although they may be seen running over the surface of refuges as they are turned over. See Oldroyd (1970) and Unwin (1981) for further details.

– Two pairs of wings (the forewings and hindwings may be linked together but can be separated gently with a pin) — 13

13 (12) Wings transparent; hindwings much smaller than forewings; wings often linked together
Hymenoptera (bees, wasps, ants, and their allies) Key R

– Not as above — 14

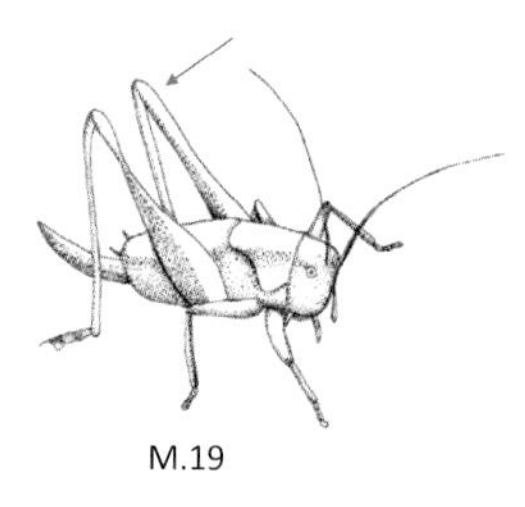
M.19

14 (13) Body grasshopper-like with large hind legs, which often overlap the top of the body (M.18 or M.19)
Orthoptera (orthopterans: grasshoppers and crickets) Key Q

– Not as above — 15

M.20

15 (14) Mouthparts a needle-like proboscis, usually resting between legs and pointing backwards (view from the side – as here – or from below) (M.20)
Hemiptera (true bugs)

Hemiptera are not often found under logs and stones but may be associated with vegetation close by. See Southwood & Leston (1959) and Unwin (2001) for further information.

– Not as above — 16

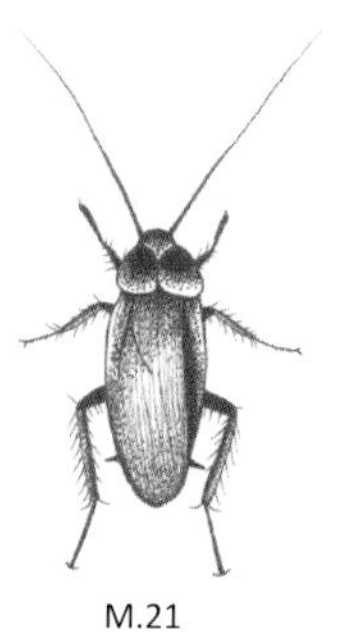
M.21

16 (15) Forewings leathery; legs spiny; two short prongs at tip of abdomen; body flattened (M.21)
Dictyoptera (cockroaches) Key P

– Not as above other groups

These include:

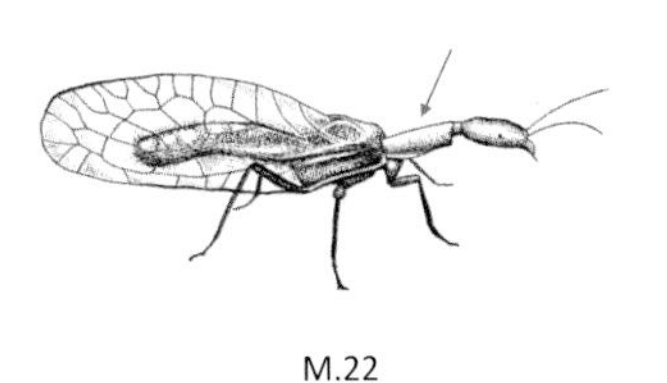

M.22

Raphidioptera – Raphidiidae (snakeflies) – these have a long 'neck' (M.22) and two pairs of similar wings. Under logs, it is mostly the larvae, pupae and newly emerged adults that might be found. See Plant (1997) for further information.

Lepidoptera – adult moths – wings covered in powdery scales. Adult moths are rarely seen under logs or stones, but some micro-moths may be found associated with vegetation close by. See Sterling *et al.* (2018) and Waring *et al.* (2018) for further information.

## Key N Springtails

Springtails are small, and detailed examination is necessary for accurate identification of species. This often means preparing specimens carefully for examination using a high-powered microscope – see Hopkin (2007) for further details. Previously, springtails were considered to comprise a single order, the Collembola. This has now been designated as a subclass and there are currently thought to be four British orders of springtails containing 19 families between them. Since the status of some springtail species in several of these families is currently rather uncertain, revisions to their taxonomy are being made fairly often. There are several useful references for help with identification, of which the most comprehensive is Hopkin (2007); this is supported by an excellent website on UK Collembola maintained by Peter Shaw (see Chapter 6). A number of other keys provide support when tackling this rather tricky group, including Fjellberg (1998 and 2007), Dallimore & Shaw (2013), as well as several useful websites (see Chapter 6). This key will identify springtails to order and then family, with some common species being highlighted. The four orders should be relatively easy to identify using a hand lens or low magnification microscope. Identifying some of the characteristics required to separate the families will need higher-magnification microscopy and a good light source. See N.1 for the general morphology of springtails. Where lengths (L) are given, they are of mature specimens from the front of the head to the end of the abdomen (ignoring any appendages such as legs or antennae). Many species are rather difficult to separate and in several places this key gives examples and descriptions of common species. Where a specimen does not fully match the description or where there are a number of similar species, consult Hopkin (2007). See Section 3.5 for further information on springtails.

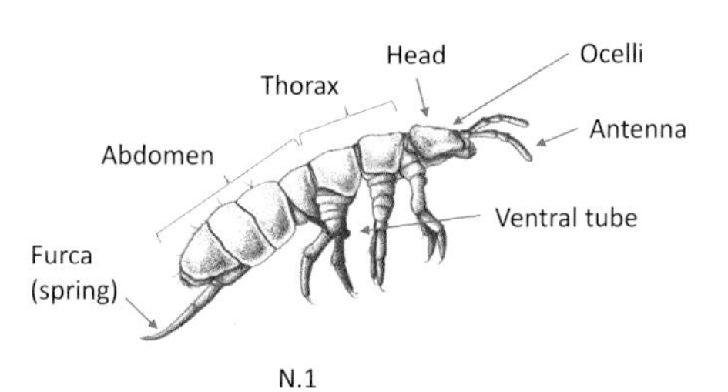


N.1

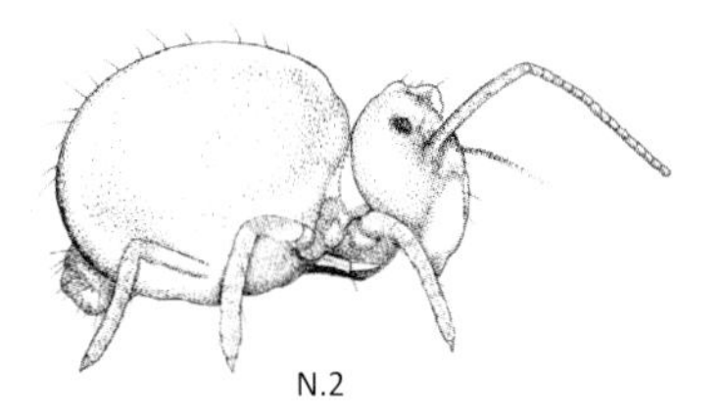
N.2

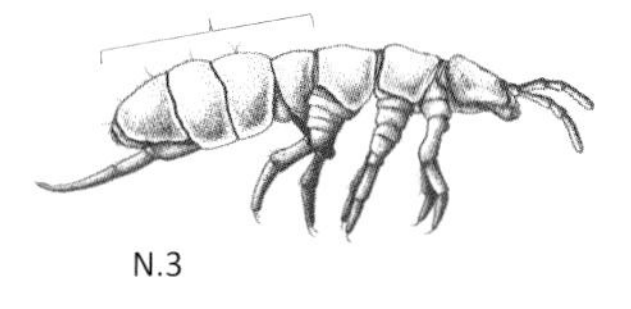
N.3

1 Body globular; abdominal segments more or less fused (N.2) Symphypleona 2

Members of this order are very fast jumpers and may be very difficult to collect when searching under logs and stones.

The order Neelipleona also keys out here. This order includes one family, the Neelidae, which contains three British species of which one is rare and one is found mainly in caves. All of these are small (<1 mm) with poorly or non-pigmented globular bodies, with short antennae (shorter than the size of the head) compared to the Symphypleona which have longer antennae (at least as long as size of the head) and no eyes (compared to the Symphypleona where each eye is made up of at least one ocellus, often eight). One very common species is:

- *Megalothorax minimus* – L:0.4–0.5 mm; body white; widespread and common, in soil.

– Body elongate; abdominal segments distinctly visible (e.g. N.3) 7

2 (1) Eyespots consist of a single ocellus Arrhopalitidae

There is one genus containing four species, of which two are scarce. The two common species, which live mainly in caves, belong to the genus *Arrhopalites* – L:1.0–1.2 mm; body white or pinkish; common.

– Eyespots consist of dark clusters of 8 ocelli (it may be difficult to count them all, but there are clearly more than 1) 3

3 (2) Terminal (4th) antennal segment much shorter than penultimate (3rd) Dicyrtomidae

There are four genera containing eight species, of which four are rare or imports. One common species that typically has distinctive markings is *Dicyrtomina saundersi* – L:3.0 mm; body brown with green patches and a pattern of pigment at the rear of the abdomen in a multibranched cross-shape (N.4), first two segments of antennae paler than the rest; very common.

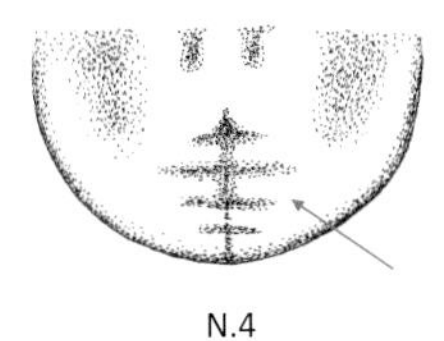
N.4

– Terminal (4th) antennal segment at least as long as penultimate (3rd) 4

4 (3) Terminal (4th) antennal segment subdivided into at least six subsegments 5

– Terminal (4th) antennal segment not subdivided 6

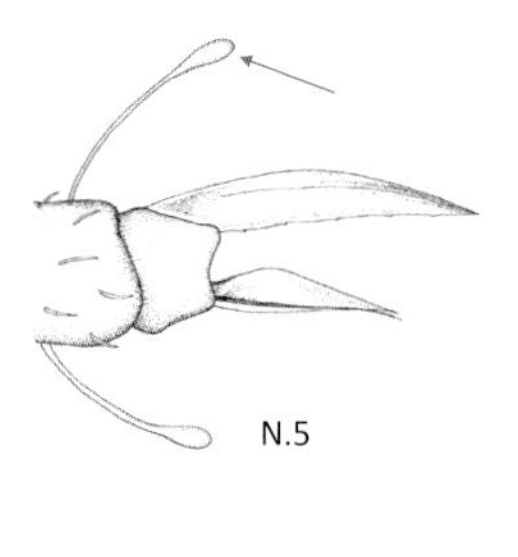
N.5

5 (4) Hairs near foot clubbed (N.5) Bourletiellidae

There are three genera containing ten species, of which three are rare and four are only moderately common. The most common species are:

- *Bourletiella arvalis* – L:1.5 mm; body yellowish; widespread and common.
- *B. hortensis* – garden springtail – L:1.3 mm; body bluish black; widespread and common.
- *Deuterosminthurus pallipes* – L:1.0 mm; body uniformly yellow or purple; widespread and common.

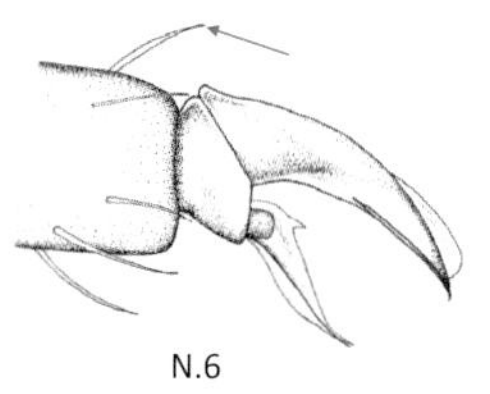
N.6

– Hairs near foot pointed (not clubbed) (N.6) Sminthuridae

There are five genera containing seven species, of which four are rare or scarce. The more common species are:

- *Allacma fusca* – L:3.5 mm; body dark brown; widespread and common.
- *Lipothrix lubbocki* – L:2.0 mm; body yellowish red with dark patches; moderately common, in ancient woodland.
- *Sminthurus viridis* – clover springtail – L:3.0 mm; body greenish or yellowish brown; widespread and very common.

6 (4) Hairs near foot clubbed (N.5) Katiannidae

There are five genera containing 17 species, of which 14 are hot-house species, or are rare or scarce. Common species are:

- *Sminthurinus aureus* – L:1.0 mm; body black, brown or greyish yellow; widespread, in leaf litter.
- *S. elegans* – L:0.7 mm; body yellow with two or four bluish stripes; widespread, and common.
- *S. niger* – L:1.0 mm; body blue-black, head with white dots behind the eye; widespread, moss and damp leaf litter.

– Hairs near foot pointed (not clubbed) (N.6) Sminthurididae

There are four genera containing 11 species, of which four are rare and two are moderately common. Males have distinctive claspers on their antennae for grasping females during mating. There are four common/very common species of the genus *Sminthurides* – L:0.5–1.0 mm; body yellow with some blue markings;

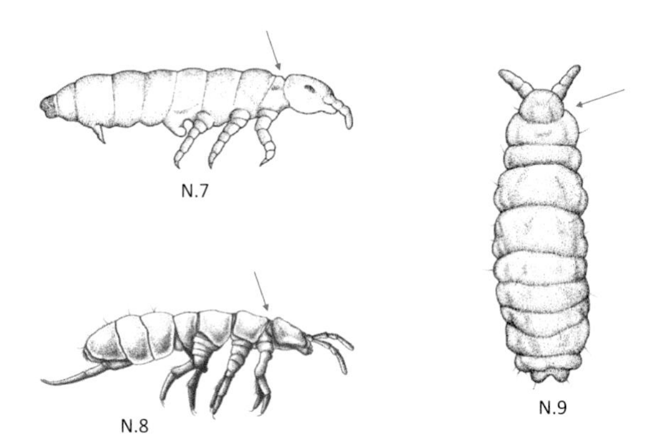


often found on the surface of fresh water. A common species frequently encountered under logs is:

- *Sphaeridia pumilis* – L:0.25 mm; body pinkish or greyish yellow; widespread and very common, mainly in soil.

7 (1) First segment of thorax easily seen when viewed from above (N.7) Poduromorpha 8

– First segment of thorax very small (may not be visible when viewed from above – e.g. N.8) Entomobryomorpha 10

8 (7) Head not distinct from body (N.9) Neanuridae

There are ten genera containing 23 species. Three of these genera (comprising seven species) have a furca (spring) although this is difficult to see unless under high magnification. The commonest of these are *Friesea* species which have three short, curved spines at the end of the abdomen.

The remaining seven genera (containing 16 species) either do not have a furca, or it is highly reduced. The most common of these species are:

- *Anurida granaria* – L:2.2 mm; body white or yellowish, body setae short, no ocelli (blind); widespread and very common, especially in waterlogged sites. Sometimes luminous when alive.
- *A. maritima* – L:3.4 mm; body dark blue, each eye composed of five ocelli; very common, around the coastline including in the subtidal zone.
- *Neanura muscorum* – L:3.5 mm; body bluish grey, each eye composed of three ocelli; widespread and very common, often under logs.

– Head distinct from body 9

9 (8) Body (with at least some) pigmentation (i.e. if white, has bluish markings) Hypogastruridae

There are seven genera containing 33 species, of which 17 are rare, scarce or only moderately common. See the second part of this couplet for the unpigmented genus *Willemia*). The more common are:

- *Ceratophysella* species – L:1.3–2.0 mm; body grey, brown, greyish blue or blue-black (*C. bengtssoni* may be pink), eight ocelli in each eye, spines at the end of the abdomen large; widespread and common, in damp or wet habitats.
- *Hypogastrura* species – L:1.5–2.4 mm; body reddish grey/brown, bluish grey, or dark blue/black, eight ocelli in each eye, spines at the end of the abdomen barely visible; widespread and common in organic material.
- *Schaefferia emucronata* – L:1.9 mm; body white with blue spots, four (or fewer) ocelli in each eye, two large spines at the end of the abdomen, furca present; widespread and very common, often in caves, also in moss and rotting wood.
- *Xenylla boerneri* – L:0.7 mm; body bluish grey, five ocelli in each eye, furca reduced to very small bumps (appears absent) unlike the other *Xenylla* species here, which all have a furca; widespread and moderately common, especially in moss.
- *X. grisea* – L:1.2 mm; body bluish grey, five ocelli in each eye, furca present; widespread and common, among organic debris. This species may be difficult to separate from *X. humicola* – L:2.0 mm; body blue-black, five ocelli in each eye, furca present; widespread and common, especially on seashores.

Four other families (at least in part) also key out here:

Brachystomellidae – one species (*Brachystomella parvula*) – L:1.0 mm; body bluish or reddish violet; widespread and common, in soil. Needs high magnification to confirm this species.

Odontellidae – two British genera each with one species, which are blue-grey, have conical antennae, and are rare.

Poduridae – one species (*Podura aquatica* – water springtail), only found on the surface and margins of lakes and ponds.

One species of Onychiuridae (also see the second part of this couplet) also keys out here:

- *Protaphorura aurantiaca* – L:2.5 mm; body orange or yellow, without ocelli; widespread and very common, in rotting wood and under stones.

– Body white (not pigmented) Onychiuridae and Tullbergiidae

Many species within these two families can only be separated by close inspection under high magnification. In all these species the furca is absent or very reduced, and eyes are absent.

Onychiuridae – 14 genera containing 17 species (one of which keys out in the first part of this couplet), of which eight are found in glasshouses or caves, or are rare or scarce and two are only moderately common. There are a number of white species with two spines at the end of the abdomen that are difficult to separate. More distinctive species include:

- *Deuteraphorura cebennaria* – L:2.3 mm; body white, no spines at the end of the abdomen (compared to two spines in the other species listed here), furca absent; widespread and common, in leaf litter and soil.
- *Kalaphorura burmeisteri* – L:3.0 mm; body white, distinctive in having very rounded segments ('Michelin-man' shape), furca reduced to two small bumps; common mainly in southern England and Wales, under logs and stones.
- *Supraphorura furcifera* – L:1.9 mm; body white, furca reduced to two pointed bumps; widespread and common.

Tullbergiidae – six genera containing 13 species (plus up to another eight species described as uncertain), of which six are rare or scarce, and two are only moderately common. All of the common species are L:0.7–1.5 mm with white bodies. Two common *Stenaphorura* species (*S. denisi* and *S. quadrispina*) can be distinguished by having four spines at the end of the abdomen (compared to two in the other common species).

One unpigmented genus of the Hypogastruridae (*Willemia*) containing six species (five of which are rare or scarce) also keys out here (see also the first part of this couplet). These species are difficult to separate from some species of Onychiuridae and Tullbergiidae – L:0.5–0.7 mm; body white, with two spines at the end of the abdomen (except for *W. denisi* which has no such spines); found in soil and moss.

10 (7) Last antennal segment less than half as long as penultimate segment — Tomoceridae

There are two genera containing five species, of which two are rare or scarce. Common species are:

- *Pogonognathellus longicornis* – L:6.2 mm; body greyish brown with bluish femurs; widespread and very common, especially in gardens.
- *Tomocerus* species – L:4.0–4.5 mm; body greyish brown/black; widespread and very common.

The family Oncopoduridae will also key out here. This family contains a single white species which is scarce.

– Last antennal segment at least half as long as penultimate segment 11

11 (10) Eyes absent Cyphoderinae

There is one genus containing two species (both found in ants' nests), one of which is very rare. The common species is:

- *Cyphoderus albinus* – L:1.6 mm; body white, the last segment of the furca is long with two teeth at the tip; common, in or near ants' nests. White springtails associated with ants are nearly always *C. albinus* (however confirmation of this species requires close examination at high magnification).

Two species of Isotomidae (see also couplet 12) also key out here, one of which is rare. The common species is:

- *Isotomodes productus* – L:0.9 mm; body white; widespread and common, in soil. This species has a shorter furca (about the same length as the head) than *C. albinus* where the furca is longer than the length of the head.

– Eyes present 12

12 (11) Abdomen with the last three segments (4, 5 and 6) fused giving the appearance of only having four abdominal segments (note – start counting abdominal segments at the first body segment after the last pair of legs – abdominal segment 1) Isotomidae (part)

The genus *Folsomia* keys out here (see couplet 13 for the remaining Isotomidae species). There are 14 species, of which six are rare or scarce, and three are moderately common or locally common. Of the remaining five species, three are grey/black in colour and are difficult to separate, while two are white, lack eyes and may be separated by their size: *F. candida* (L:3.0 mm; often found in compost and flowerpots, also widely cultured in laboratories) and the smaller *F. fimetaria* (L:1.4 mm; often found in compost).

– Abdomen with the last three segments (4, 5 and 6) unfused, i.e. with six segments 13

13 (12) Fourth abdominal segment over twice as long as third abdominal segment Entomobryidae

There are 10 genera with 35 species, of which many are rare, scarce or uncommon, or are found mainly climbing vegetation and trees, or in buildings and caves. The most common species under logs and stones are:

- *Entomobrya* species – L:2.0–2.5 mm; body yellow with some dark colouration to the abdomen, antennae with four segments; common and widespread.
- *Heteromurus nitidus* – L:3.0 mm; body very pale brown, antennae with five segments (the terminal one being subdivided into many subsegments); widespread and common.
- *Lepidocyrtus* species – L:2.0–3.5 mm; body white or pale brown with some blue colouration especially on the head and the bases of the legs, antennae with four segments; common and widespread. Another very common and widespread species (*L. cyaneus* – L:1.5 mm) has a bluish violet body with a silvery blue appearance.
- *Orchesella cincta* – L:4.0 mm; body reddish brown to blackish with a pale second segment to the abdomen (giving a belted appearance – *cincta* comes from the Latin *cinctum* meaning girdle or belt), antennae with six segments; widespread and extremely common, in a wide range of habitats.
- *O. villosa* – L:5.5 mm; body mid-brown with darker markings, very hairy, antennae with six segments; widespread and very common, under dead wood and stones in a wide range of habitats including gardens.
- *Pseudosinella* species – L:1.0–2.8 mm; body white, antennae with four segments; very common and widespread.

– Fourth abdominal segment about as long as third abdominal segment Isotomidae

There are 23 genera containing 57 species (see couplet 12 for the remaining Isotomidae species). Nine of these species are completely white, of which six are rare or scarce. The three more common species are difficult to separate (L:0.9–1.8 mm). Forty-eight species have some pigmentation, the majority of which are found either climbing trees, or in montane or boggy areas, or are rare, scarce or have a local distribution. Many of the remaining widespread and common species are greyish brown or greyish blue and are difficult to separate (L:1.0–2.5 mm). Several common species are typically green(ish) including:

- *Agrenia bidenticulata* – L:2.3 mm; common in wet habitats, often under stones alongside and in mountain streams.
- *Isotoma viridis* – varied (or green) springtail – L:4 mm; common and widespread, across a range of habitats.
- *Pseudisotoma sensibilis* – L:1.8 mm; body dark blue; one of the most common UK springtails, found in a range of habitats.

## Key O Earwigs

Four British species of earwigs key out here. Note that a number of introduced species including those restricted to hot-houses are not covered by this key. The lengths (L) given are for adult specimens from the front of the head to the end of the forceps at the end of the abdomen, ignoring any appendages (e.g. antennae and legs). See O.1 for the general morphology of earwigs. If your specimen does not exactly match the descriptions, then consult Hincks (1956) or Marshall & Haes (1988). Marshall & Ovenden (1999) provide images of British species. A checklist, species descriptions and distribution data are given on the Orthoptera & Allied Insects website (see Chapter 6). The latest distribution atlas was published by Haes & Harding (1997). See Section 3.5 for further information on earwigs.

1 Folded hindwings visible (just projecting beyond the hardened forewings – tegmina) (O.2) 2

– Hindwings absent or not visible (i.e. not projecting beyond the hardened tegmina) (O.3) 3

2 (1) Large (adults >10 mm); second tarsal segment lobed (O.4) *Forficula auricularia*

- *F. auricularia* – common earwig – L:10–15 mm; body dark brown with robust pincers; widespread and very common in a wide range of habitats including woodland, scrub, hedgerows, grassland, heathland and marsh.

– Small (adults <8 mm); second tarsal segment not lobed (O.5) *Labia minor*

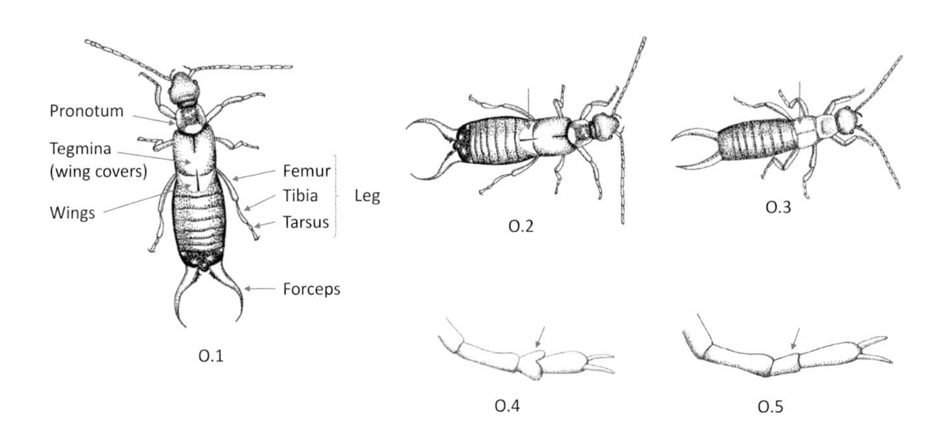


O.1

O.2

O.3

O.4

O.5

- *L. minor* – lesser earwig – L:4–7 mm; body yellow-brown with slender pincers; found in compost, rubbish tips and dung piles.

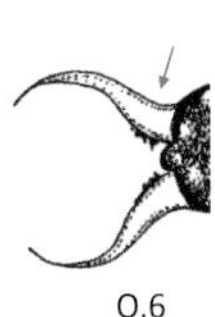

O.6

3 (1) Light brown; male forceps thickened at base (O.6) *Forficula lesnei*

- *F. lesnei* – Lesne's earwig – L:6–10 mm; body light brown; found in woodland, scrub, hedgerows and open areas including coastal grassland.

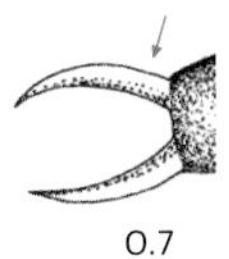

O.7

– Reddish brown with lighter wing-cases; male forceps not thickened at base (O.7) *Apterygida media*

- *A. media* – short-winged (or hop-garden) earwig – L:6–11 mm; body reddish brown with lighter brown wing-cases; a local species found in wood edges, scrub, and hedgerows in eastern England.

## Key P Cockroaches

There are three species of native British cockroaches that can be found outside buildings, all of which have a scattered distribution in the south of England. This key covers adult cockroaches and all lengths (L) given are for adult specimens from the front of the head to the end of the abdomen, ignoring any appendages (e.g. antennae, legs or tails – cerci). Nymphs can be recognised by their wing buds (which are separate, i.e. they do not overlap each other, unlike in adults where the left wing overlaps the right one). There are several species of cockroach associated with buildings that are not covered by this key. Note that several invasive species have also been recorded outdoors on occasions, including *Ectobius vittiventris* (garden cockroach) and *E. montanus* (Italian cockroach), as well as those belonging to the *Planuncus tingitanus* species group (variable cockroach). These, and indeed possibly other species, may be spreading across continental Europe and into the UK where at least some seem to be establishing. Some of these additional species are not well described and in at least one case may be an aggregate of several species (e.g. *P. tingitanus*). As such these are not covered in the key below. If your specimen does not match the descriptions given in the key, then consult Hincks (1956) or Marshall & Haes (1988). Marshall & Ovenden (1999) provide images of British species. A checklist, species descriptions and distribution data are given on the Orthoptera & Allied Insects website (see Chapter 6). The latest distribution atlas was published by Haes & Harding (1997). See Section 3.5 for further information on cockroaches.

| | | |
|---|---|---|
| 1 | Wings fully developed, reaching the end of the abdomen (e.g. P.1) | 2 |
| – | Wings not fully developed, not reaching the end of the abdomen (P.2 or P.3) | 4 |

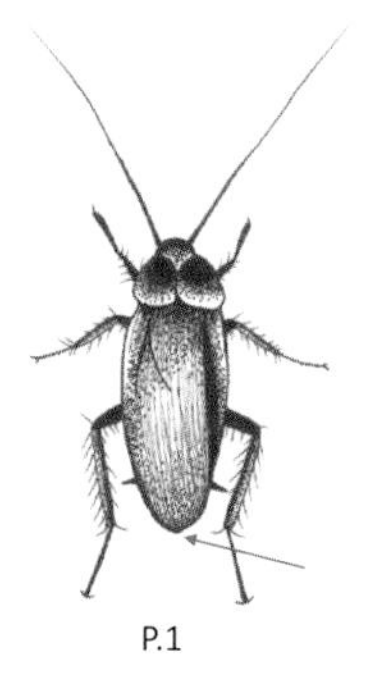
P.1

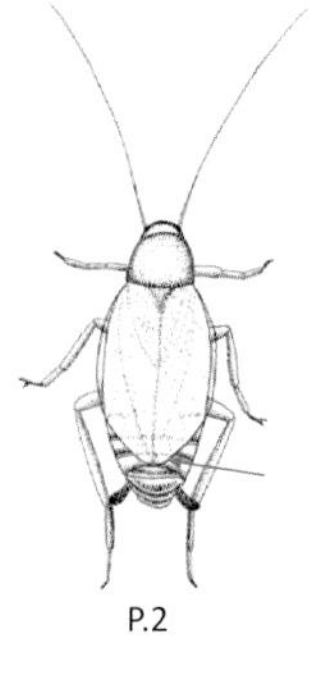
P.2

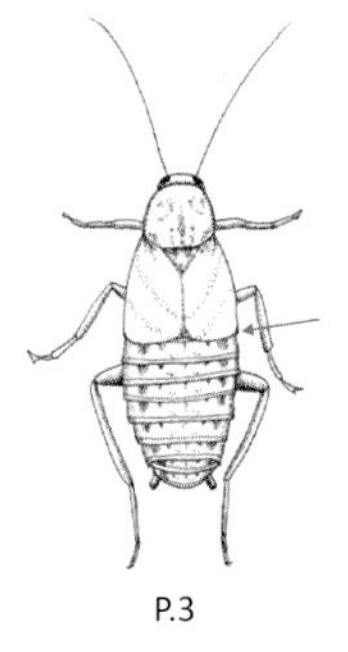
P.3

2 (1) Golden yellowish brown *Ectobius pallidus* (males and females)

- *E. pallidus* – tawny cockroach – L:8–9.5 mm; body golden-yellowish brown all over; in woodland clearings, grassland, heathland and dunes.

– Darker brown (at least on pronotum) 3

3 (2) Dark brown with lighter flecks on the pronotum *Capraiellus panzeri* (males)

- *C. panzeri* – lesser cockroach – L:5–8 mm; body darkish brown with speckled pronotum; mainly coastal sites, often under scrub, on dunes or shingle beaches.

– Light brown with darker brown pronotum *Ectobius lapponicus* (males)

- *E. lapponicus* – dusky cockroach – L:7–11 mm; body light-brown, sometimes greyish, usually with darker pronotum in males; in woods, scrub and heathland.

4 (1) Wings nearly reach the end of the abdomen (P.2) *Ectobius lapponicus* (females)

See couplet 3 for description

– Wings reach less than halfway down the abdomen (P.3) *Capraiellus panzeri* (females)

See couplet 3 for description

## Key Q Orthopterans (crickets and allied insects)

Orthopterans such as grasshoppers, groundhoppers and bush-crickets are usually seen near to or on logs and stones rather than underneath them. Mole crickets usually live underground but may occasionally be found under logs and stones, while true crickets, as well as being seen in burrows and on the ground, may also be found under refuges. Mole crickets and all species of true crickets are restricted to southern England. All measurements given here are adult body lengths (L) taken from the front of the head to the end of the abdomen, ignoring any appendages (e.g. antennae, legs or tails – cerci). This key enables the identification of orthopterans to family, with mole crickets and true crickets being taken to species. See Q.1 for the general morphology of orthoperans. Where animals do not fully fit the descriptions given, consult Marshall & Haes (1988), Marshall & Ovenden (1999), or Evans & Edmundson (2007). Some older texts may also be useful, including Hincks (1956), Bellman (1988), and Brown (1990). The Grasshoppers and Related Insects Recording Scheme website also provides support for identification as well as species descriptions and distribution maps (see Chapter 6). The latest atlas was published by Haes & Harding (1997). See Section 3.5 for further information on orthopterans.

1 Front legs relatively massive and used for digging (Q.2) Gryllotalpidae (mole crickets)

The single representative of this family is a relatively rare species, which may be found under logs and stones:

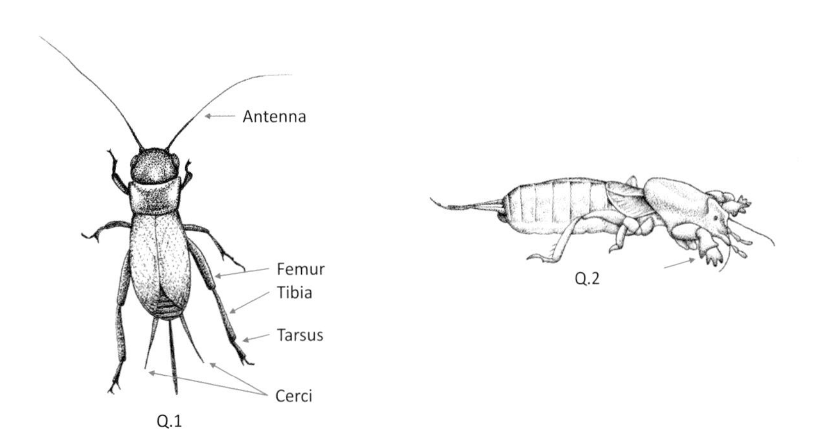

Q.1

Q.2

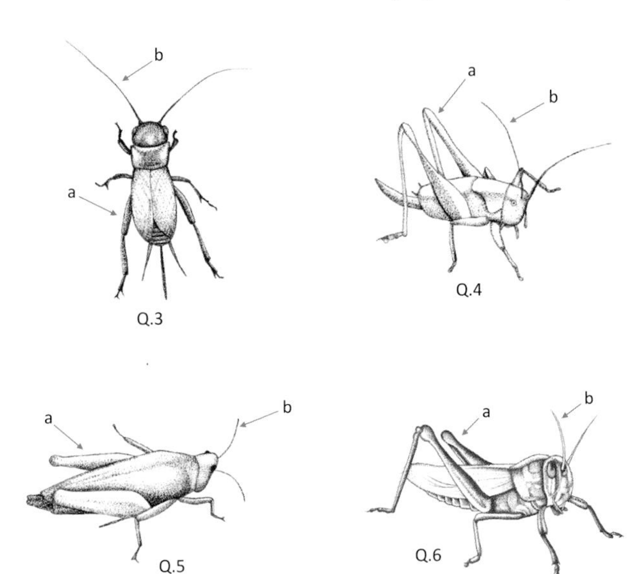


- *Gryllotalpa gryllotalpa* – mole cricket – L:35–46 mm; brown with yellow-brown patches on the thorax and abdomen, fully winged; damp grasslands.

– Front legs not so massive; hind legs large and used for jumping (Q.3a, Q.4a, Q.5a, Q.6a) 2

2 (1) Antennae longer than body (Q.3b or Q.4b) 3

– Antennae shorter than body (Q.5b or Q.6b) 4

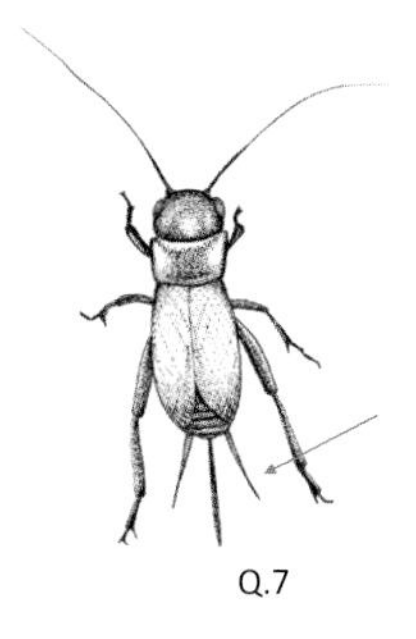
Q.7

3 (2) Paired prongs (cerci) at hind end of abdomen longer than upper surface of thorax (pronotum) (Q.7); tarsi with three segments Gryllidae (true crickets)

There are four native species, and two occasionally introduced species (one that is only seen outdoors in summer and another that is restricted to hot-houses). The four native species are:

- *Acheta domesticus* – house cricket – L:16–20 mm; body yellow-brown, fully winged; in buildings, sometimes on rubbish tips.
- *Gryllus campestris* – field cricket – L:20–26 mm; body shiny black, fully winged; in grassland.
- *Nemobius sylvestris* – wood cricket – L:7–10 mm; body dark brown, short winged; in woodland leaf litter.
- *Pseudomogoplistes vicentae* – scaly cricket – L:8–13 mm; body greyish brown, wingless; on shingle beaches.

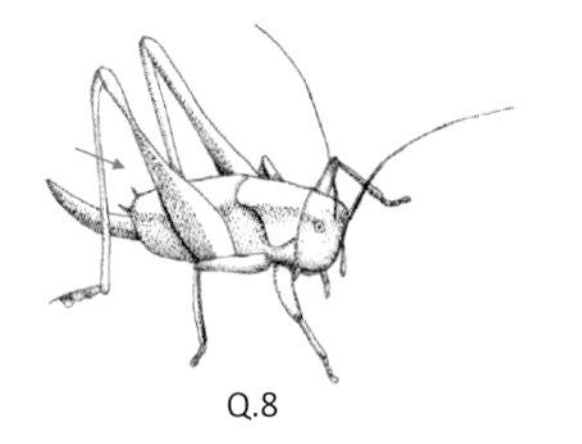
Q.8

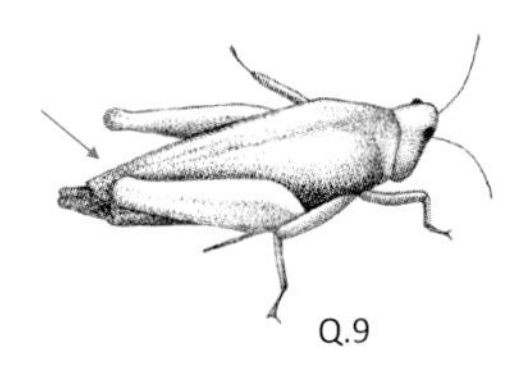
Q.9

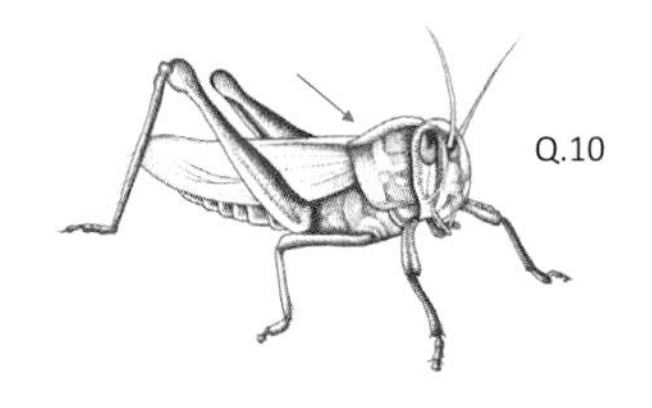
Q.10

– Paired prongs at hind end of abdomen shorter than upper surface of thorax; tarsi with four segments (Q.8)
Tettigoniidae (bush crickets)

Often found on vegetation above the ground in woods or grasslands.

4 (2) Upper surface of thorax (pronotum) extended backwards to cover most of the abdomen (Q.9)
Tetrigidae (groundhoppers)

On vegetation but can be found on the ground especially in damp areas with open ground

– Upper surface of thorax not extended backwards over most of abdomen (Q.10) Acrididae (grasshoppers)

On vegetation but can be found on the ground especially in grassy areas.

## Key R Ants (workers)

This key can be used to identify the common workers of ant species likely to be found under logs and stones. Most non-ant Hymenoptera (bees, wasps, etc.) will usually be seen around or on logs and stones, rather than under them. Some will be using or creating burrows in the soil. The first couplet of the key below should separate these from ants. Non-ant Hymenoptera identification is covered by various Handbooks for the Identification of British Insects (volumes 6, 7 and 8: available from the Royal Entomological Society of London website – see Chapter 6) and a variety of field guides/handbooks (e.g. Wilmer, 1985; Prŷs-Jones & Corbet, 2014; Yeo & Corbet, 2015; Benton, 2017; and Falk & Lewington, 2018).

The majority of ants encountered under logs and stones are workers. In some cases, nests can be found in these micro-habitats; turning over the refuge will disturb large numbers of ants. Queens and males may also be present on occasions, especially when a nest is disturbed. Only workers are keyed out in this key. Note that queens are generally larger and may be either winged or have the remnants of wings present. The males of many species will also be winged, and in males the genitalia are usually obvious, being paired and projecting from the end of the abdomen. For the identification of males or queens, see Bolton & Collingwood (1975) or Skinner & Jarman (2024). Lengths (L) given are from the front of the head to the end of the abdomen, ignoring any appendages (e.g. antennae and legs). See R.1 for the general morphology of ants. Where specimens do not exactly fit the descriptions and illustrations, or where species are difficult to separate, or for species that are rare under logs and stones that have not been keyed out here to species level, consult Lebas *et al.*

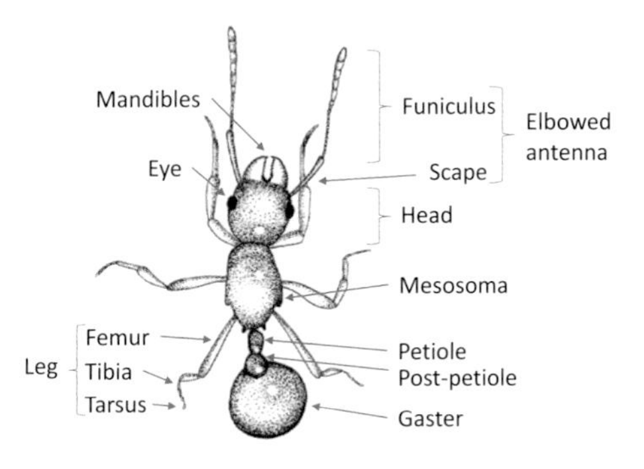


R.1

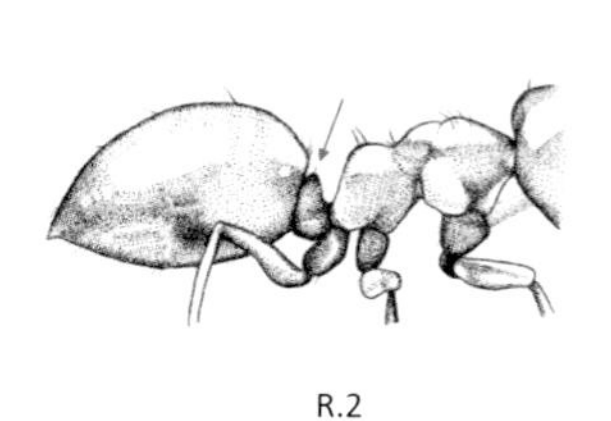
R.2

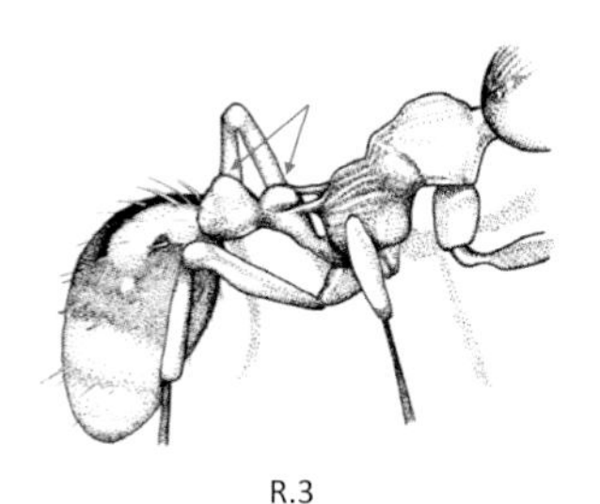
R.3

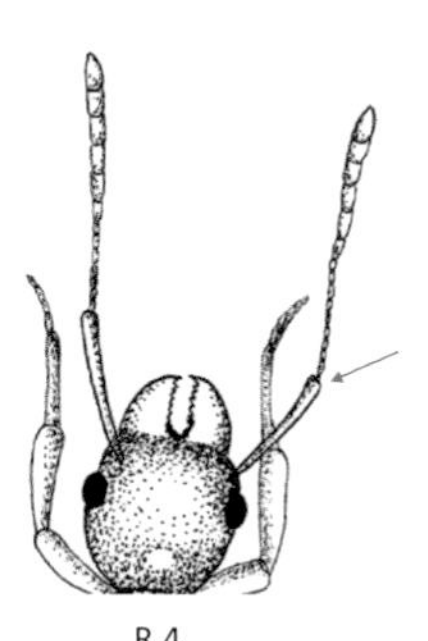
R.4

(2019) or Skinner & Jarman (2024). For further descriptions of the species found, including their distribution, see Barrett (1979), Lebas *et al.* (2019), and the BWARS (Bees, Wasps & Ants Recording Society) and AntWiki websites (see Chapter 6). See Section 3.5 for further information on ants.

1 One or two segments of abdomen forming a waist (comprising the petiole and, where present, the postpetiole) which is small and scale-like (R.2 or R.3); antennae elbowed with fewer than 16 segments (R.4) — Formicidae (ants) 2

Note that any ants with wings or wing remnants, or obvious paired genitalia protruding from the hind end of the abdomen will be queens or males and are not covered by this key.

– Not as above — other Hymenoptera (bees, wasps, parasitic wasps, etc.)

2 (1) Waist comprising one segment – petiole (R.2) — 3

– Waist comprising two segments – the petiole and postpetiole (R.3) — Myrmicinae 6

3 (2) Segment comprising waist clearly visible from above — 4

– Segment comprising waist not visible from above (petiole hidden by the first segment of the gaster) — Dolichoderinae

Two species of *Tapinoma*, which are difficult to separate, key out here. The most widespread, common British species is:

- *Tapinoma erraticum* – erratic ant – L:2–4.2 mm; black and shiny; nests under stones.

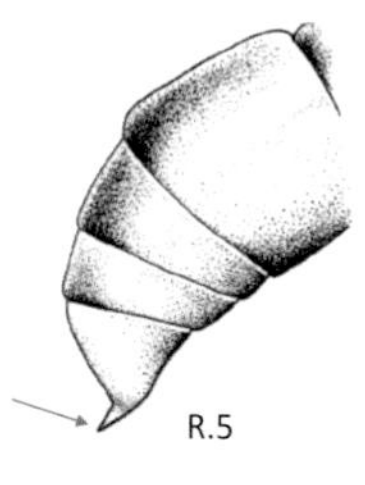
R.5

4 (3) Sting present at end of gaster (R.5) — Ponerinae

Two species of *Ponera*, which are difficult to separate, key out here. The most common British species is:

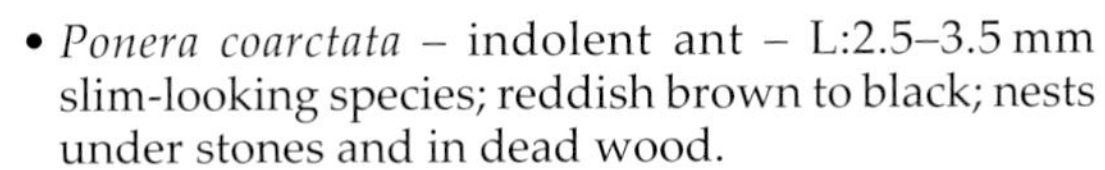

- *Ponera coarctata* – indolent ant – L:2.5–3.5 mm slim-looking species; reddish brown to black; nests under stones and in dead wood.

– No sting at end of gaster **Formicinae 5**

The introduced species *Linepithema humile* – the Argentine ant, a member of the subfamily Dolichoderinae, also keys out here. This has only rarely been found in Britain (mainly in buildings and glasshouses) and can be distinguished from the Formicinae by having a hairless anal slit, rather than a round anal opening fringed with hairs.

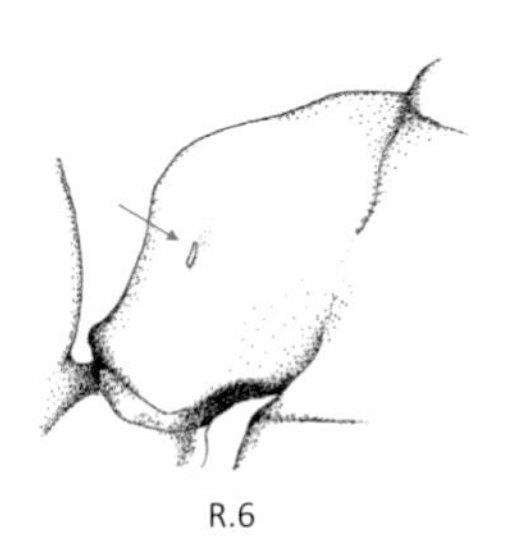

R.6

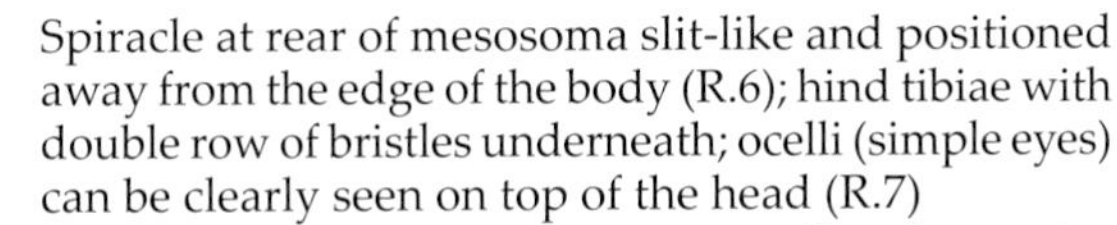

5 (4) Spiracle at rear of mesosoma slit-like and positioned away from the edge of the body (R.6); hind tibiae with double row of bristles underneath; ocelli (simple eyes) can be clearly seen on top of the head (R.7) ***Formica* species**

There are 11 British species, which require close examination to separate them. Common species under logs and stones include:

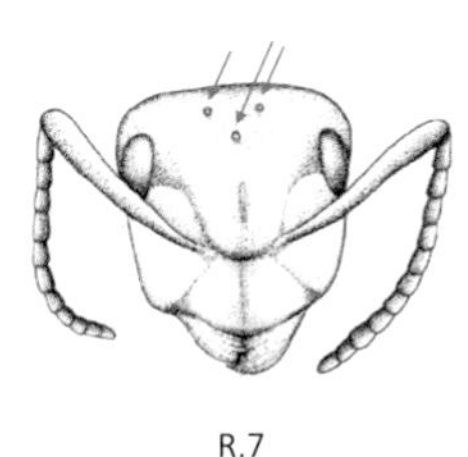

R.7

- *F. fusca* – common black ant – L:4.5–7 mm; black with fairly densely hairy gaster; common, especially in the south, nests under stones or rotten logs. This species is difficult to separate from *F. lemani* (common black ant – L:4.5–7 mm; nests under stones and dead wood), which is more northerly in distribution.
- *F. rufa* – red (or southern) wood ant – L:4.5–9 mm; dark gaster, head and mesosoma with dark brown patches; nests in mounds made of fragments of vegetation, occasionally found on and under dead wood in areas where nests are common.
- *F. sanguinea* – blood-red slave-maker ant – L:6–9 mm; dark gaster with reddish head and paler reddish mesosoma; nests in old tree trunks and sometimes under stones.

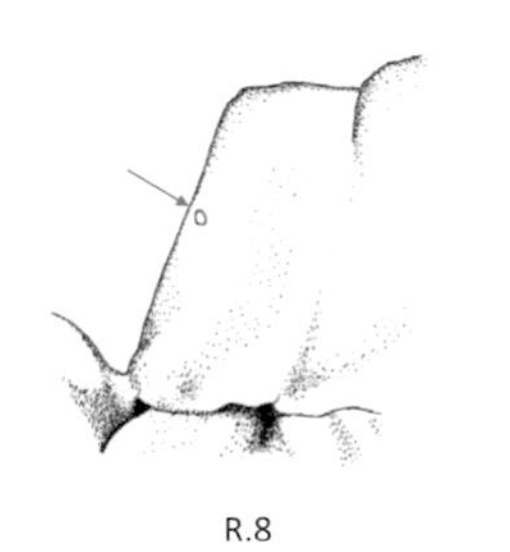

R.8

– Spiracle at rear of mesosoma almost as wide as long and positioned near to the edge of the body (R.8); hind tibiae with bristles only at tip (no double row underneath); ocelli not obvious on head (except for *L. fuliginosus* – jet black ant – which has a shiny, black heart-shaped head when viewed from the front) ***Lasius* species**

There are 14 British species of *Lasius* species (moisture ants), including one very locally distributed introduced species (*L. neglectus* – invasive garden ant), and one restricted to the Channel Islands (*L. myops*), which require close examination to separate. Common species under logs and stones include:

- *L. flavus* – yellow meadow ant – L:2.2–4.8 mm; yellowish, larger individuals with darker head; nests in mounds of earth and also under stones.

- *L. mixtus* – 3.5–4.5 mm; entirely yellow; sometimes nests under stones.
- *L. niger* – black garden ant – L:2.5–5 mm; mesosoma and hind body uniformly dark brown; nests in small mounds of soil and under stones (including paving stones), plant pots and other refuges.
- *L. platythorax* – L:2.5–5 mm; mainly black; nests in tree stumps or dead wood on the ground.
- *L. umbratus* – yellow lawn ant – L:3.8–5.5 mm; entirely yellow; sometimes nests under stones or dead wood.

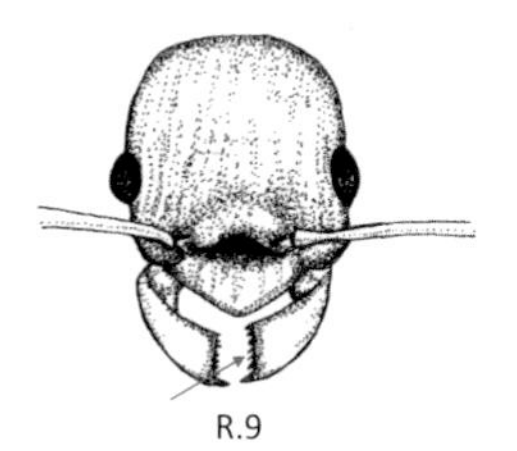

R.9

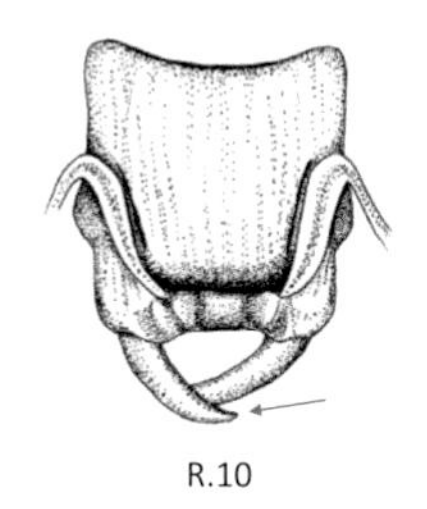

R.10

6 (2) Mandibles with many teeth (R.9) 7

– Mandibles long, slender and toothless (R.10) *Strongylognathus testaceus*

- *S. testaceus* – L:2–3.6 mm; body pale yellow, head darker in larger individuals; found in nests of another ant (*Tetramorium caespitum* – see couplet 12) under stones.

7 (6) Funiculus of antennae with 10 or 11 segments 8

– Funiculus of antennae with 9 segments *Solenopsis fugax*

- *S. fugax* – thief ant – L:1.5–2.2 mm; entirely yellow, larger individuals with darker head; nests under stones.

8 (7) Funiculus of antennae with 10 segments 9

– Funiculus of antennae with 11 segments 10

9 (8) Hind segment of waist with downward-pointing projection (R.11) *Formicoxenus nitidulus*

- *F. nitidulus* – shining guest ant – L:2.8–3.4 mm; shiny orange-red body; nests with ants of *Formica* species (see couplet 5), sometimes found under stones when they form part of the dome of the nest.

– Hind segment of waist without downward-pointing projection (R.12) *Leptothorax acervorum*

- *L. acervorum* – slender ant – L:3–4 mm; head and gaster dark brown, mesosoma reddish; nests in dead wood and under bark, branches and tree stumps near the ground.

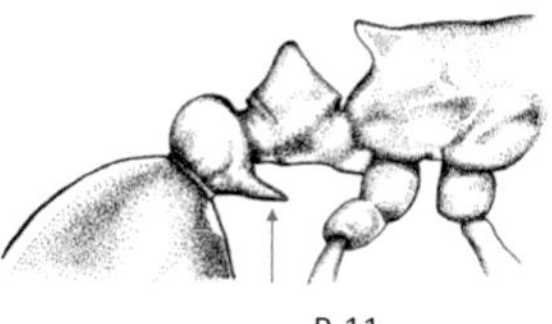

R.11

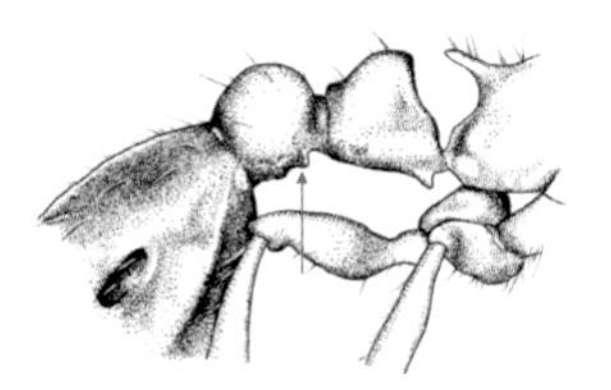

R.12

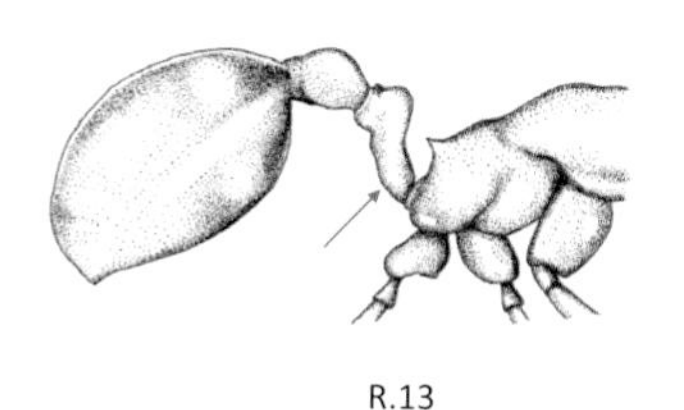
R.13

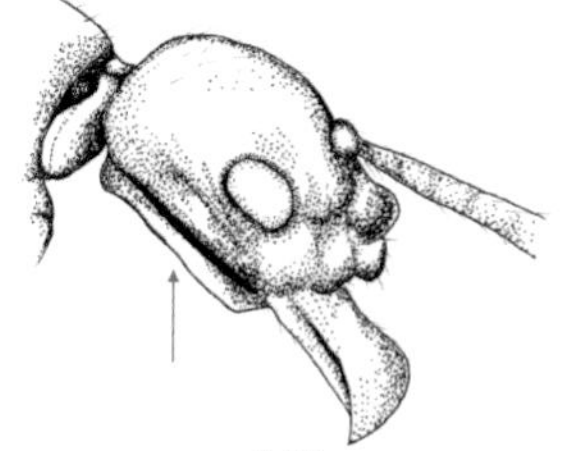
R.14

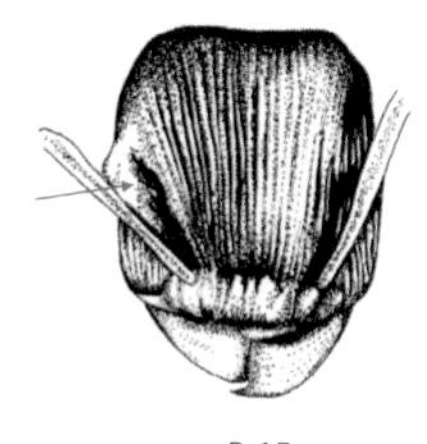
R.15

10 (8) Petiole elongated into a long stalk (R.13) *Stenamma* species

There are two *Stenamma* species in Britain, of which the commonest is:

- *S. debile* – L:3–4 mm; entirely reddish; nests under stones.

– Petiole not elongated 11

11 (10) Underside of head with a ridge either side (R.14) *Myrmecina graminicola*

- *M. graminicola* – grass ant (or woodlouse ant) – L:3.5–5 mm; black or reddish with pale legs and a yellowish tip to the gaster, cube-shaped petiole; nests under large stones, rolls into a ball when disturbed.

– Underside of head without ridges running each side 12

12 (11) Ridge in front of antennal insertion on top of head (R.15) *Tetramorium caespitum*

- *T. caespitum* – pavement ant – L:2.5–4 mm; entirely black; nests under stones, sometimes in dead wood.

Introduced species of this genus are associated with hot-houses and have the ridges on the head extending behind the level of the eyes.

– No ridge in front of antennal insertion on top of head 13

13 (12) Mandibles with at least six teeth (e.g. R.9), last three antennal segments less than half the length of the funiculus (L:3.5–6 mm) *Myrmica* species

There are 12 British species, which require close examination to separate them. Common species under logs and stones include:

- *M. rubra* – common red ant (or European fire ant) – L:3.5–5 mm; entirely red; nests under stones, plant pots and dead wood on the ground.
- *M. ruginodis* – red ant – L:4–6 mm; entirely red; nests under stones and in dead wood on the ground.
- *M. sabuleti* – L:4–5 mm; red body; nests under stones.
- *M. scabrinodis* – elbowed red ant – L:4–5 mm; entirely red; nests under stones and bark of dead wood.

– Not as above *Temnothorax* species

There are three British species, which are difficult to separate. The commonest species under logs and stones is:

- *T. albipennis* – turf ant – L:2–3 mm; yellow body, darker tips to antennae, dark band on first segment of gaster; nests under a variety of natural refuges including stones and branches on the ground.

## Key S Beetles

Ground beetles (Carabidae) and rove beetles (Staphylinidae) are the commonest families living under logs and stones (and as such, further keys to common species of these two families are given later in this book – Key T and key U respectively). However, some members of other families may be found occasionally, including when overwintering. This key allows the identification of adult beetles belonging to those families described by Dibb (1948) as Lapidicoles (i.e. those found under logs, stones and other materials on the ground), as opposed to wood-boring beetles (which live, either as adults or as larvae, under the bark and within the body of the dead wood). The identification of the latter is beyond the scope of this book and those who are interested should consult works such as Bevan (1987) and Alexander (2002). Lengths given (L) are from the front of the head to the end of the abdomen, ignoring any appendages (e.g. antennae and legs). Note that gravid females may be larger than the lengths given if their abdomen is swollen with eggs. See S.1a for general beetle morphology. The number of segments in the tarsi of the front, mid and hind leg can be important in identifying some families (count from the end of the tibiae ignoring the claw – see S.1b). More comprehensive keys to British beetle families in general can be found in Crowson (1956) and Unwin (1984). A variety of images covering a range of species within the families identified below can be found in Harde & Hammond (2000). For identification guides to families other than the Carabidae and Staphylinidae, see the Royal Entomological Society Handbooks for the Identification of British Insects series volumes 4 and 5 (available from

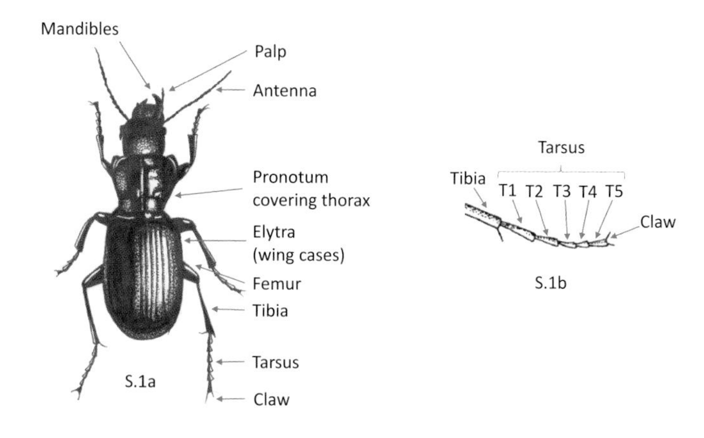

the Royal Entomological Society website – see Chapter 6). Further keys and information can be found on the websites for Mike Hackston's Insect Keys, UK Beetles, and UK Beetle Recording (see Chapter 6). Duff (2018) provides a checklist to the beetle species of the British Isles. See Section 3.6 for more information on beetles.

S.2

1 Elytra (wing-cases) short leaving 3–6 segments of abdomen exposed (S.2 or S.3); (be careful when examining preserved specimens – for example most carabid beetles do not normally have any abdominal segments exposed, but in specimens that have fallen into pitfall traps containing fluid the abdomen may swell and protrude backwards and sideways beyond the elytra) 2

– Elytra covering the whole (or almost the whole) of the abdomen (at most 1–2 segments exposed) 4

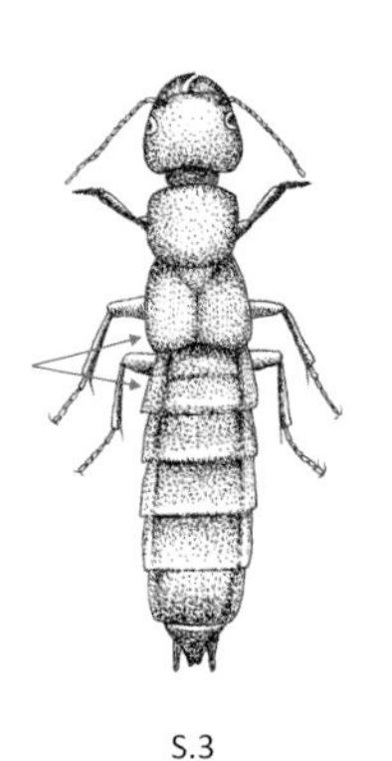

S.3

2 (1) Medium to large beetles (L:9–30 mm); antennae with distinct club comprising the last four segments (S.4) or gradually thickened (S.5) from the base to the tip; at most only three or four segments of the abdomen exposed by the short wing-cases, which are at least as long as the exposed part of the abdomen (S.2); all tarsi with five segments (count as in S.1b) Silphidae

Silphidae – burying (or carrion) beetles – seven genera containing 21 species; occasionally found under logs and stones, especially when these are near to carcases. Relatively common examples include *Nicrophorus* species (sexton beetles – L:10–25 mm, black, often with orange bands on the elytra, antennae with clubs comprising four segments) and *Silpha* species (snail beetles – L:11–20 mm; usually black, quite convex; antennae gradually thickened from the base to the tip). The Silphidae Recording Scheme website and Mike Hackston's Insect Keys website have keys for British Silphidae.

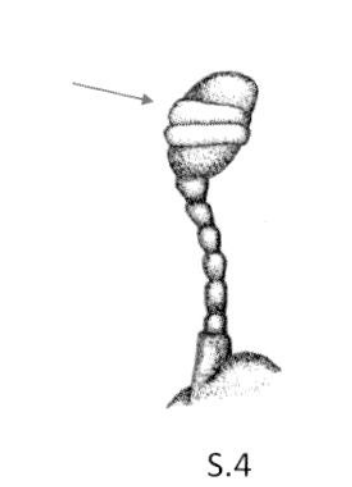

S.4

– Not as above 3

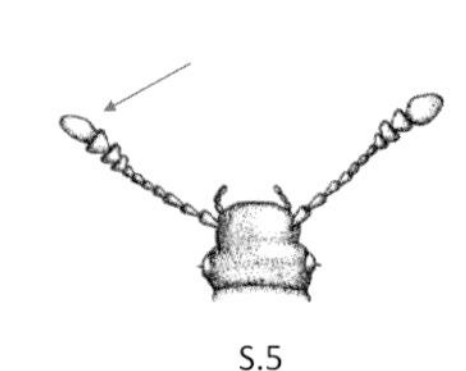

S.5

3 (2) Wing-cases, which are shorter than the exposed part of the abdomen, meet in the midline (in some species these may overlap at the midline) and expose at least 3, and often 6 segments of the abdomen (S.3); antennae comprising 9–11 segments (S.6), not clubbed or with a small club comprising the last three segments Staphylinidae (rove beetles – part – see also couplet 9) Key U

– Not as above other families

One genus (*Claviger*) of Staphylinidae, from the subfamily Pselaphinae (ant-loving beetles), will key out here (see also Key U, couplet 3).

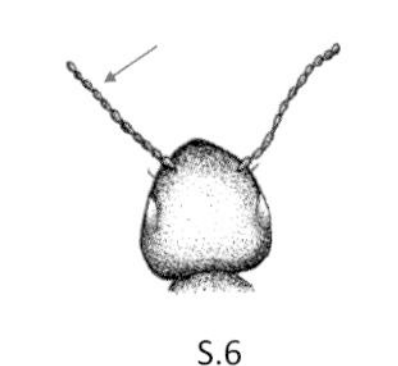

S.6

There are a number of other families of beetles that are not covered by this key which will key out here (see Crowson 1956, or Unwin 1984 for further identification). Most of these are not common under logs and stones.

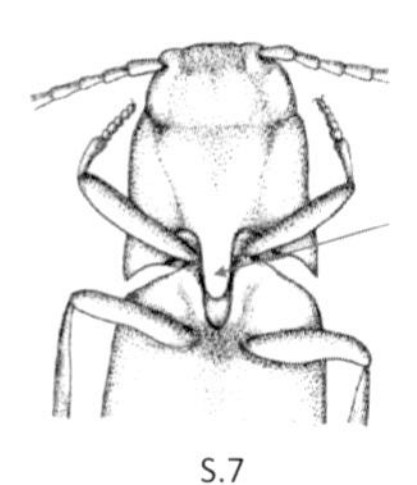

S.7

4 (1) Body long and thin; body outline relatively smooth between thorax and abdomen (no 'waist') and hind angles of upper side of thorax point down past front angles of elytra; underside of thorax projects back as a peg (S.7); tarsi all with 5 segments
Elateridae, Eucnemidae and Throscidae

Elateridae (click beetles) – 38 genera containing 73 species; occasional under logs and stones. When disturbed, these animals are able to jump away (when lying on their back) with an audible click by pressing a process on the underside of the thorax firmly into a groove in the abdomen. Mike Hackston's Insect Keys website has keys to British Elateridae. Common species include:

- *Agriotes lineatus* – lined click beetle – L:7–10 mm; various shades of brown, showing as stripes of alternating light and dark brown down the elytra, covered with grey or yellow hairs; common, especially in the south of England, sometimes under stones in woods, grassland, waste ground and gardens.
- *Hypnoidus riparius* – L:5.5–7.0 mm; shiny black with a metallic bronze sheen, pale brown legs; widespread, under stones near to streams.
- *Prosternon tessellatum* – chequered click beetle – L:9–11 mm; brown to black covered with creamy-yellow hairs; may be found under stones, in lowland areas often near rivers or in scrub.
- *Selatosomus aeneus* – brilliant click beetle – L:10–15 mm; metallic green, blue or bronze-black with dark antennae and reddish brown to black legs; locally common, in woodland, grassland and farmland, under debris and leaf litter especially early in the year.
- *Zorochros minimus* – L:2.5–3.8 mm; black with four yellowish marks on the elytra (two at the front edge and two at the rear); under stones at the sides of streams.

Eucnemidae (false click beetles) – six genera containing seven species, which are usually found in standing dead or decaying wood, although adults may be found in the soil around the roots of decaying trees. These species are able to jump (click) away when disturbed, although not as strongly as Elateridae species. All UK species are scarce or have a local distribution. Mike Hackston's Insect Keys website has a key to British Eucnemidae.

Throscidae (small false click beetles) – two genera containing five species, three of which are locally distributed and restricted in range. They are generally small beetles (L:1.5–4.0 mm) and have antennae with a three- to five-segmented club. These species may occasionally be seen to jump (click) away when disturbed but are much more likely to lie still with their legs and antennae tucked in so that they resemble seeds. Mike Hackston's Insect Keys website has a key to British Throscidae. The two common species are:

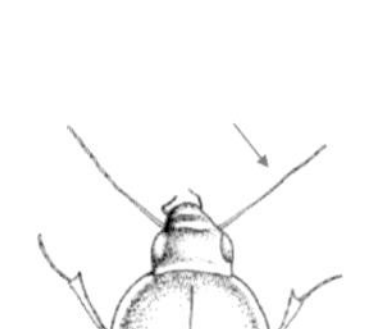

S.8

- *Trixagus dermestoides* – L:2.5–3.3 mm; body brown to dark reddish brown; widespread and common throughout England and Wales as well as southern Scotland, in leaf litter and on tree trunks. A very similar species *T. carinifrons* – L:2.5–3.0 mm is common throughout England and Wales, in leaf litter and on tree trunks. The two can be distinguished by the position of the notch in the lower margin of the eye – in *T. dermestoides* this is at about the midline of the eye, in *T. carinifrons* it is behind the midline of the eye.

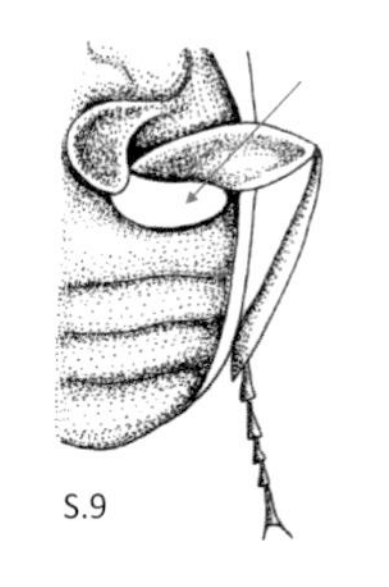

S.9

– Not as above 5

5 (4) Antennae always long and thread-like (S.8); thorax broader than head; all tarsi with five segments; elytra hard; behind the base of each hind femur is a large projection (trochanter) (S.9), which is longer than the diameter of the femur

Carabidae (ground beetles) Key T

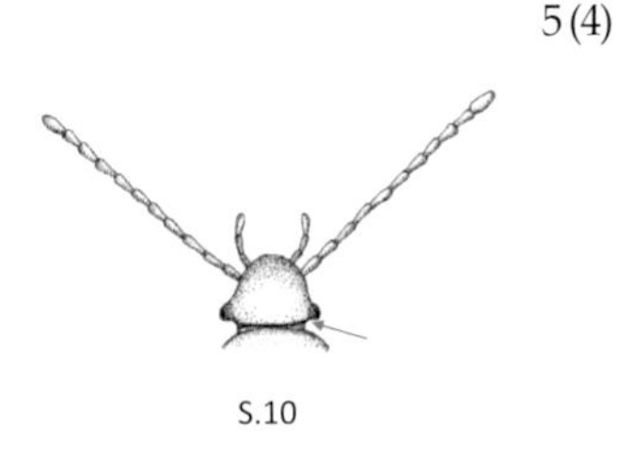

S.10

Some species of the family Leiodidae (fungus beetles) that have thread-like antennae may also key out here (see also couplet 9). These are small beetles (e.g. *Choleva* species – L:5–7 mm), which are widely found in decaying vegetable matter. They may be distinguished from Carabidae by the shape of the head, which is sharply angled behind the eyes and has an obvious ridge running along the back of the head where it is separated from the neck (S.10). In addition, Carabidae usually have short scutellary stria (i.e. the groove nearest to the join between the two elytra is shorter than the rest – S.11), which are absent in Leiodidae. Mike Hackston's Insect Keys website has keys to British Leiodidae.

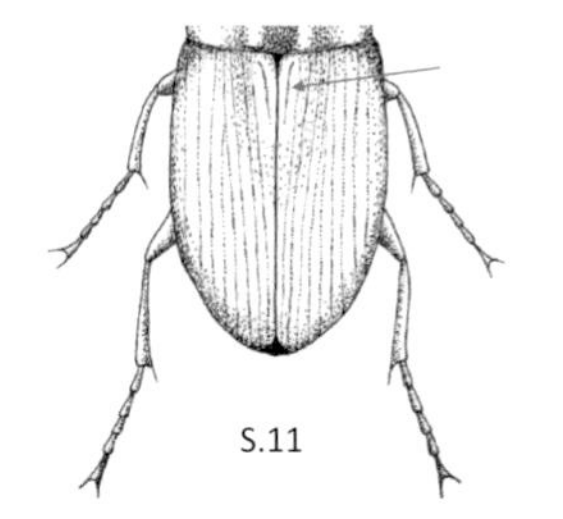

S.11

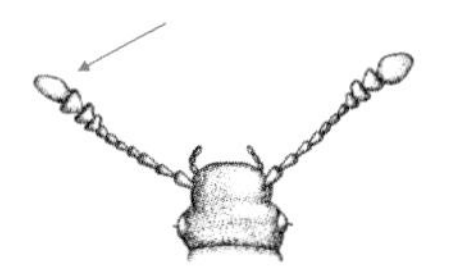
S.12

– Antennae may be long and thread-like (S.8), or thickened at the end (S.12), or clubbed (S.13 or S.14); hind femurs either without trochanters, or with trochanters smaller than diameter of femur 6

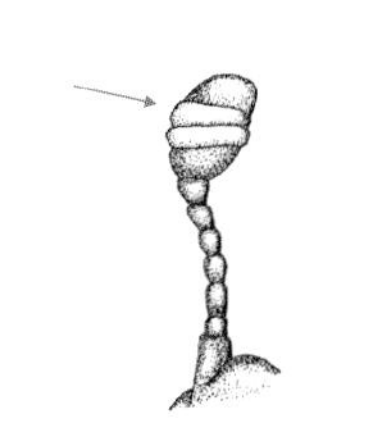
S.13

6 (5) Antennae with a club composed of plates (S.14); legs broad with large spines; all tarsi with five segments dung beetles 7

Note that the term dung beetle does not refer to those from one particular family, rather they come from two families – Geotrupidae and Scarabaeidae. Dung beetles are occasionally found under logs and stones, especially when these are near to dung

– Not as above 8

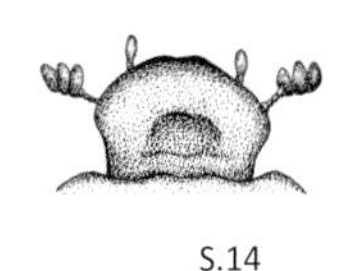
S.14

7 (6) Antennae with 11 segments (including a three-segmented club) Geotrupidae

Geotrupidae – dor beetles – five genera containing eight species; occasionally found under logs and stones. Jessop (1986), and the UK Beetles and Mike Hackston's Insect Keys websites have keys to UK Geotrupidae (see Chapter 6). One common example is:

- *Geotrupes stercorarius* – dor beetle (or lousy watchman, or dumbledore) – L:16–25 mm, black (sometimes with a bluish tint), often covered with mites, and with short antennae; found associated with dung, including digging tunnels under dung.

– Antennae with 9 or 10 segments (including an asymmetric club) Scarabaeidae

Two genera of dung beetles from the Scarabaeidae (*Aphodius* and *Onthophagus*) key out here. The Dung Beetle UK Mapping Project website provides keys to species for these two genera, while the remaining genera are covered by Jessop (1986) and Mike Hackston's Insect Keys website (see Chapter 6). Common examples of these two genera are:

*Aphodius* species (42 species – note that this genus is undergoing a revision that is likely to lead to many species being separated into other genera) – one common species is:

- *A. rufipes* – L:4–12 mm; dark brown to black with broad and spiny front legs; found associated with animal dung, but also in gardens where they are attracted to lights.

*Onthophagus* species (eight species) – one common example is:

- *O. similis* – L:4.5–7.0 mm; pronotum dark with metallic sheen, elytra pale brown with darker mottling; widespread and common especially in Wales and central and southern England, in and around dung.

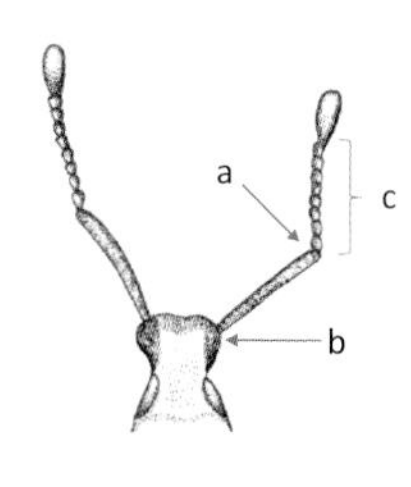

S.15

8 (6) Antennae clubbed and with an elbow (S.15a); head prolonged into short snout (S.15b); all tarsi with four segments — Curculionidae

Curculionidae – weevils – 146 genera containing 476 species; occasional under logs and stones. Morris (1997–2012) and Mike Hackston's Insect Keys website have keys to British Curculionidae. Common species include:

- *Barynotus moerens* – L:8–9 mm; mottled appearance caused by covering of brown and bronze scales, antennae brown with darker club; locally common especially in the south of England, under logs in open woodland and wood edges.
- *B. obscurus* – ground weevil – L:8–9.5 mm; light brown, elytra with mottling of pale patches; under stones, in woods.
- *Gronops lunatus* – L:3.0–3.9 mm; body covered in pale and dark grey scales giving the appearance of two pale bands across the body, elytra broad at base with pointed ends; ground-active in coastal habitats such as salt marshes and dunes.
- *Mitoplinthus caliginosus* – hop root weevil – L:6–9 mm; body dull black, antennae and legs reddish brown; under logs and stones in woods.
- *Otiorhynchus clavipes* – red-legged weevil – L:8.6–12.5 mm; body shiny black with very long black antennae, and reddish legs; under stones in woods and scrub.
- *O. ligneus* – L:4.2–6.5 mm; body black to reddish brown with rows of yellowish curved hairs on the pronotum and elytra, antennae and legs reddish brown to black; under stones, especially on sandy soils.
- *O. rugifrons* – L:4.3–6.1 mm; body dull black, antennae and legs black, segments comprising the funicle (filamentous part of the antenna – S.15c) wider than long; under stones and debris at the coast.
- *O. rugosostriatus* – (rough) strawberry root weevil – L:5.5–7.6 mm; dark brown (pronotum darker than the elytra), both pronotum and elytra covered in small bumps, antennae long and thin with long, narrow club; found in ground vegetation and leaf litter during the day, wood edges, heathland and moorland.

– Antennae without an elbow — 9

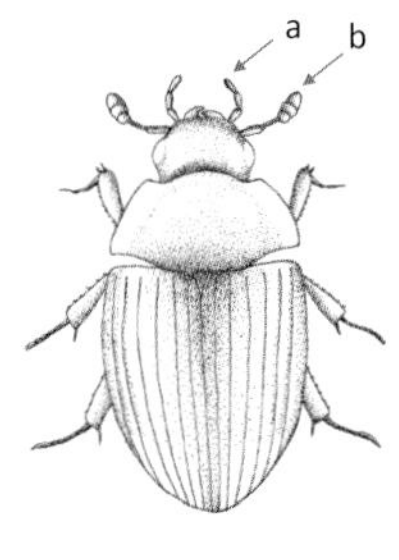

S.16

9 (8) Palps long, at least ⅔ length of antennae (S.16a); antennae with a 3–5-segmented club (S.16b) — Hydrophilidae

Water scavenger beetles – 18 genera containing 70 species; occasionally found under logs and stones, especially when these are near water. These species have all tarsi with five segments. Mike Hackston's Insect Keys website has keys to British Hydrophilidae (see Chapter 6). Common examples include:

- *Cercyon marinus* – L:2.2–3.4 mm; body oval in shape, shiny black with pale (yellowish) margin to the extreme rear of the elytra; margins of freshwater habitats such as lakesides.
- *C. ustulatus* – L:2.6–3.4 mm; body oval in shape, shiny black with red or brown ends to the elytra which appear as separate patches and are more extensive than in *C. marinus*; under debris in wetlands and the margins of freshwater habitats.

Members of the Hydraenidae (minute moss beetles) key out here – five genera containing 31 species; although they have long palps and clubbed antennae, they have four rather than five segments on all tarsi. Mike Hackston's Insect Keys website has keys to the British Hydraenidae.

– Not as above other families

Some Coccinellidae (ladybirds) may key out here – 30 genera containing 53 species. See Roy *et al.* (2013) and Mike Hackston's Insect Keys website for further identification of British Coccinellidae. Occasionally species may be found overwintering under refuges at ground level, one example being:

- *Hyperaspis pseudopustulata* – false-spotted ladybird – L:3–4 mm; body black with orangey-red sides to the pronotum and two orangey-red spots at the hind end of the elytra; overwinters in leaf litter, and sometimes under logs and stones.

Some Leiodidae (fungus beetles) will key out here (see also couplet 5) – 20 genera containing 93 species. These small beetles (L:1.3–6 mm) are found associated with fungi in decaying vegetation. They have antennae ending with clubs formed from the last 3–5 segments, and in most species antennal segment 8 is smaller than either segments 7 or 9. Mike Hackston's Insect Keys website has keys to British Leiodidae. One of the more easily identifiable species is:

- *Ptomaphagus subvillosus* – 2.2–3.7 mm; body brown with the microscopic sculpture of the thorax and elytra forming transverse lines; widespread, often associated with carrion.

Some Tenebrionidae (darkling beetles) may also be found under stones – 33 genera containing 47 species (many of which are pests of stored products). In these species the fore- and mid-tarsi have five segments, while the hind-tarsi have four. Mike Hackston's Insect

Keys website has keys to British Tenebrionidae. One common species found under refuges is:

- *Phylan gibbus* – L:7–9 mm; body black, legs and antennae reddish, front tibiae broad; under debris on sandy and gravel beaches.

Two subfamilies of Staphylinidae (Scydmaeninae and Scaphidiinae) will also key out here (see also couplet 3 and Key U) – these have longer elytra than the majority of Staphylinidae with at most the last 1–2 abdominal segments being visible:

Scydmaeninae (ant-like stone beetles) – nine genera containing 32 species; typically having a distinct 'waist' between the thorax and abdomen, and the elytra completely covering the abdomen. All the tarsi have five segments. See Key U, couplet 2 for some common examples.

Scaphidiinae (shining fungus beetles) – three genera containing five species; typically without a distinct 'waist' between the thorax and abdomen, and only the final segment or two of the abdomen showing beyond the slightly truncated elytra. All of the tarsi have five segments. See Key U, couplet 2 for a common example.

## Key T Ground beetles

Carabidae (ground beetles) are some of the most frequent and obvious beetles found under refuges such as logs and stones. Many of the over 360 UK species can be encountered in such microhabitats. Four subfamilies are listed in the latest checklist to beetles (Duff, 2018). The current key does not cover three of these subfamilies: Omophroninae, Cicindelinae and Brachininae. The single UK example of the Omophroninae is a disc-shaped species living in burrows along sandy margins of fresh water (and is rare in Britain). The five species of Cicindelinae (tiger beetles) are long-legged active hunters usually seen running over the ground (rather than seeking refuge under logs and stones). Four of these have very local distributions, the other (*Cicindela campestris* – the green tiger beetle) is widespread and common in open grassland and heathland. The two species of Brachininae (bombardier beetles) include one that is very rare, while the other is very locally distributed and declining in the UK. Both species use a mixture of chemicals to expel fluid at 100 °C from their anus as a defence against predation and are typified by having extra abdominal segments (seven in the female and eight in the male) compared to the six in the remaining Carabidae family (the Carabinae).

This key enables the identification of most of the common genera and species of Carabinae, especially those that are larger and more obvious, but does not include all those that may be found. See T.1 for the general morphology of carabid beetles (T.1a – full body, T.1b underside of head). Lengths (L) given are from the front of the head to the end of the abdomen, ignoring any appendages (e.g. antennae and legs). Note that gravid females may be larger than the lengths given if their abdomen

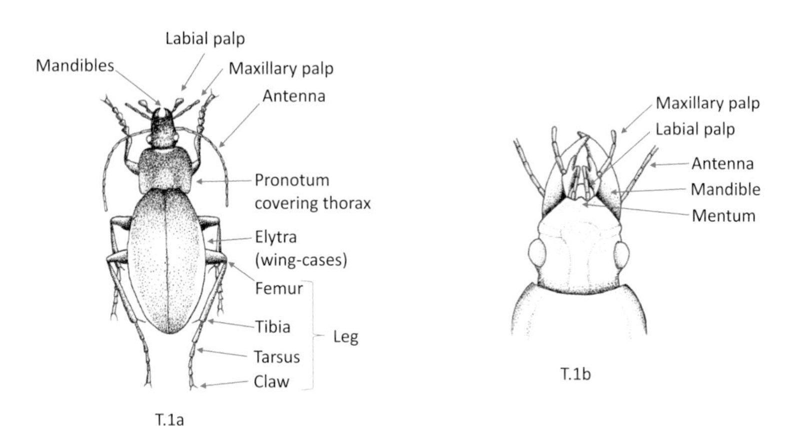


T.1a

T.1b

is swollen with eggs. Examples of common species are given, but where specimens do not exactly match the description, consult Forsythe (2000) or Luff (2007). Distribution data can be found in Luff (1998) or the UK Beetle Recording website (see Chapter 6). This site also provides descriptions and images of many of the British species and a checklist of the British species (see also Duff, 2018). Further keys and information can be found on the websites for Mike Hackston's Insect Keys, UK Beetles, UK Beetle Recording, and John Walters and Mark Telfer's Guides to British Beetles website (see Chapter 6). See Section 3.5 for further information on ground beetles.

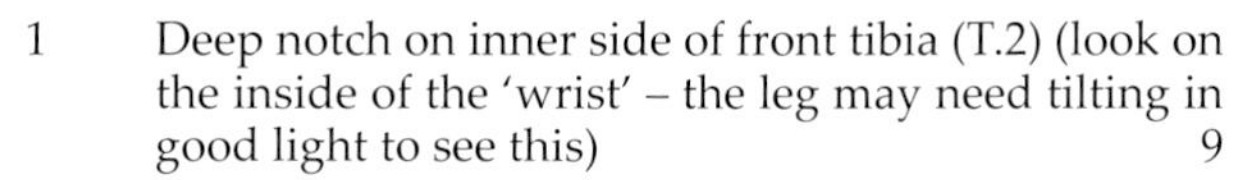

1 Deep notch on inner side of front tibia (T.2) (look on the inside of the 'wrist' – the leg may need tilting in good light to see this) 9
– No notch on front tibia 2

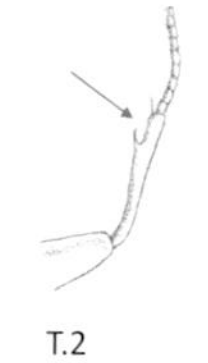

T.2

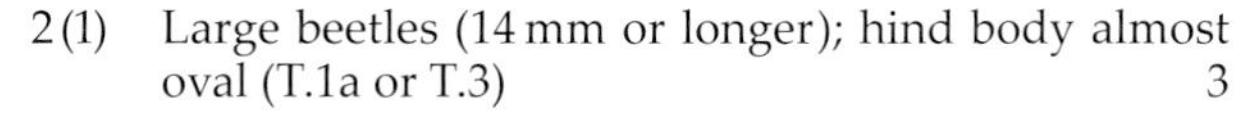

2 (1) Large beetles (14 mm or longer); hind body almost oval (T.1a or T.3) 3
– Less than 14 mm; hind body not oval 4

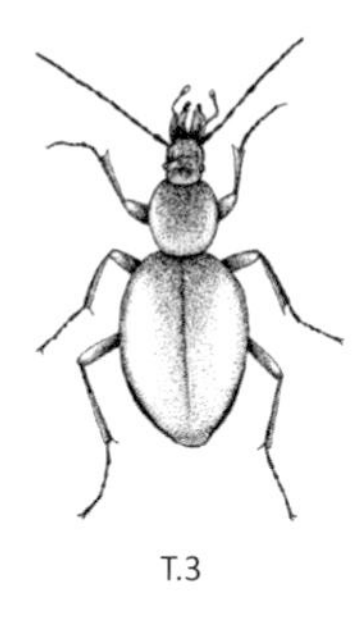

T.3

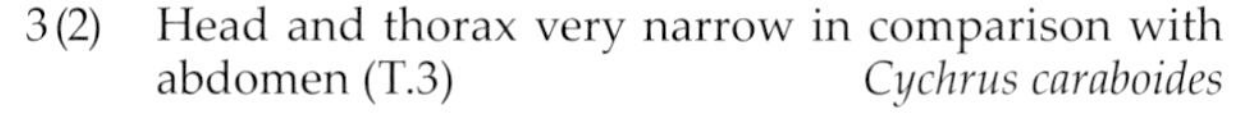

3 (2) Head and thorax very narrow in comparison with abdomen (T.3) *Cychrus caraboides*

- *C. caraboides* – buzzing snail-hunter – L:14–19 mm; body dull black with black appendages and long mandibles, prominent spoon-shaped palps, may make a buzzing sound if picked up; woods, upland grassland and moorland.

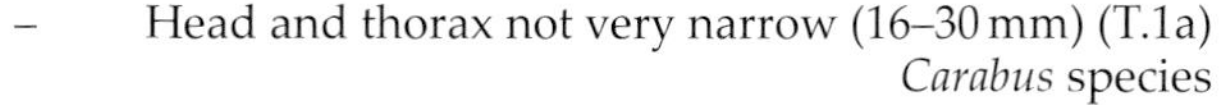

– Head and thorax not very narrow (16–30 mm) (T.1a) *Carabus* species

There are 11 British species, of which five are local or scarce. The more common species are:

- *C. arvensis* – moorland ground beetle – L:16–20 mm; elytra usually coppery (can be green, purple, blue, brown or black), each with three rows of elongated bumps; in moorland, heathland, also wood edges.
- *C. glabratus* – smooth ground beetle – L:22–30 mm; smooth black elytra (may have metallic blue tint); moorland.
- *C. granulatus* – sausage ground beetle – L:16–23 mm; elytra coppery (may have green or blue tint) with continuous raised ridges and elongated (sausage-shaped) bumps between these; in wetlands such as bogs and marshes.
- *C. nemoralis* – bronze (or green) ground beetle – L:20–26 mm; elytra green or purple bronze, rough in texture with fine ridges although without raised bumps, pronotum much wider than long; in a wide range of habitats including woods, grassland, farmland and gardens.

- *C. problematicus* – ridged (or grainy) violet ground beetle – L:20–28 mm, elytra black or blue-black (with metallic violet, blue, green or coppery sheen, especially at the margins) with fine ridges and rough in texture although without raised ridges or bumps, pronotum much wider than long and widest behind the middle; in woods, grassland and moorland.
- *C. violaceus* – violet ground beetle – L:20–30 mm; elytra black or blue-black (with metallic violet, blue, green or coppery sheen, especially at the margins), smooth in texture without either fine ridges or raised ridges or bumps (specimens of *C. problematicus* and *C. violaceus* can often be confused – one trick is to rub a fingernail across the elytra – in *C. problematicus* you will feel the ridges quite clearly, whereas *C. violaceus* feels smooth), unmetallic specimens may be confused with *C. glabratus* which has a wider and more convex body; wide range of habitats including woods, grassland, moorland, parks and gardens

Another genus (*Calosoma*) containing two species will also key out here – these have a straight hind edge to the pronotum, whereas *Carabus* species have the hind angles projecting backwards (as in T.1a). One of these species has a very local distribution, while the other is an occasional immigrant. Neither is regularly found under logs and stones, being active hunters in tree canopies.

4 (2) Elytra (wing-cases) with shining patches ('mirrors') *Elaphrus* species

There are four British species, of which two are scarce or local. More common species are:

- *E. cupreus* – copper peacock (or copper dimple-back) – L:8–9.5 mm; coppery-black with a purple or green sheen, both elytra with numerous purplish mirrors, legs black with blueish metallic sheen (middle of tibiae pale and not metallic); wet grassland, marshes and moorland.
- *E. riparius* – green-socks peacock (or green dimple-back) – L:6.5–8 mm; olive green (occasionally brown) with coppery sheen, elytra with numerous purplish mirrors, legs metallic green (middle of tibiae pale and not metallic); margins of freshwater habitats.

– Elytra without such mirrors (although scattered large, unmetallic punctures may be present) 5

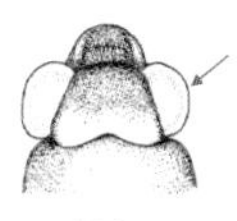

T.4

5 (4) Eyes very large and bulging (T.4); body with metallic sheen (coppery or bronze) *Notiophilus* species

Eight British species, of which two are scarce or local. If the specimen is under 4.5 mm, double check for the presence of a tibial notch (couplet 1) since *Asaphidion* species (couplet 11) may key out here if this feature has been missed. The more common *Notiophilus* species can be difficult to separate:

- *N. aquaticus* – black-legged springtail-stalker – L:5–6 mm; body bronze to black, elytra without pale hind spots, legs black; open habitats including grassland, moorland, dunes and near rivers.
- *N. biguttatus* – common springtail-stalker – L:5–6 mm; body bronze or coppery, elytra with yellowish spots at the tip, legs dark (tibiae reddish); widespread and very common, woods, grassland, fields and gardens. Difficult to separate from *N. substriatus* (frosted springtail-stalker – L:4.5–5.5 mm; most intervals between elytral stria rather dull (these are shining in *N. biguttatus*); open dry habitats).
- *N. germinyi* – heath springtail-stalker – L:4.5–5.5 mm; bright bronze or coppery to almost black (may have blue or purple sheen especially to the head), elytra without pale hind spot, legs dark (tibiae paler); grassland, heathland, and moorland. Difficult to separate from *N. palustris* (rough-necked springtail-stalker – L:4.5–5.5 mm; body dark copper or bronze to black, elytra without pale hind spots, legs dark (tibiae paler); woods and damp grassland).
- *N. rufipes* – red-legged springtail-stalker – L:5.5–6.5 mm, body brassy, elytra without pale hind spots, legs reddish; woods and gardens.

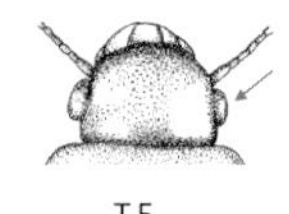

T.5

– Eyes not so large (e.g. T.5) 6

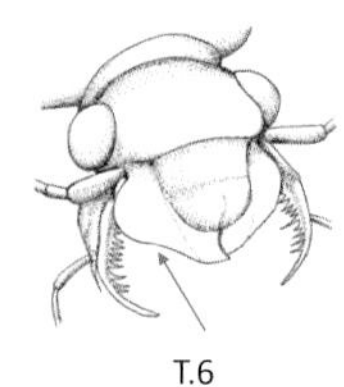

T.6

6 (5) Sides of mandibles expanded into side plates and palps with inwardly projecting spines (T.6) *Leistus* species

There are six British species, of which one is scarce. The more common species are:

- *L. ferrugineus* – rusty plate-jaw (or rusty flange-mouth) – L:6–8 mm; body (including head) pale to mid orangey or reddish brown; open habitats such as fields and gardens.
- *L. fulvibarbis* – blueish plate-jaw (or bluish flange-mouth) – L:6.5–8.5 mm; body dark brown to black, elytra with a faintly blue metallic sheen, legs reddish; in woods.
- *L. spinibarbis* – Prussian plate-jaw (or blue flange-mouth) – L:8–10.5 mm; body blue-black, elytra strongly metallic, legs black with paler tarsi; wide range of habitats including woods, grasslands and gardens.
- *L. rufomarginatus* – red-rimmed plate-jaw (or colonising flange-mouth) – L:8–9.5 mm; body

brownish with paler margins to the pronotum and elytra; woods, hedgerows and gardens.

- *L. terminatus* – black-headed plate-jaw (or black-headed flange-mouth) – L:6–8 mm, body pale to mid brown, head and tips of elytra black; damp woods, grassland, moorland and gardens, as well as near to freshwater habitats.

– Sides of mandibles without side plates **7**

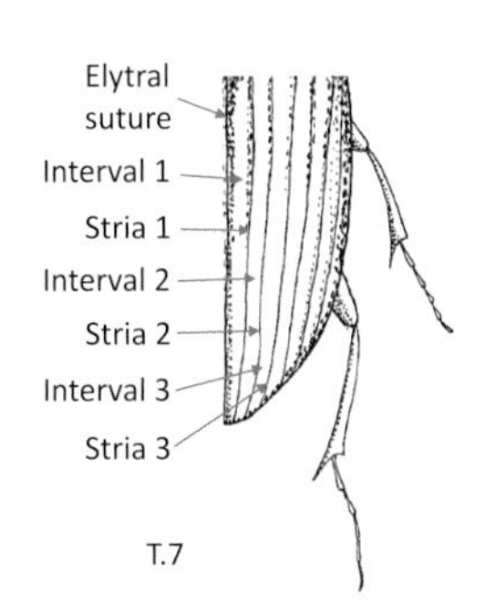


T.7

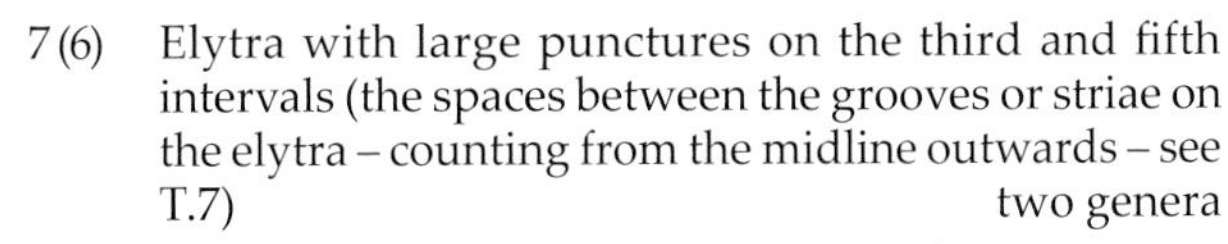

7 (6) Elytra with large punctures on the third and fifth intervals (the spaces between the grooves or striae on the elytra – counting from the midline outwards – see T.7) **two genera**

These two genera each contain a single species, one of which (*Pelophila borealis* – ten-lined dimple-back – L:9–12 mm) is restricted to Orkney, Shetland and north and west Ireland. The more common species is:

- *Blethisa multipunctata* – the many-dimpled – L:10.5–13.5 mm; black with a coppery sheen and a number of deep punctures on the elytra, head with two punctures containing setae at inner edge of each eye (one near the middle and the other near the back of the eye); widespread but local in Britain and Ireland, often near freshwater habitats such as fens and marshes.

– Elytra not heavily punctured **8**

8 (7) Elytra the same colour all over; body blackish brown **_Nebria_ species**

There are five British species, of which two are scarce. The more common are:

- *N. brevicollis* – common heart-shield – L:10–14 mm; elytra black, antennae and palps paler, top surface of hind tarsi with fine hairs (high magnification and a good light source are needed to distinguish between this and *N. salina* – note that these hairs may rub off, so it is worth examining both hind legs to be sure); widespread and abundant, in many habitats including woods, grasslands and gardens.
- *N. rufescens* – upland heart-shield – L:9–11 mm; black or reddish elytra, antennae mainly black; moorland, river and stream margins.
- *N. salina* – bare-footed heart-shield – L:11–14 mm; elytra black, antennae and palps pale, top of hind tarsi smooth; heathland, moorland and coastal habitats such as dunes.

– Not as above **other genera and species**

One distinctive species that will key out here is *Eurynebria complanata*:

- *E. complanata* – the beachcomber (or strandline beetle) – L:16–23 mm; sandy to golden brown with two ragged dark patches running across the elytra; sandy beaches above the high tide line, very local to south Wales and south-west England.

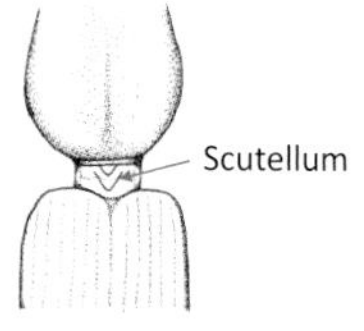


T.8

9 (1) A short waist visible between thorax and elytra (animals caught in traps containing fluid may expand giving the illusion of a waist, but a true waist has the scutellum on it – T.8) four genera

These four genera include one (*Miscodera*) that is rather scarce. Two genera (*Clivina* and *Dyschirius*) have broad front legs with strong spines adapted for digging, while the other two (*Broscus* and *Miscodera*) do not.

*Clivina* – two British species:

- *C. collaris* – two-tone tunneller – L:5–6 mm; head and pronotum dark brown, elytra paler brown; margins of freshwater habitats.
- *C. fossor* – common tunneller – L:6–6.8 mm; head, pronotum and elytra blackish brown; open habitats such as grasslands, fields and gardens.

*Dyschirius* – 11 British species, of which seven are local or scarce. More common species are:

- *D. globosus* – least mole-beetle – 2.3–2.9 mm; very rounded shiny brown to black body (with rounded margins to the pronotum and elytra), appendages brown; in damp habitats including bogs, coastal dunes and near saline pools.
- *D. luedersi* – marsh mole-beetle – L:3.3–3.9 mm; body shiny brassy black, appendages black; near water both fresh and saline. Difficult to separate from *D. salinus* – clay-coast mole-beetle – L:3.5–4.5 mm; body black with a brassy sheen (elytra may be reddish), appendages brown or black; saltmarshes.
- *D. thoracicus* – sandy-coast mole-beetle – L:3.5–4.5 mm; body shiny brassy black, femora and most of antennae black, base of antennae, tibiae and tarsi reddish; saltmarshes.

*Broscus* – one British species:

- *B. cephalotes* – strandline burrower – L:17–22 mm; dull black; sandy beaches and dunes.

– No such waist visible between thorax and elytra (at least part of the scutellum projects onto the elytra) 10

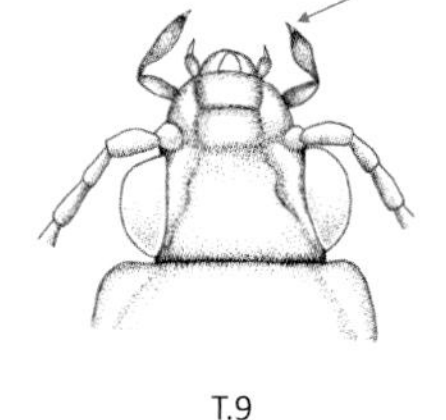

T.9

10 (9) Last joint of maxillary palp very short (T.9), less than half the length of the next-to-last segment 11

– Last joint of maxillary palp not so short, at least half the length of the next-to-last segment 12

Note that this couplet refers to the length rather than the width of the last segment of the palp – species with thinner than normal last segments should key out as normal length (go to couplet 12).

11 (10) Elytra covered with fine hairs *Asaphidion* species

Four British species, of which two are local or scarce:

- *A. curtum* – common velvetback (or shorter lesser hairy pin-palp) – L:3.8–4.5 mm; body bronze, legs pale yellow, antennae usually pale but darkening towards the tips; damp woodland and wetlands.
- *A. flavipes* – yellow-legged velvetback (or yellow-legged lesser hairy pin-palp) – L:3.9–4.7 mm; dark bronze, legs yellow or pale brown with darker joints between the femora and tibiae, last half of antennae brown; open habitats near fresh water.

– Elytra smooth six genera

These six genera include four which are scarce or local: *Bracteon* – two species; *Cillenus* – one species; *Elaphropus* – two species; and *Tachys* – four species. The other two genera are:

*Bembidion* – the largest carabid genus with 54 species, many of which (33) are local, scarce or rare. Several species can be difficult to separate. A few of the more common species are:

- *B. dentellum* – L:5.2–6.0 mm; body with a bronze sheen, elytra with paler brown marks especially at the shoulders and towards the tips, legs brown, antennae black with brown basal segments; damp and marshy sites near fresh water.
- *B. lampros* – common gleaming pin-palp – L:3–4 mm; shiny brassy or bronze, legs reddish brown, antennae dark with paler basal segments; dry habitats including fields and gardens.
- *B. quadrimaculatum* – common four-dotted pin-palp – L:2.8–3.4 mm; elytra dark (brown to black) with four pale marks at the corners; dry open sites such as fields and gardens.
- *B. tetracolum* – common four-spotted pin-palp – L:5–6 mm; metallic green head and pronotum, elytra metallic dark brown with paler marks at the shoulders and towards the tips, legs pale; open habitats especially near fresh water.
- *B. tibiale* – L:5.5–6.5 mm; body dark metallic green, legs brown with black bases to the femora and tibiae, antennae black (first segment brown); gravel and shingle banks of rivers and streams.

*Ocys* – two species, one of which is very scarce. The more common species is:

- *O. harpaloides* – L:4.2–5.8 mm; body shiny reddish brown with yellow to reddish brown appendages; in damp areas in woods.

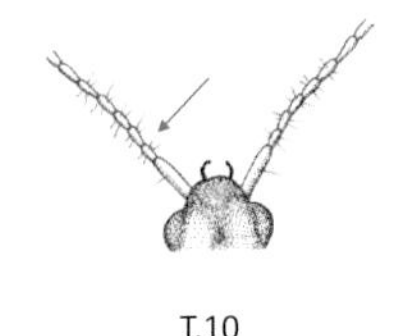
T.10

12 (10) Antennae with very long hairs on segments 2–5 (T.10); body with metallic sheen (brassy) *Loricera pilicornis*

- *L. pilicornis* – hair-trap ground beetle – L:6–8 mm; metallic black or dark brown with a green or copper sheen, elytra with distinct punctures; woods, grassland, fields, and gardens.

– Not as above 13

13 (12) Elytra truncate revealing at least part of the last segment of the abdomen (note that animals caught in traps containing fluid may expand, giving the illusion that the abdomen naturally extends past the elytra; similarly, gravid females may have the abdomen expanded with eggs – in both of these cases, the abdomen expands sideways as well as to the rear); elytra often flattish at the tips 14 genera

Of these 14 genera, six are scarce or local: *Cymindis* – three species; *Lionychus* – one species; *Masoreus* – one species; *Odacantha* – one species; *Polistichus* – one species; and *Somotrichus* – one species. In three of the other genera, the common species are more often found climbing plants including trees than under refuges on the ground: *Calodromius* – one species; *Dromius* – four species; and *Paradromius* – two species. The other five genera are:

*Demetrias* – three species, of which two are local or scarce (and more commonly climb plants). The more common species is:

- *D. atricapillus* – hairy-templed thatcher – L:4.5–5.5 mm; head black, pronotum orange-red, elytra pale yellow with a darker central mark which broadens towards the front edge, appendages pale, tarsi with bilobed penultimate segments (T.11); grasslands, fields and dunes.

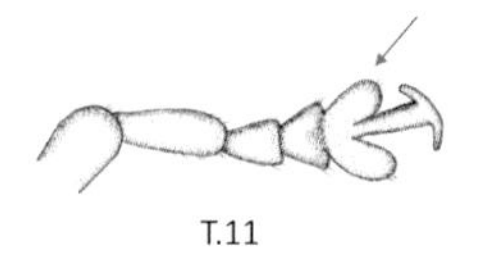
T.11

*Lebia* – flat ground beetles – three species, of which two are local and very scarce. The more common species is:

- *L. chlorocephala* – L:5.8–8.0 mm; head and elytra bright metallic green (with a golden or bluish sheen), pronotum reddish yellow; grasslands.

*Microlestes* – two species:

- *M. maurus* – least flimsy – L:2.3–2.8 mm; body shiny black, sides of elytra rather rounded, appendages black; dry sandy soils, sometimes at the coast.
- *M. minutulus* – overlooked flimsy – L:2.8–3.7 mm; body somewhat shiny black, sides of elytra quite straight, appendages black; open sites on sandy or gravelly soils.

*Philorhizus* – five species, of which three are local and very scarce. The more common species are:

- *P. melanocephalus* – black-headed stem-runner – L:2.8–3.5 mm; head black, pronotum bright orange-red, elytra pale yellow with darker central line which broadens towards the front edge, appendages pale, tarsi with unlobed penultimate segments; dry grassland and dunes.
- *P. notatus* – dark stem-runner – L:3.0–3.7 mm; head black, pronotum dark reddish brown to nearly black, elytra dark with pale yellow blotches on shoulders and near the tips, appendages pale; dry grassland and dunes.

*Syntomus* – three species, of which one has a rather local distribution and is scarce. The more common species are:

- *S. foveatus* – L:3.2–3.8 mm; body dark bronze, sides of elytra slightly rounded, appendages black; grasslands, dry heathland, fields, waste ground and dunes.
- *S. obscuroguttatus* – L:2.8–3.3 mm; body quite shiny black, sides of elytra straight, antennae black, legs brown (femora darker); damp habitats among litter and moss.

– Elytra not truncate, covering all of the abdominal segments; elytra relatively smoothly rounded at tips 14

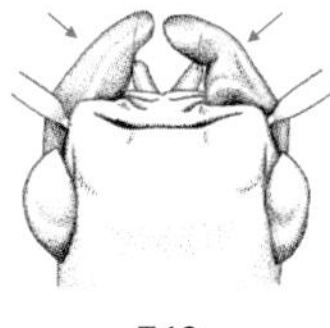

T.12

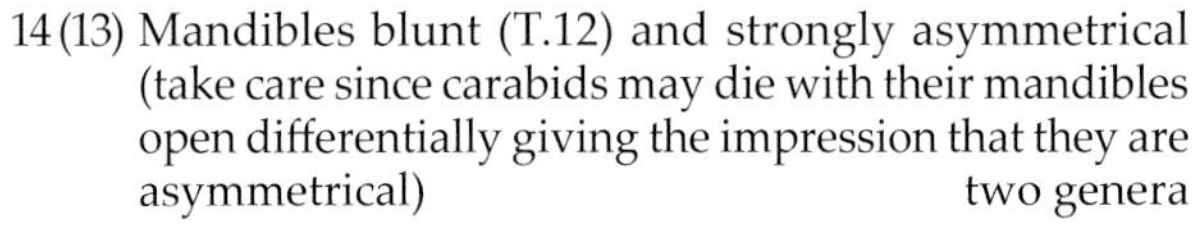

14 (13) Mandibles blunt (T.12) and strongly asymmetrical (take care since carabids may die with their mandibles open differentially giving the impression that they are asymmetrical) two genera

The two species in one of these two genera (*Licinus*) are scarce and local. The other genus (*Badister*) has seven species, of which five are local or scarce. The more common species are:

- *B. bullatus* – L:4.8–6.3 mm; head black, pronotum bright red-brown, elytra red with a black 'C'-shaped mark towards the tip of the left-hand elytron (singular of elytra) and its mirror image on the right-hand one, first antennal segment pale; dry open habitats such as grassland, heathland and dunes.
- *B. sodalis* – L:3.8–4.6 mm; head black, pronotum dark brown or black with paler margins, elytra mid to dark brown with a pale patch on each shoulder; damp woods.

– Mandibles not blunt; mandibles symmetrical 15

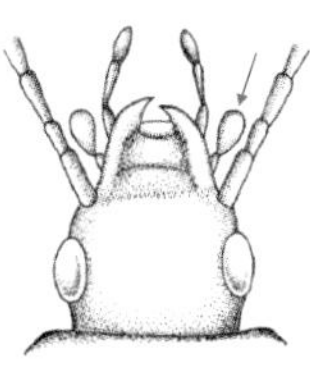

T.13

15 (14) Last segment of palps (especially the maxillary palps) broad at the end (T.13) two genera

The two species in one of these two genera (*Panagaeus*) are scarce and local. The other genus (*Synuchus*) contains a single species:

- *S. vivalis* – pear-palp ground beetle – L:6.0–8.5 mm; body mid to dark brown with light reddish brown appendages; woods, grassland, fields, and gardens.

– Last segment of palps not broad at end 16

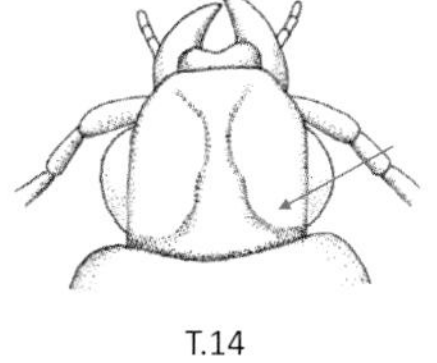

T.14

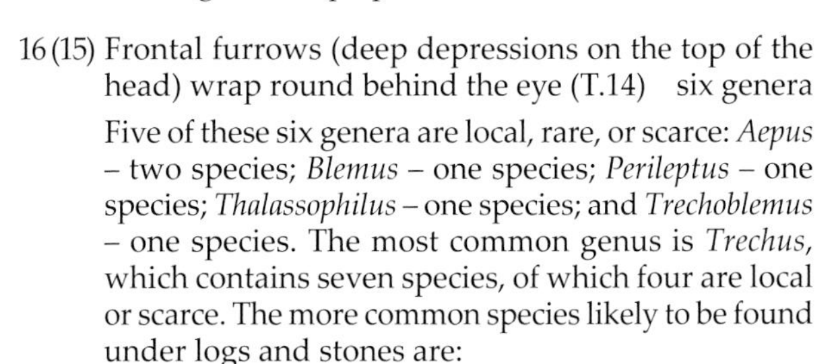

16 (15) Frontal furrows (deep depressions on the top of the head) wrap round behind the eye (T.14) six genera

Five of these six genera are local, rare, or scarce: *Aepus* – two species; *Blemus* – one species; *Perileptus* – one species; *Thalassophilus* – one species; and *Trechoblemus* – one species. The most common genus is *Trechus*, which contains seven species, of which four are local or scarce. The more common species likely to be found under logs and stones are:

- *T. obtusus* – short-winged spear-palp – L:3.6–4.0 mm; body reddish brown (head usually darker), appendages paler; heathland, moorland, fields and gardens.
- *T. quadristriatus* – long-winged spear-palp – L:3.6–4.1 mm; body reddish brown (head may be darker), appendages paler; in a wide range of habitats including fields, gardens, and human-disturbed sites.
- *T. secalis* – inland spear-palp – L:3.5–4.0 mm; body pale to mid reddish brown with distinct constriction between pronotum and elytra, appendages yellow; woods, damp grassland and fields.

– Frontal furrows (deep depressions on the top of the head) do not wrap round behind the eye, usually ending by the side of (not behind) the eye 17

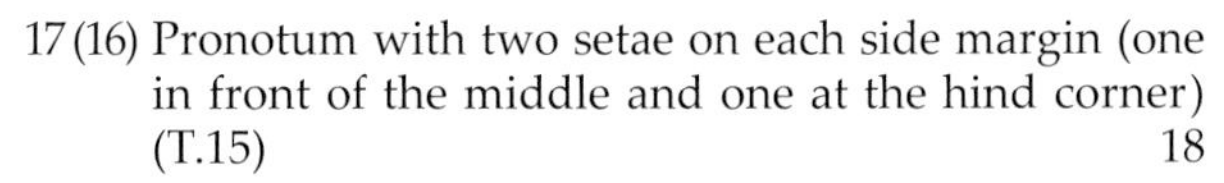

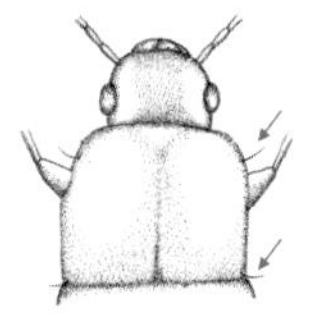

T.15

17 (16) Pronotum with two setae on each side margin (one in front of the middle and one at the hind corner) (T.15) 18

– Pronotum with a single seta on each side margin (at or near the middle) (T.16) – a few species have an additional seta at the front corner (not the hind corner) of the pronotum 11 genera

Three of these 11 closely related genera are scarce or local introductions (with one species each): *Anthracus*, *Diachromus* and *Scybalicus*. Several of the remaining species can be difficult to separate. The more commonly found species likely to be found under logs and stones are:

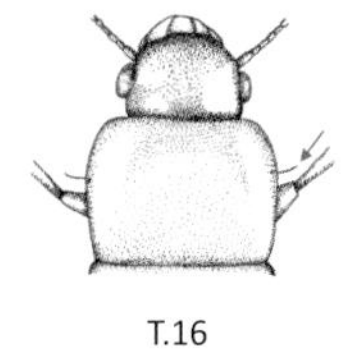

T.16

*Acupalpus* – eight species, of which five are local or scarce. The more common are:

- *A. dubius* – L:2.5–2.8 mm; pale to mid reddish brown, pronotum paler than head, antennae dark brown with basal and terminal segments paler, legs yellow or pale brown; near fresh water.

- *A. meridianus* – L:3.0–3.7 mm; body dark brown or black with ends and central part of elytra pale yellow-brown, antennae dark (basal segments paler), legs pale to mid brown (femora may be darker); open habitats such as fields and gardens.
- *A. parvulus* – L:3.1–3.9 mm; head black, rest of body variable – pronotum red (may have dark mark in the middle, elytra red-brown (may have dark mark in middle towards the rear), antennae mid brown (basal segments paler), legs pale (tibiae and femora may be darker in places); damp vegetation.

*Anisodactylus* – three species, of which two are local in distribution. The more common species is:

- *A. binotatus* – common shortspur – L:10–13 mm; body black, head with pale reddish spot between the eyes, legs usually black; damp grassland, marshes and fields.

*Bradycellus* – seven species, three of which are scarce. The more common species are:

- *B. harpalinus* – common brownling – L:3.8–5.0 mm; body mid reddish brown with pronotum and the margins and central line of the elytra paler, appendages pale, eyes large and rounded, hind angles of pronotum rounded; dry sites including woods, grassland, heathland, fields and gardens.
- *B. ruficollis* – heather brownling – L:2.5–3.4 mm; body dark brown to nearly black, pronotum and central line of elytra paler, appendages pale, eyes large but flattish, hind angles of pronotum sharp; drier heathland and moorland.
- *B. sharpi* – L:4.0–4.8 mm; body dark red-brown to nearly black (elytra darker than rest of body), appendages pale, hind angles of pronotum sharp; woods and damp grassland.
- *B. verbasci* – ruderal brownling – L:4–5 mm; body pale red-brown, elytra darker towards the tips, appendages pale, hind angles of pronotum sharp; open sites including fields and waste ground.

*Dicheirotrichus* – two species, one of which is local in distribution. The more common species is:

- *D. gustavii* – lesser saltling – L:5.5–7.5 mm; body variable (can be pale yellow-brown to nearly black with pale margins to head, pronotum and base of elytra), appendages yellow; in saltmarshes and at the shoreline.

*Harpalus* – 20 species, of which 14 are local or scarce. The more common species are:

- *H. affinis* – sunshining seed-eater – L:9–12 mm; body variable (can be metallic golden-green, reddish copper, bluish purple or black with a metallic sheen), elytra with outer three intervals (gaps between the grooves or striae on the elytra – see T.7) hairy; dry open sites such as fields, gardens and waste ground.

- *H. anxius* – heath seed-eater – L:6.5–8.0 mm; body black, antennae pale (sometimes except for segments 2 and 3), legs dark (except tarsi); dunes and sandy heathland.
- *H. latus* – wide-headed seed-eater – L:8.5–10.5 mm; body black or dark brown, margins of pronotum pale, appendages yellow or pale brown; dry grassland and upland heathland. *H. latus* can be difficult to separate from *H. rubripes* – L:9.0–11.5 mm; body variable (males shiny black with blue-green sheen; females dull black with metallic sheen to elytra), antennae pale reddish brown, legs brown with darker femora or mainly dark brown or black; dry open sandy and chalky sites.
- *H. rufipes* – golden-haired seed-eater – L:11–16 mm; body dullish black, appendages reddish brown, elytra with short golden hairs; dry open sites such as fields.
- *H. tardus* – common seed-eater – L:8.5–11 mm; body shiny black (in females the elytra are dull black), antennae pale, legs black with brown tarsi; dry open sites including grassland, fields, gardens and dunes.

*Ophonus* – 13 species, of which 11 are local or scarce. The more common species are:

- *O. puncticeps* – L:6.5–9.0 mm; body mid to dark brown, appendages light yellowish brown, sides of pronotum with single long seta near the middle; dry open sites, including fields.
- *O. rufibarbis* – L:6.5–9.5 mm; body mid to dark brown, elytra may be darker than rest of body, appendages yellow to pale brown, pronotum with two long setae on each side (one near the middle and the other at the shoulder); dry partially vegetated sites.

*Stenolophus* – three species, of which two are rather local in distribution. The more common species is:

- *S. mixtus* – L:5–6 mm; head and pronotum black (margins of pronotum brown), elytra dark brown (slightly iridescent) with margins, central line and shoulders paler, antennae dark (except for first segment), legs pale yellow-brown; marshes and edges of fresh water.

*Trichocellus* – two species:

- *T. cognatus* – northern hairy-eye – L:4.0–5.5 mm; head and pronotum black with paler margins, elytra brown with a black patch in the centre and towards the front, antennae black with first segment (counting from the head) reddish brown; widespread especially in the north of Britain, on moorland and heathland.

- *T. placidus* – southern hairy-eye – L:4.0–5.5 mm; head and pronotum brown with dark central patch, elytra light to mid brown with a long dark patch along intervals 2 to 4 (gaps between the grooves or striae on the elytra – see T.7); widespread especially in the east of Britain, damp woods and grassland and marshes.

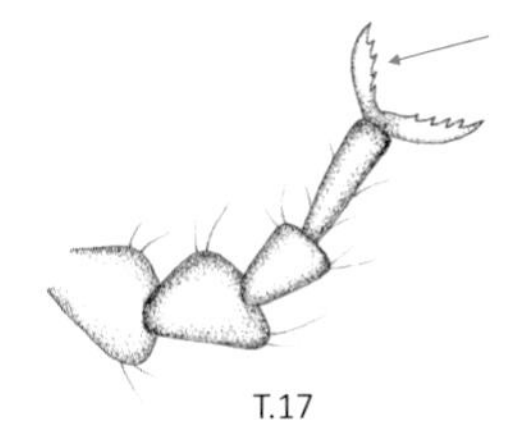

T.17

18 (17) Claws at end of tarsi toothed (T.17) *Calathus* species

There are eight British species, of which two are local or scarce. The common species are:

- *C. erratus* – common heathland raker – L:8.5–11.5 mm; body black, margins of pronotum paler, appendages reddish, hind angles of pronotum sharp; dry open sites including dunes.
- *C. fuscipes* – common raker – L:10–14 mm; body black, antennae brown, legs light brown with darker tarsi; grasslands, fields and gardens.
- *C. melanocephalus* – common bi-coloured raker – L:6.0–8.5 mm; pronotum reddish, head and elytra black, appendages reddish or yellow-brown; grassland, heathland, fields and gardens.
- *C. micropterus* – northern raker – L:6.5–8.5 mm; body black, margins of pronotum paler and widened at the base, hind angles of pronotum rounded, appendages pale yellow-brown; upland areas such as woods and moorland.
- *C. mollis* – dune raker – L:6.5–9.0 mm; pale to mid reddish brown, appendages pale yellow-brown; dunes and some sand pits inland.
- *C. rotundicollis* – round angled (or pitchy, or forest) raker – L:8.5–10.5 mm; body mid to dark brown, margins of pronotum and elytra paler, legs and antennae reddish, hind angles of pronotum rounded; woods, gardens and dunes. Confusion is possible between this species and *Synuchus vivalis* which is much less common and has broad last segments to the labial palps (see couplet 15), whereas the last segments of the labial palps in *C. rotundicollis* are cylindrical (like the other segments).

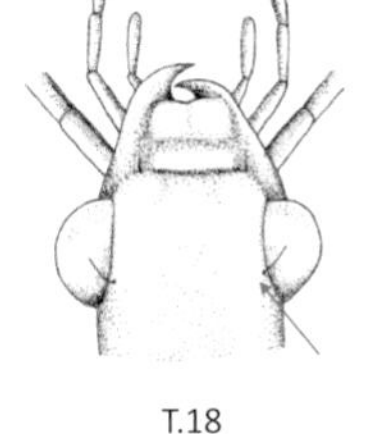

T.18

– Claws at end of tarsi simple (not toothed) 19

19 (18) Head by the inner edge of the eye with a single puncture containing a seta (near the middle of the eye) (T.18) 20

– Head by the inner edge of the eye with two punctures each containing a seta (one towards the front edge of the eye and the other near the rear or behind the eye) (T.19) 21

T.19

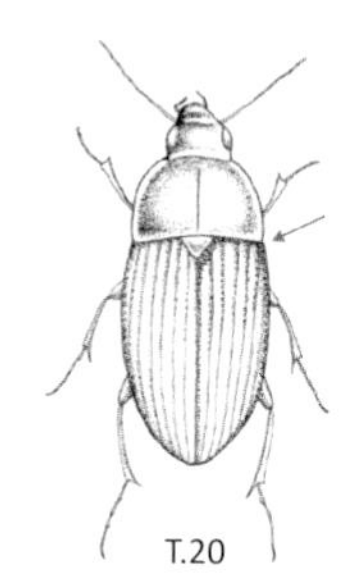
T.20

20 (19) Beetle oval, with head, thorax and abdomen almost continuous in outline (T.20) two genera

These two genera (*Oodes* and *Zabrus*) both contain a single species which are rather local in distribution.

– Not as above three genera

Of these three genera, two (with one species each) are very scarce: *Callistus* and *Drypta*. The more common genus is *Chlaenius* containing four species, of which two are local and scarce. The more common species that may be found under logs and stones are:

- *C. nigricornis* – green night-runner – L:10–12 mm; head coppery or metallic green, pronotum reddish copper, elytra metallic green, antennae mainly dark; damp grasslands and marshes.
- *C. vestitus* – yellow-bordered night-runner – L:9–11 mm; body bright metallic green (may have golden or bluish sheen), sides and tips of elytra pale yellow, appendages pale; muddy areas near water.

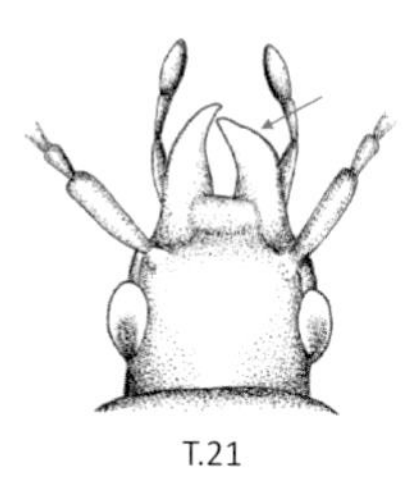
T.21

21 (19) Mandibles large, straight and protruding – as long as rest of the head (T.21) *Stomis pumicatus*

- *S. pumicatus* – long-jaw ground-beetle – L:6.5–8.5 mm; body dark reddish brown to black, appendages reddish brown; widespread but often scarce, woods, grassland, and disturbed areas.

A much smaller and paler species with a rather local distribution may also key out here due to its mandibles protruding strongly forwards: *Perigona nigriceps* – compost ground beetle – L:2–2.5 mm.

– Mandibles not as above 22

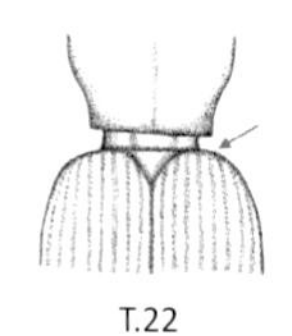
T.22

22 (21) Groove (stria) along shoulder of elytra absent (T.22); head strongly constricted behind the eyes *Patrobus* species

There are three British species, of which one is scarce. The more common species are:

- *P. assimilis* – northern (or upland) pinch-neck – L:6.5–9 mm; body dark reddish brown, appendages reddish, third antennal segment about the same length as the first; woods, grassland and moorland.
- *P. atrorufus* – common pinch-neck – L:7.5–9.5 mm; body mid to dark reddish brown, appendages reddish, third antennal segment longer than the first; woods and damp grassland.

– Groove along shoulder of elytra present (T.23) and/or head not strongly constricted behind the eyes 23

T.23

*Batenus livens* has a constricted neck but also a groove along the shoulder of the elytra (see couplet 26).

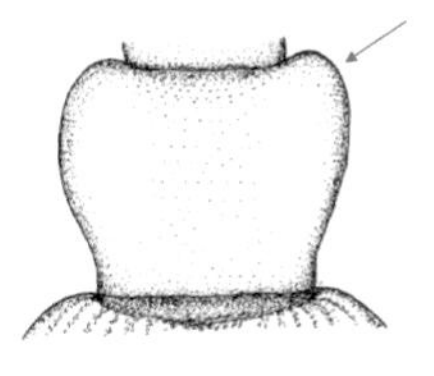
T.24

T.25

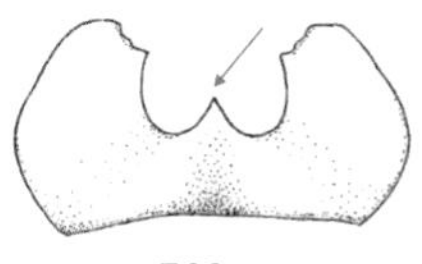
T.26

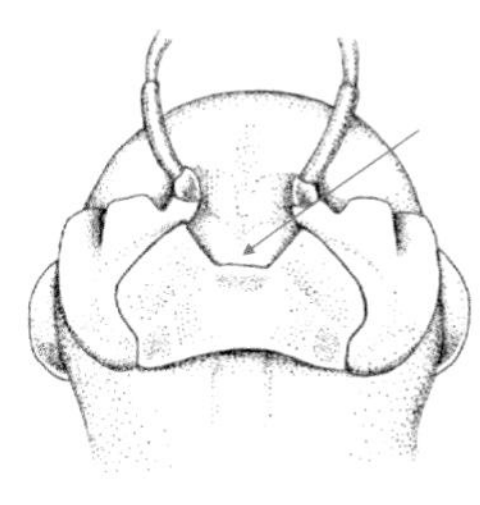
T.27

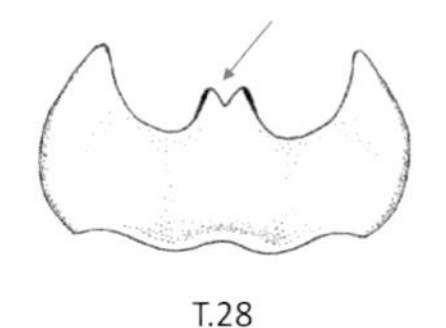
T.28

23 (22) Pronotum heart-shaped with base much narrower than elytra (T.24) six genera

Two of these six genera are scarce: *Laemostenus* – two species; and *Sphodrus* – one species. The more common genera (each with a single species) that are more likely to be found under logs and stones are:

- *Anchomenus dorsalis* – L:6–8 mm; head and pronotum metallic green or blue, elytra reddish brown with hind end green or blue; fields, gardens and waste ground.
- *Oxypselaphus obscurus* – L:5.0–6.5 mm; body mid to dark brown, head black, legs pale, antennae dark with paler basal segments; damp woods and marshes.
- *Paranchus albipes* – L:6.5–8.8 mm; body mid to dark brown, appendages yellow; at the margins of fresh water.
- *Platynus assimilis* – L:9.0–12.5 mm; body shiny black, legs mid to dark brown (bases of femora black), antennae brown with black basal segments; woods especially near water.

– Pronotum not heart-shaped with base not much narrower than elytra 24

24 (23) Pronotum with central projection forming a double curve on front edge (T.25) *Platyderus depressus*

- *P. depressus* – L:6.0–8.5 mm; body reddish brown, appendages pale red-brown; scarce, fields, waste ground and gardens.

– No central projection on front of pronotum 25

25 (24) Mentum (underside of mouthparts – examine from underneath) with tooth (T.26) 26

– Mentum without tooth (T.27) 27

26 (25) Tooth on mentum (underside of mouthparts – examine from underneath) with notch (T.28) *Pogonus* species

There are three British species, two of which are scarce. The most common species is:

- *P. chalceus* – punctate driftliner – L:5.5–7.0 mm; body coppery or metallic green, legs mid-brown with darker bases to the femora; coastal in saltmarshes and tidal debris.

– Tooth on mentum without notch (T.26) three genera

Two of these genera are local or scarce (with one species each): *Batenus* and *Sericoda*. The most common genus is *Agonum* – 16 species, of which nine are local or scarce. The more common species are:

- *A. emarginatum* – L:7.5–9.0 mm; body shiny black with black appendages (bases of antennae, tibiae and tarsi may be brown); marshes and near fresh water.
- *A. fuliginosum* – L:5.5–7 mm; head and pronotum black, elytra black or dark reddish brown, legs mid to dark brown, antennae black, pronotum wider than long; damp grassland, marshes and moorland.
- *A. gracile* – L:5.8–7.1 mm; body black (may have a metallic sheen), appendages black, pronotum wider than long; upland grasslands, marshes, bogs and moorland.
- *A. marginatum* – yellow-sided agonum – L:8.8–10.4 mm; body bright coppery-green, elytra with pale yellow or cream margins, appendages dark metallic brown (tibiae and first antennal segment paler); wet lowland sites especially near water, also at the coast.
- *A. muelleri* – L:7–9 mm; head and pronotum dark coppery-green, elytra dark copper or brassy, appendages dark brown or black (tibiae and first antennal segments paler); woods, damp grassland, fields, gardens and dune slacks.
- *A. thoreyi* – L:6–8 mm; head black, pronotum brown or black with paler margins, elytra yellowish brown or (very) dark brown, legs light to dark brown, antennae dark with paler basal segments; wetlands such as fens and marshes.
- *A. viduum* – L:7.5–9.0 mm; black with green or brassy metallic sheen especially on elytra, appendages black; in marshes and near water.

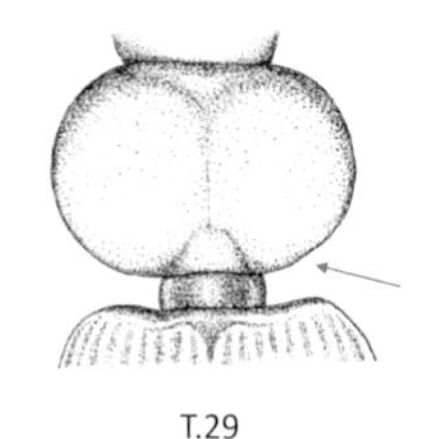
T.29

27 (25) Base of pronotum completely rounded (hind angles not distinct) (T.29) *Olisthopus rotundatus*

- *O. rotundatus* – L:6.5–8.0 mm; body reddish brown with a bronze sheen, legs yellow with darker tarsi, antennae dark with basal segments yellow; damp woods and marshes.

– Base of pronotum not completely rounded – at most hind angles (which are distinct) are rounded (e.g. T.30) 28

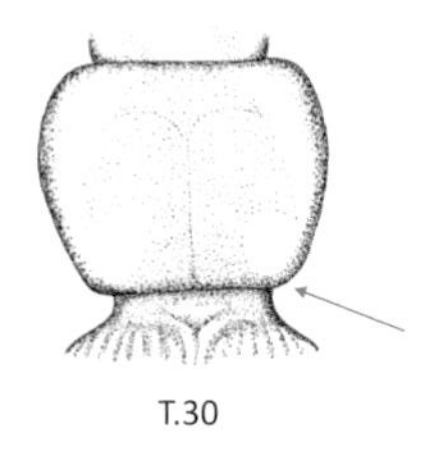
T.30

28 (27) Third elytral interval (the gap between the grooves or striae on the elytra – see T.7) with at least one puncture (needs to be viewed using high magnification and good lighting) 29

– Third elytral interval with no punctures 30

29 (28) Beetle with a strong metallic reflection *Poecilus* species

There are four British species, of which two are scarce. The more common species that may be found under logs and stones are:

- *P. cupreus* – copper greenclock – L:11–13 mm; body bright coppery-green (occasionally bluish, purple or black), appendages black except for bright reddish basal two antennal segments, head with many small pits between the eyes; widespread especially in the south, dry habitats including fields.
- *P. versicolor* – rainbow (or vari-coloured) greenclock – L:10.5–12.5 mm; body shiny green, copper, bluish or occasionally black, appendages black except for bright reddish basal two antennal segments, head without small pits between the eyes; damp grassland, moorland and fields.

– Without a strong metallic reflection (with at most a slight metallic sheen) *Pterostichus* species

There are 19 British species, of which nine are rather local or scarce. The more common species are:

- *P. adstrictus* – upland blackclock – L:10–13 mm; body black, elytra often with a dark bronze sheen, appendages black, pronotum with sharp hind angles; upland moorland and heathland.
- *P. madidus* – common blackclock – L:14–18 mm; body black, legs black or reddish (two forms), hind angles of pronotum rounded (T.30); widespread and very common, woods, grasslands and gardens.
- *P. melanarius* – rain beetle (or common black ground beetle, or strawberry ground beetle) – L:13–17 mm; body and appendages black, pronotum with a protruding tooth at the sharp hind angles and two depressions on either side at the base; widespread and very common, grassland, fields and gardens.
- *P. minor* – lesser (or small) blackclock – L:6.8–7.9 mm; body black, legs dark reddish brown, pronotum with sharp hind angles and two depressions on each side at the base; wet grassland and marshes.
- *P. niger* – great blackclock – L:16–21 mm; body and appendages black, pronotum with sharp hind angles and two depressions on each side at the base; woods, damp grassland and moorland.
- *P. nigrita* – mitten blackclock – L:9–12 mm; body and appendages usually black, pronotum with a protruding tooth at the sharp hind angles; widespread and common especially in damp lowland sites, particularly near water. *P. nigrita* is difficult to separate from *P. rhaeticus* (pincer blackclock – L:9–11.5 mm) with which it overlaps in habitats, although the latter species is mainly found in wet grasslands and moorland at higher altitudes.
- *P. strenuus* – rough-chested blackclock – L:6.0–7.2 mm; body black, legs and base of antennae reddish brown; widespread and common in a wide range of habitats, especially grasslands. *P. strenuus*

is difficult to separate from *P. diligens* (smooth-chested blackclock – L:5.5–7.0 mm; damp grassland, marshes and moorland). In *P. strenuus* the sides of the prosternum (underside of the pronotum) are rough and punctured, whereas in *P. diligens* they are smooth and relatively unpunctured. *P. vernalis* (spring blackclock – L:6.0–7.5 mm; damp lowland habitats especially grasslands) is another similar species which can be distinguished by the lack of the short striae (or groove) near to the scutellum that can be seen in the other two species.

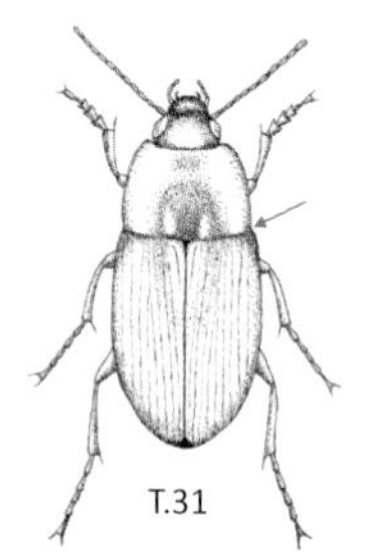
T.31

30 (28) Beetle oval, with head, thorax and abdomen almost continuous in outline (T.31) *Amara* species

There are 27 British species, of which 16 are local or scarce. The more common species are:

- *A. aenea* – common sunshiner – L:6.5–8.8 mm; body shiny brassy or coppery, antennae black (basal three segments pale), legs black (tibiae mid to dark brown); dry grasslands, gardens, waste ground and dunes.
- *A. apricaria* – lesser brown sunshiner – L:6.5–8.5 mm; body mid to dark reddish brown with a metallic sheen, appendages mid brown; open sites including fields.
- *A. bifrons* – pale moonshiner – L:5.5–7.3 mm; body mid to dark brown with a brassy sheen, appendages pale yellow-brown; dry open sites.
- *A. communis* – interrupted sunshiner – L:6–8 mm; body shiny brassy black, antennae black with the basal two segments and the basal half of segment 3 pale, legs black (tibiae a little paler); grassland and moorland.
- *A. eurynota* – rumple-backed sunshiner – L:9.6–12.5 mm; body coppery or bronze, appendages black except the basal three antennal segments; open dry habitats such as fields and dunes.
- *A. familiaris* – red-legged sunshiner – L:5.5–7.3 mm; body dark brass or copper, antennae black with basal three and part of fourth segment pale, legs pale to mid reddish brown; grassland, heathland and dunes.
- *A. lunicollis* – mesophile sunshiner – L:7.5–9.0 mm; body shiny black or with coppery sheen, appendages black except for first one or two segments of antennae; open habitats.
- *A. ovata* – broad-dimpled sunshiner – L:8.0–9.5 mm; body black often with a brassy sheen, appendages black with the basal three antennal segments pale; dry fields and gardens.
- *A. plebeja* – lesser trident-spurred sunshiner – L:6.0–7.8 mm; body black with a metallic sheen, antennae black (basal three segments red), legs black (tibiae red or light brown), front tibiae with a

three-pronged spur as opposed to a single-pronged spur in all other *Amara* species except for the very scarce and larger *A. strenua* (great trident-spurred sunshiner – L:8.0–9.6 mm) which also has a three-pronged spur; damp grasslands and fields.

- *A. similata* – narrow-dimpled sunshiner – L:8.0–9.5 mm; body brassy or coppery, antennae black except basal three segments pale, legs black (tibia mid to dark brown); open fields and gardens.
- *A. tibialis* – least sunshiner – L:4.5–5.9 mm; body black often with a metallic sheen, antennae black (basal three segments pale), legs dark with brown tibiae; dry open ground including heaths and dunes.

– Beetle not oval, distinct pinch point between pronotum and elytra 31

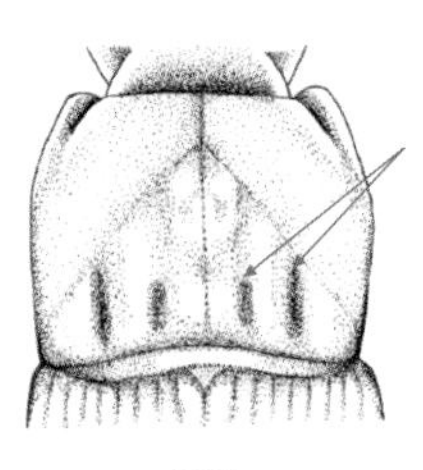

T.32

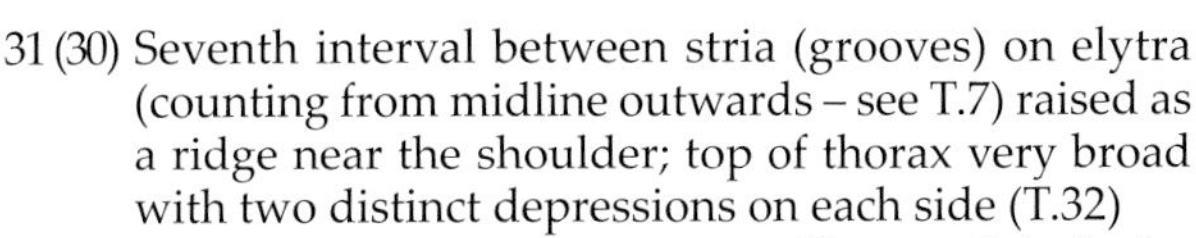

31 (30) Seventh interval between stria (grooves) on elytra (counting from midline outwards – see T.7) raised as a ridge near the shoulder; top of thorax very broad with two distinct depressions on each side (T.32) *Abax parallelepipedus*

- *A. parallelepipedus* – common shoulder-blade – L:17–22 mm; body wide and shiny black, pronotum with sharp hind angles and two depressions on each side at the base; woods and moorland.

– No such ridge near the shoulder *Curtonotus* species

There are three British species, one of which is scarce. The more common species are:

- *C. aulicus* – common stem-climber – L:11–14 mm; body black or dark reddish brown often with a brassy sheen, appendages mid brown; dry open habitats.
- *C. convexiusculus* – saltmarsh stem-climber – L:11–12.8 mm; body bronze or slightly metallic dark brown, appendages light to mid brown; lowland grasslands and coastal dunes and saltmarshes.

## Key U Rove beetles

This is one of the largest families of beetles, comprising 19 subfamilies containing over 1,000 British species. Many of them are difficult to identify with certainty – even for experts – and, especially with smaller species, expert help should be sought in order to confirm identification. See U.1 for the general morphology of rove beetles (U.1a – full body, U.1b underside of head). Lengths (L) given are from the front of the head to the end of the abdomen, ignoring any appendages (e.g. antennae and legs). It is worth noting that dead specimens may alter in length with those drying out shrinking somewhat, while those kept in preservative may expand due to the preserving fluid. Gravid females may also increase in size if their abdomen is swollen with eggs. This key enables the identification of rove beetles to subfamily. Examples of species are listed to give some idea of the types of animals that may be found. These should be taken as indications only and not as definitive identifications. Where the specimen does not fully match the description given, consult Joy (1932), Tottenham (1954), Hodge & Jones (1995), Lott (2009) and Lott & Anderson (2011). The latter two texts cover species identification for nine of the 19 subfamilies found in the UK and, together with Tronquet (2006), provide illustrations of many of the species found in Britain. Further keys and information can be seen on the websites for Mike Hackston's Insect Keys, UK Beetles, and UK Beetle Recording (see Chapter 6). Duff (2018) offers a checklist of the British beetle fauna including staphylinids. A national distribution atlas has not been published as yet, although a few may be available at county level – e.g. Marsh (2016) has produced a provisional atlas to many of the Staphylinidae

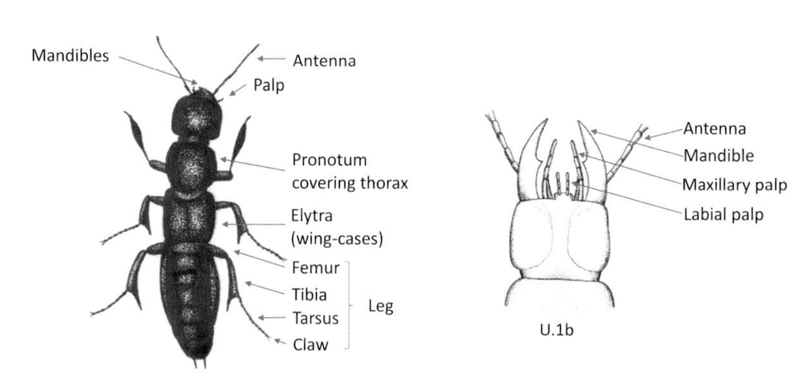


U.1a

U.1b

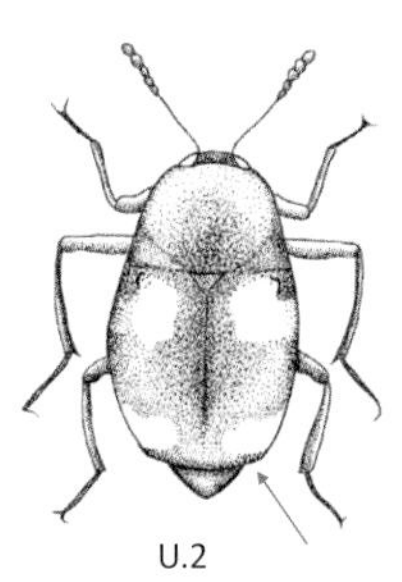
U.2

for the historical vice-counties of Yorkshire. Databases that include distribution data for Staphylinidae can be found at the NBN Atlas website (see Chapter 6). See Section 3.5 for further information on rove beetles.

1 Elytra (wing-cases) covering the whole or most of the abdomen (at most 1–2 segments exposed) (U.2, U.3 or U.4) 2

– Elytra (wing-cases) short leaving 3–6 segments of the abdomen exposed (U.5) 3

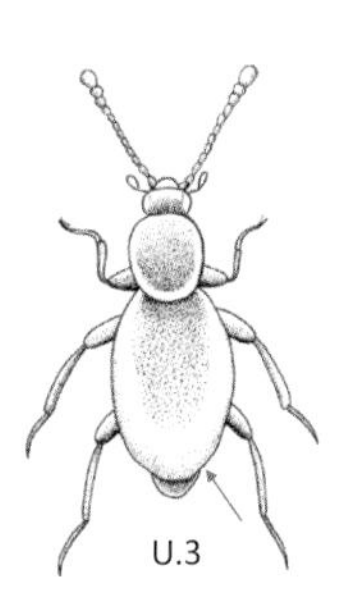
U.3

2 (1) Outline of body smooth – no 'waist' between the thorax and abdomen (often oval in shape – U.2); upper surface shiny; jaws point downwards Scaphidiinae

Scaphidiinae (shining fungus beetles) comprise three genera containing five species, which tend to be associated with fungi in rotting heartwood. One common example is:

- *Scaphidium quadrimaculatum* – orange-spotted scaphidium – L:4.5–6.0 mm; body shiny black with two reddish patches on each elytron; deciduous woods under logs, sometimes in aggregations.

– Outline of body not smooth – distinct 'waist' between thorax and abdomen (not oval in shape – U.3); upper surface with at least some hairs (often quite hairy) Scydmaeninae

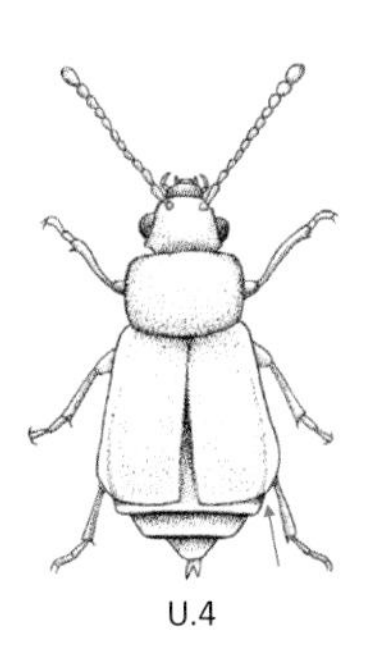
U.4

Scydmaeninae (ant-like stone beetles) include nine genera containing 32 species. Common examples include:

- *Scydmaenus tarsatus* – L:2.0–2.1 mm; body shiny dark reddish brown covered with pale hairs; may be found in decaying wood and ants' nests, also overwinter in leaf litter and under logs and stones.
- *Stenichnus collaris* – L:1.5–1.7 mm; body shiny black with some yellowish hairs, more elongate than *S. scutellaris* (elytra twice as long as broad – examine at high magnification using an eyepiece micrometer), antennae and legs reddish brown; leaf litter and rotting wood.
- *Stenichnus scutellaris* – L:1.4–1.5 mm; body shiny black with some yellowish hairs, squatter than *S. collaris* (elytra less than twice as long as broad – examine at high magnification using an eyepiece micrometer), antennae and legs reddish; mainly woods and wood edges in leaf litter and in association with ants.

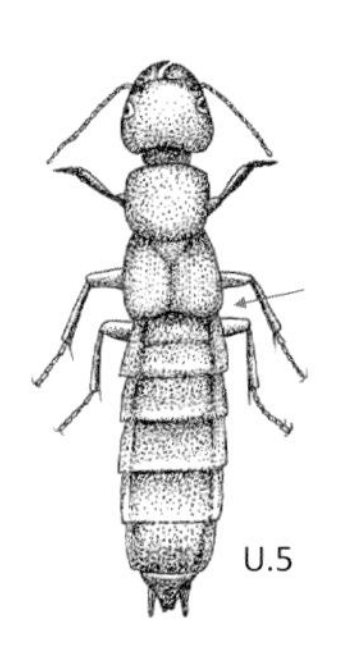
U.5

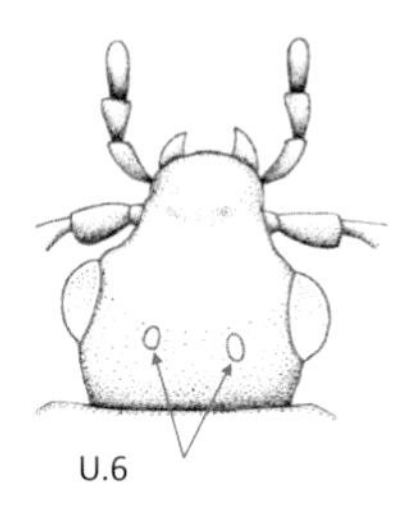

U.6

Six species of Omaliinae (ocellate rove beetles – see also couplet 11) in the genus *Eusphalerum* key out here. These can be recognised by a pair of ocelli (small simple eyes) on the top of the head behind the eyes (U.6). A common species is:

- *Eusphalerum luteum* – L:2–3 mm; body pale brown to yellowish, slightly triangular with the pronotum wider than the head and the elytra wider than the pronotum; more often on flowerheads but can be found on the ground in wood edges and hedgerows.

3 (1) Tarsi each with three segments; elytra broader than pronotum; abdomen fairly inflexible; antennae with club formed from the final segment; body < 3 mm long — Pselaphinae

Pselaphinae (ant-loving beetles) contain 54 species from 19 genera, many of which are associated with ants' nests (especially those of *Lasius* species). One distinctive UK genus (*Claviger*) contains two species – L:2.0–2.7 mm; pale brown with fine hairs, eyes absent, antennae thick with only six segments (compared to 11 in the other genera); found (sometimes in large numbers) in ants' nests, especially of *Lasius* species (see Key R – Ants).

– Not as above — 4

4 (3) Antennae with nine segments (last segment clubbed) — Micropeplinae

There are two genera containing five species, all of which are under 2.5 mm long. All the UK species are dark, with oval bodies, strongly ridged pronotums (with four or five longitudinal ridges), and 11-segmented antennae without a club. One very common example is:

- *Micropeplus staphylinoides* – L:2 mm; body black with paler margins to the pronotum, and paler legs and antennae, elytra with longitudinal ridges between which are many small punctures, abdominal segments with strong longitudinal ridges and a ridge along the hind edge; widespread and common throughout England, Wales and Scotland (where it is mainly concentrated in the east), in woodland, damp grassland and wetlands.

– Antennae with 10 or 11 segments — 5

5 (4) Antennae long and thin with whorls of hairs on each segment — Habrocerinae/Trichophyinae

Each of these subfamilies has a single UK species, both of which are rather local in distribution. The two families are distinguishable by the position of the antennae: in Trichophyinae (L:3.5 mm) they are fixed to the upper surface of the head, compared to Habrocerinae (L:2.5–3.5 mm) where they are fixed to the side of the front of the head.

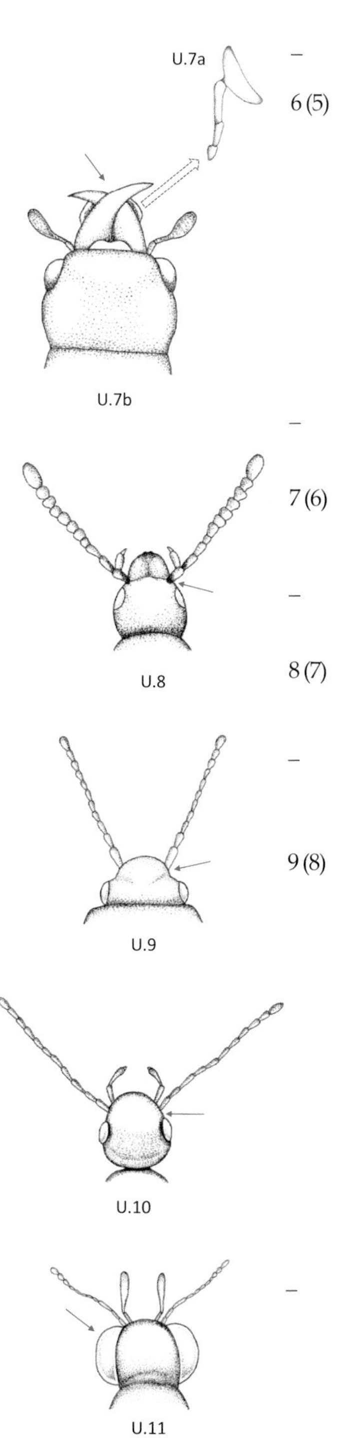


– Antennae without whorls of hairs on each segment 6

6 (5) Final segment of labial palps large and crescent shaped (U.7a); mandibles long and crossed over in front of head (U.7b) Oxyporinae

Oxyporinae (cross-toothed rove beetles) containing a single, quite distinctive species:

- *Oxyporus rufus* – L:6–8 mm; head large (wider and longer than pronotum) and shiny black, pronotum and anterior sections of the elytra and abdomen are orange (the remainder being black), orange legs and antennae; found in woods and parkland often associated with fungi, sometimes on dead wood.

– Final segment of labial palps similar to other segments 7

7 (6) Both pronotum and elytra with distinct ridges Pseudopsinae

A single, rare UK species.

– Pronotum or elytra, or both, without distinct ridges 8

8 (7) Antennae originate on top of the head near the front of the eyes (on the 'forehead' – U.8) – origination point clearly visible from above 9

– Antennae originate on the front of the head (U.9), or on the side of the head (U.10) – origination point often not clearly visible from above 10

9 (8) Antennae originate near to the front of the eyes; head large and broad, wider than thorax; eyes large and bulging (U.11) Steninae

Steninae (water skaters) are a difficult subfamily to separate, comprising many similar species (75 species in two genera). Two common examples are:

- *Stenus bimaculatus* – L:6–7 mm; body black, elytra each with an orange spot, yellowish red legs with dark joints; often found active on soil and within low vegetation especially in wetlands and damp woods, adults may overwinter under logs and stones.
- *S. clavicornis* – L:5–6.5 mm; black body, yellowish red legs with dark joints; open habitats including wetlands and grassland, often within low vegetation or plant debris.

– Head smallish; eyes normal; antennae with 11 (rarely 10) segments; last segment of labial palp at least as long as broad Aleocharinae

This subfamily comprises a large number of difficult to identify, often small species (463 species in 126 genera). Common genera include *Oxypoda* (L:1.5–4 mm, all

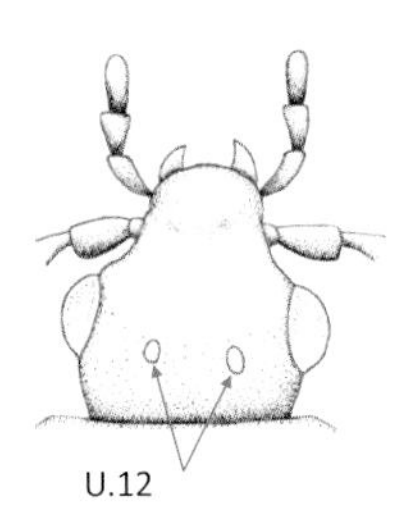

U.12

tarsi with five joints) and *Atheta* (vexing rove beetles – L:1–4 mm, front tarsi with 4 joints).

10 (8) Top of head with one or two simple eyes (ocelli) (U.12 or U.13) – these show as small, round contrastingly coloured areas on the top of the head in line with the hind margins of the eyes (they may need to be examined at high magnification under good light conditions, especially for species with heavily pitted heads) 11

– Top of head with no ocelli 12

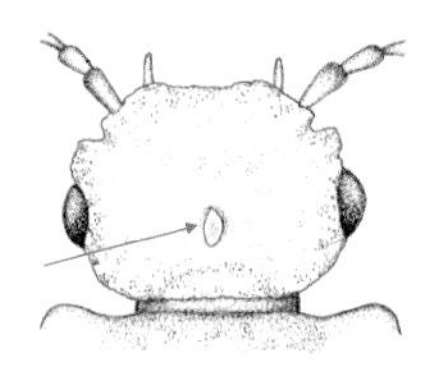

U.13

11 (10) Top of head behind eyes with pair of small simple eyes (ocelli) (U.12) Omaliinae

Omaliinae (ocellate rove beetles) comprise 27 genera containing 64 species (see also couplet 2 for the genus *Eusphalerum*). Common species include:

- *Lesteva sicula* – L:3.5–4 mm; hairy body which is dark brown to black, orange-brown legs, last segment of maxillary palp four times the length of the previous one; found in damp areas in vegetation debris and under logs and stones.
- *Olophrum piceum* – L:4.1–6.0 mm; shiny, dark brown body with paler legs and antennae, thorax broader than long, all segments of palps elongate; woods, heaths and moorland, often under heather litter.

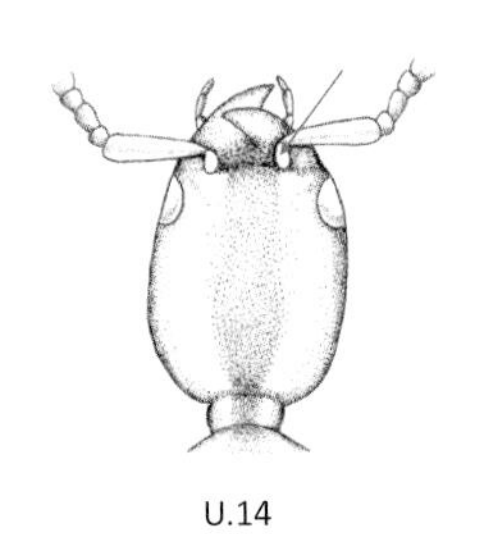

U.14

– Top of head behind eyes with a single simple eye (ocellus) (U.13) Proteininae

A single genus containing one species keys out here (see also couplet 20):

- *Metopsia clypeata* – L:2.5–3.0 mm; body pale brown or yellowish, head somewhat hexagonal in shape, head and pronotum highly punctured and whole upper surface covered in short hairs; found associated with fungi, dung and damp vegetation.

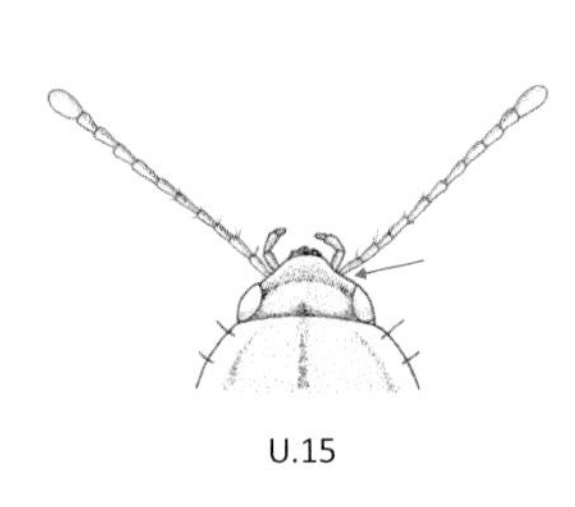

U.15

12 (10) Antennae originate inside the base of the mandibles (U.14) 13

– Antennae originate outside the base of the mandibles (U.15) 16

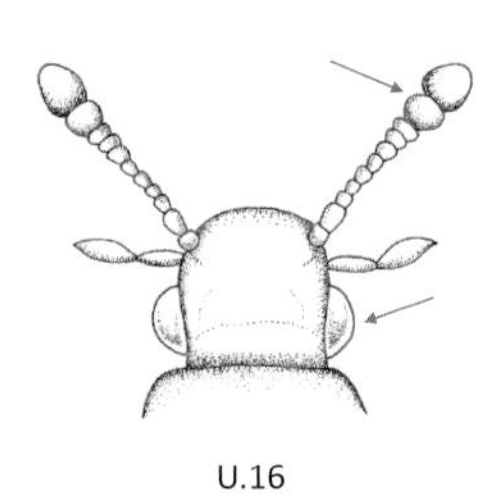

U.16

13 (12) Antennae with club comprising last two segments; eyes positioned towards the back of the head (U.16) Euaesthetinae

There are two genera containing four species, all of which are fairly local in distribution. The most common is:

- *Euaesthetus ruficapillus* – L:1.4–1.7 mm; body dark, eyes small, pronotum strongly pitted; marshy areas under wet vegetation debris.

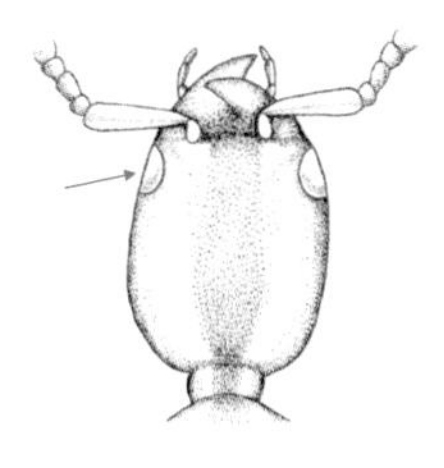

U.17

– Antennae without a club; eyes set forward of the back of the head (U.17); head not much narrower than the pronotum, which is never less than half the length of the elytra Staphylininae 14

There are 33 genera of Staphylininae (large rove beetles) containing 185 species which can be divided into three tribes (see couplets 14 and 15).

14 (13) Origination points of antennae closer to eyes than to each other (easier to see under high magnification using an eyepiece micrometer) Staphylinini

This is the largest tribe in the Staphylininae comprising 23 genera and 158 species, many of which are very similar and may be difficult to separate. A few common (and more striking) species include:

- *Ocypus olens* – Devil's coach-horse beetle (or cocktail) – L:23–32 mm (the largest UK species in the family); body dull black, often seen with the tail raised over the back; found in woods, grassland, and urban areas.
- *Philonthus decorus* – L:10–14 mm; body dull black with coppery front parts, pronotum with 4 pits on each side; widespread, especially in shaded habitats.
- *Quedius levicollis* – L:11–13 mm; body black with red or dark red appendages (legs, antennae and maxillary palps), quite large eyes; widespread especially in open areas.

– Origination points of antennae as close (or closer) to each other as to the eyes (easier to see under high magnification using an eyepiece micrometer) 15

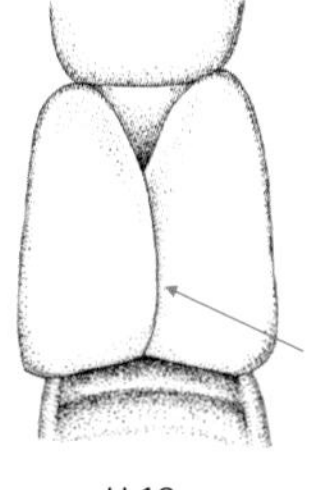

U.18

15 (14) Elytra overlap at midline (U.18); distinct narrow neck (less than half the width of the head); head same width as thorax, which is longer than broad) Xantholinini

This tribe contains eight genera and 21 species, several of which are quite similar. A common example is:

- *Xantholinus linearis* – L:7–8 mm; head black, pronotum dark red to black with a slightly metallic appearance, elytra brown to dark brown, yellow or reddish legs, antennae and maxillary palps; very common and widespread especially in open and dry areas.

– Elytra meet at midline without overlapping; neck wider (at least half as wide as head) Othiini

This tribe contains two genera and six species, one of which has only been recorded once. The largest common species is:

- *Othius punctulatus* – L:9–13 mm; head black, pronotum dark red or black, elytra red; widespread and common in damp woods and grassland.

16 (12) Tarsi with three segments — Oxytelinae

Oxytelinae (spiny-legged rove beetles) include 11 genera containing 87 species that key out here (see also couplet 21), many of which are difficult to separate. One of the most common species is:

- *Anotylus sculpturatus* – L:3.5–4.5 mm; body black, with ridges along the top of the pronotum, yellow legs (which have rows of spines along the outer edges of the tibiae) and dark antennae; widespread in dung and leaf litter.

– Tarsi with four or five segments — 17

U.19

17 (16) Abdomen strongly narrowed to the tip (U.19) — Tachyporinae

Tachyporinae (crab-like rove beetles) comprise 13 genera containing 67 species, many of which are difficult to separate. Common species include:

- *Tachinus rufipes* – L:4.5–6.5 mm; body black with paler reddish hind margins to the pronotum and elytra; occur under logs and other debris in damp wood margins, grassland and parkland.
- *Tachyporus nitidulus* – L:1.1–1.4 mm; body reddish with darker head and abdomen; found among fungi and mosses in woods, especially in autumn and winter.
- *Tachyporus hypnorum* – L:3–4 mm; head black, pronotum dark with paler yellowish hind and side margins, elytra yellowish with darker front margin, antennae and legs pale; found among leaf litter in open areas.

Note that *Tachyporus* species have setae on the edges of the thorax and elytra, while *Tachinus* species do not.

– Abdomen not strongly narrowed to the tip — 18

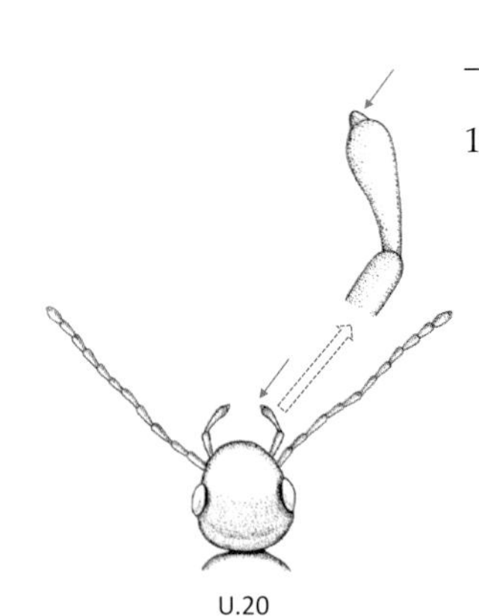
U.20

18 (17) Last segment of maxillary palp very small (U.20) — Paederinae

This subfamily contains 16 genera and 62 species, many of which are difficult to separate. The strikingly coloured genus *Paederus* (pronotum and the anterior four abdominal segments are orange, while the head and the last two abdominal segments are black, and the elytra are metallic blue) has a toxin within the haemolymph ('blood'), which can cause dermatitis in humans. A common species of Paederinae is:

- *Lathrobium brunnipes* – L:8.0–9.5 mm; body black with reddish yellow legs; found in leaf litter and vegetation debris from mires and damp woods and grassland.

– Last segment of maxillary palp not very small — 19

19 (18) Body densely covered with golden hairs Phloeocharinae

The single UK species is:

- *Phloeocharis subtilissima* – L:1.5–2.0 mm; body dark brown with reddish elytra, body entirely covered in long pale (golden) hairs; found in woods associated with fallen wood and fungi.

– Body smooth or any hairs not golden 20

20 (19) Under 3 mm in length; elytra broad (wider at hind edge than widest part of pronotum); antennae with two basal segments enlarged Proteininae

Ten species in two genera key out here (see also couplet 11). The genera can be separated as follows: in *Megarthrus* species the pronotum has a central groove and a notch in the hind angles of the pronotum, while *Proteinus* species do not have these features and are more oval in shape and covered in very fine hairs. Two common species are:

- *M. denticollis* – L:1.5–2.5 mm; body brown with paler side margins to the pronotum, legs pale brown, antennae dark with paler first segments; found in a range of habitats in decaying wood and other vegetation.
- *P. brachypterus* – L:1.5–1.9 mm; body dark brown with paler legs, antennae dark with paler first segments; found in a range of habitats, often associated with decaying fungi and leaf litter.

– Not as above (if under 3 mm in length, then body strongly metallic) 21

21 (20) Elytra and pronotum of similar width (body shape somewhat straight-sided); males with horns on side of head and mandibles Piestinae

Piestinae (flat rove beetles) include a single UK species:

- *Siagonium quadricorne* – L:4.0–5.5 mm; body dark brown to black, with paler pronotum, elytra and legs; may be common in areas with fallen timber, found under bark in damp conditions.

– Elytra wider than pronotum (body shape not straight-sided); no horns on head or mandibles Oxytelinae

Oxytelinae (spiny-legged rove beetles) include four genera (each with a single species) that key out here (see also couplet 16). Three of these are found at the edge of streams and ponds and tend to be locally distributed. The most common species is:

- *Syntomium aeneum* – L:2.2–3.0 mm; body dark with bronze-green metallic sheen; widespread on dry soils, especially in upland areas.

## Key V Insect larvae

It is almost impossible to identify many insect larvae to species. If animals can be kept until they emerge, identification of the adults may be possible. However, it is often difficult to culture specimens through pupation to emergence. Ford (1963), Oldroyd (1973) and the Amateur Entomologists' Society webpages (see Chapter 6) provide details of culturing techniques appropriate for many types of insects, while Wong (1972) reviews the early literature on rearing a wide range of insects. Here we include a quick tabular key to the major body shapes of insect larvae (Table 4.2). The main key enables the identification to order, and in some cases family, of those larvae commonly found under logs, stones and other debris. Where the specimen does not fully match the description given in the following key, consult the series of Royal Entomological Society Handbooks covering fly (Smith, 1989) and beetle larvae (Hammond *et al.*, 2019), or for a wider range of insect larvae see Chu & Cutkomp (1992) or Marshall (2017). See Section 3.5 for further information on insect larvae.

**Table 4.2** Tabular key to insect larval body shapes – five basic body shapes cover most insect larvae, which can be separated on the presence/absence of a well-developed head and/or legs/prolegs as well as their general appearance

| | Body shape | | | | |
|---|---|---|---|---|---|
| | **Campodeiform** | **Elateriform** | **Eruciform** | **Scarabaeiform** | **Vermiform** |
| Well-developed head | Present | Present | Present | Present | May be absent, sometimes present |
| Jointed legs on thorax | Long | Short | Short | Short | Absent |
| Prolegs on abdomen | Absent | Absent | Present | Absent | Absent |
| Appearance | Insect-like | Worm-like | Caterpillar-like | Grub-like | Maggot-like |
| Examples | Neuroptera (snakeflies: Raphidiidae – V.8); Coleoptera (soldier beetles: Cantharidae – V.9; ground beetles: Carabidae – V.11; rove beetles: Staphylinidae – V.12). | Coleoptera (click beetles: Elateridae – V.10). | Hymenoptera (sawflies: Symphyta – V.6); Lepidoptera (butterflies and moths – V.7). | Coleoptera (longhorn beetles: Cerambycidae; dung beetles and scarabs: Scarabaeidae; weevils: Curculionidae – V.3). | Diptera (crane flies: Tipulidae – V.1; horse flies: Tabanidae – V.2; blow flies: Calliphoridae – V.4); Hymenoptera (ants: Formicidae – V.5). |

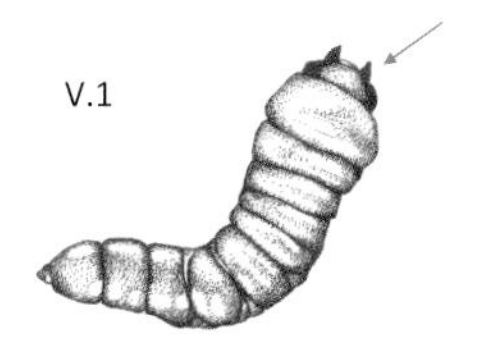


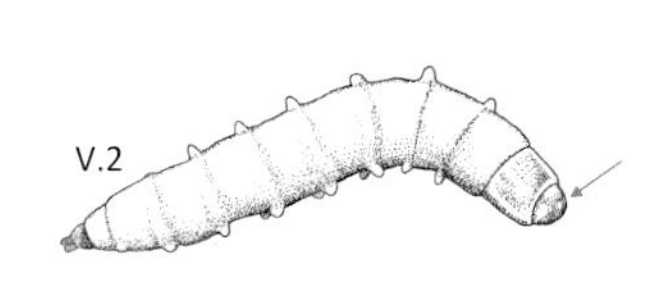


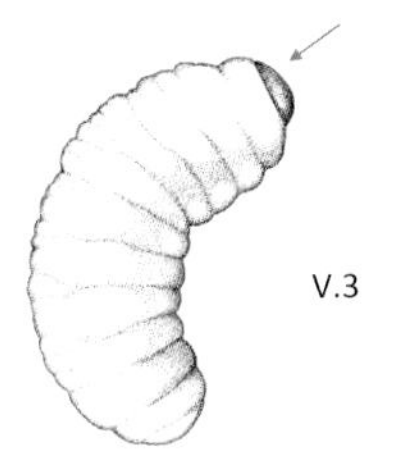


1 Without jointed legs — 2

– With jointed legs (note that these may be quite small) — 3

2 (1) Head a distinct hard capsule (may be partially retracted into the thorax), usually darker than the rest of the body (V.1, V.2 or V.3) — Diptera (true flies – part) and Coleoptera (beetles – part)

These include Diptera:

- Mycetophilidae (fungus gnats) – white or transparent bodies with dark head, may be associated with mucus threads or trails;
- Tipulidae (crane fly larvae – V.1) – with leathery skin and head partially retracted into thorax;
- Tabanidae (horse fly larvae – V.2) – cylindrical but tapering at both ends, with a small head which is retracted into the thorax;

Coleoptera:

- Curculionidae (weevil larvae – V.3) – often distinctly crescent shaped, fleshy and pale.

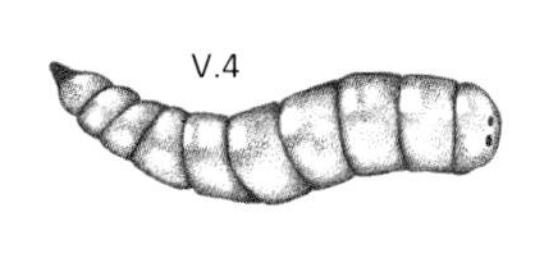


– Head not a hard capsule, not darker than the rest of the body (V.4, V.5) — Diptera (true flies – part) and Hymenoptera (bees, wasps and ants – part)

Diptera larvae that key out here include:

- Calliphoridae (blow fly larvae – V.4) – these are maggots and often pale.

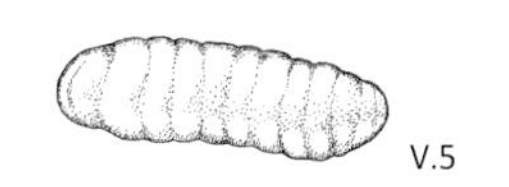


Bee, wasp and ant larvae (Hymenoptera – V.5) that key out here are not pointed at either end. Ant larvae are most likely under logs and stones and will almost always be in association with adult ants (usually in an ants' nest), often with workers carrying them to safety when disturbed.

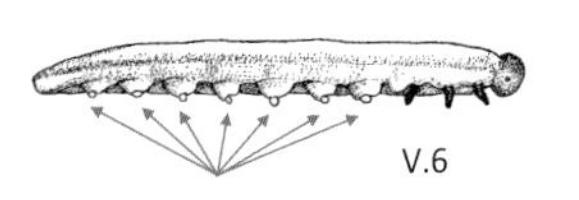


3 (1) Prolegs (fleshy – not jointed) present on abdomen in addition to the jointed legs on the thorax (V.6 or V.7) — 4

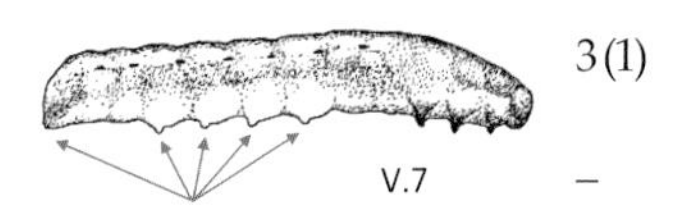


– No prolegs on abdomen — 5

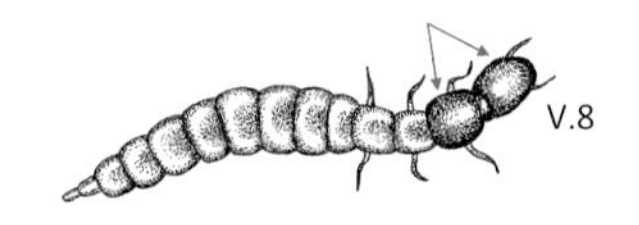


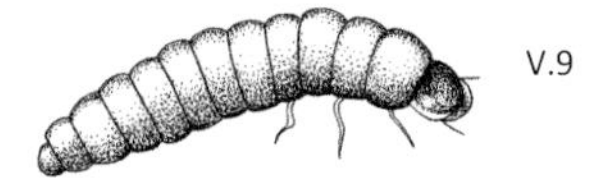


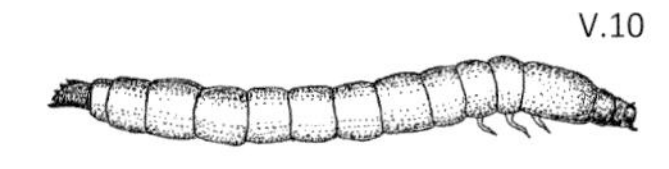


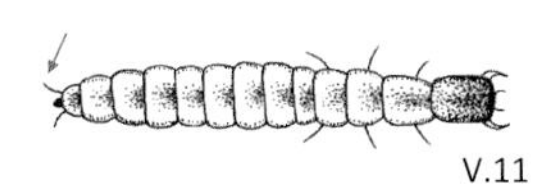


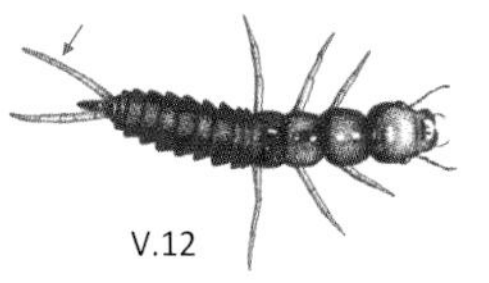


4 (3) At least six pairs of prolegs (V.6) — Hymenoptera: Symphyta (sawflies)

– At most five pairs of prolegs (V.7) — Lepidoptera (butterflies and moths)

5 (3) Head and next segment hard, dark and elongate (V.8) — Raphidioptera: Raphidiidae (snakeflies)

– Not as above — 6

6 (5) Head capsule dark and well developed — Coleoptera (beetles) 7

– Head capsule not dark or well developed — other larvae

7 (6) Skin of larva velvety; no projections (cerci) from hind end (V.9) — Cantharidae (soldier beetles)

– Skin not velvety; projections (cerci) from hind end may be present — 8

8 (7) Pale brown; surface shiny, slightly waxy; shape cylindrical (V.10) — Elateridae (click beetles)

– Not as above — 9

9 (8) Large mandibles (jaws); head often hard and dark; projections (cerci) from hind end long (e.g. V.11 or V.12) — other Coleoptera

These include:

Carabidae (ground beetle) larvae (V.11) (legs with six segments and usually two claws per leg); Staphylinidae (rove beetle) larvae (V.12) (legs with five segments and one claw per leg).

– Not as above — other larvae

Other beetle larvae may key out here including Scarabaeidae (scarab beetles such as chafers, dung beetles and some species found in rotting wood) and Cerambycidae (longhorn beetles which live in decaying wood). These are grub-like, with obvious hardened heads and jointed legs, and may be curled up into a C-shape.

## Key W Amphibians

This key covers adult amphibians, and while it does not include tadpoles, it may help to identify some juveniles. In the UK, there are two groups of native amphibian – frogs and toads (Anura), which do not have tails as adults, and newts (Caudata) which do. Tailed amphibians can be distinguished from lizards by their smooth and (usually) moist skin. Several introduced species, some of which are now naturalised, may occasionally be found, although most have a restricted distribution. Lengths (L) given are maximum full body lengths including the tail (where present). Where specimens do not correspond with the description given consult Beebee (2013) or Inns (2018), and the Froglife website (see Chapter 6). Both texts include illustrations (as does Roberts *et al.*, 2016) and distribution maps for all UK species. See Section 3.6 for further information on amphibians.

1 Has a tail — newts 2

– Does not have a tail — frogs and toads 5

2 (1) Up to 100 mm in total length (including the tail) — 3

Small specimens (and juveniles) of other species may key out here – see the descriptions in couplet 4 if specimens do not match those in couplet 3.

– Over 100 mm in total length (including the tail) — 4

3 (2) Underside of throat pink (or yellow) and unspotted; belly pale yellow with few, if any, dark spots — *Lissotriton helveticus*

- *L. helveticus* – palmate newt – L:90 mm; body brown, green or grey, breeding males with a dark filament at the end of the tail; widespread, scrub, open woodland, heathland, moorland, pasture and gardens.

– Underside of throat white with black spots; belly yellow with black spots or blotches — *Lissotriton vulgaris*

- *L. vulgaris* – smooth newt – L:100 mm; body grey or brown, breeding males have a wavy crest along the back; widespread and common, open woodland, scrub, pasture and gardens.

4 (2) Underside bright yellow/orange with well-defined black spots — *Triturus cristatus*

- *T. cristatus* – great (or northern) crested newt – L:150 mm; body dark brown or black with a rough (warty) appearance, males have a wavy crest with an obvious break between the back and tail (the crest is more pronounced during the breeding season); widespread, open woodland, pasture.

– Underside without black spots, or with ill-defined spots or blotches (i.e. with fuzzy edges) other newt species

Two introduced species will key out here – *Ichthyosaura alpestris* (Alpine newt – L:110 mm; belly unspotted) is uncommon, mainly in parks and gardens, while *Triturus carnifex* (Italian crested newt – L:160 mm; skin smooth, belly with large round and fuzzy-edged spots) is restricted to a couple of sites around Birmingham and in Surrey.

5 (1) Skin dry or dryish with warty appearance toads 6

– Skin smooth and moist (without warty appearance) frogs 8

W.1

6 (5) Eyes with horizontal pupils; head with prominent raised glands (parotoid glands) behind the eyes (W.1) 7

– Eyes with vertical pupils; head without prominent raised glands behind the eyes *Alytes obstetricans*

- *A. obstetricans* – midwife toad – L:50 mm; body greyish, belly white, underside often with grey spots; scattered distribution in England, in gardens.

7 (6) Central yellow stripe along the back *Epidalea calamita*

- *E. calamita* – natterjack toad – L:80 mm; body green or brown, iris of eye greenish; very rare, in coastal marshes, dunes and heaths.

– Without central yellow stripe along the back *Bufo bufo*

- *B. bufo* – common toad – L:90 mm: body olive-brown to brown, belly pale with dark speckles, iris of eye coppery or golden; common and widespread throughout Britain, in woodland, scrub and rough grassland.

W.2

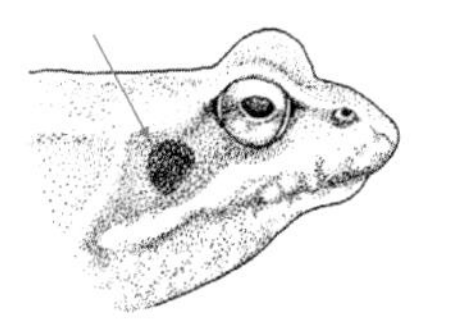

8 (5) Obvious brown patch behind the eye (W.2) *Rana temporaria*

- *R. temporaria* – common frog – L:90 mm; olive-green or brown (breeding males are grey with a pale blue throat); widespread and common, in damp woodland and grassland, marshes, moorland, parks and gardens.

– No obvious brown patch behind the eye green (or water) frogs

These include one native (pool frog) and a couple of introduced species (edible and marsh frogs) that are not found under logs and stones.

## Key X Reptiles

The UK reptile fauna is quite small, consisting of two native species of legged lizards, one species of slow worm (legless lizard) and three species of snakes. Most of these species use refuges to hide under. Legged lizards can be distinguished from amphibians with tails (newts) by their dry, scaly skin. There are a number of reptile species that have been introduced to the UK, including two lizards from the Channel Islands. This key allows the identification of adults of the most common native species, with some notes on juveniles. Lengths (L) are maximum sizes of adults, measuring from the tip of the head to the end of the tail. Note that in some lizards (including slow worms) the tail may be damaged or have been shed to avoid predation. Even if the tail regrows following shedding, it will usually be rather stubbier than the original size. Therefore, it is important to take account of any loss of tail when measuring body lengths (where relevant the proportion of the total body length contributed by the tail is given in the species descriptions). Where specimens do not correspond with the description given, consult Beebee (2013) or Inns (2018), or the Froglife website (see Chapter 6). Both texts include illustrations (as does Roberts *et al.*, 2016) and distribution maps for all UK species. See Section 3.6 for further information on reptiles.

| | | |
|---|---|---|
| 1 | With legs | 2 |
| – | No legs (snake-like) | 3 |

2(1) Body greyish or greenish brown, mottled with rows of dark and white flecks running along the back *Zootoca vivipara*

- *Z. vivipara* – viviparous (common) lizard – L:150 mm (tail contributes half to two-thirds of this length); body greyish or greenish brown with rows of dark and white flecks, female often has paler back, darker sides, a dark central line and a yellowish underside; widespread throughout the UK, in open woodland, scrub, heaths, moors and dunes.

A similar looking, if slightly larger, introduced species will also key out here: *Podarcis muralis* (wall lizard) – L:180 mm (tail contributes up to two-thirds of this length). As its common name suggests, this species is more likely to be found high on rocks and walls rather than under refuges; mainly in the Channel Islands, where it is native, and the south of England).

A larger, bright green species may also be found in coastal heaths and dunes in Jersey, where it is native, and on south-facing cliffs around Bournemouth, where it has been introduced: *Lacerta bilineata* (green lizard) – L:400 mm (tail contributes up to two-thirds of this length).

– Body grey-brown marked with eyespots (dark spots with paler centres) *Lacerta agilis*

X.1

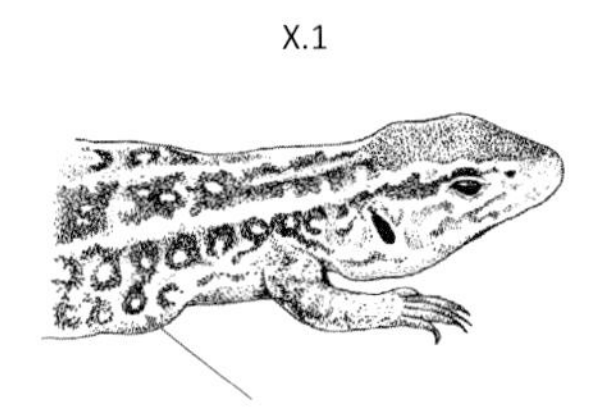

- *L. agilis* – sand lizard – L:190 mm (tail contributes about half of this length); body greyish or brownish mottled with darker patches which have paler (white or cream) centres (X.1), breeding males have bright green flanks; with a scattered and local distribution, mainly in the south and coastal west of England, in dry heaths and sand dunes.

3 (1) Body up to 450 mm long; blunt end to the tail *Anguis fragilis*

- *A. fragilis* – slow worm – L:400–450 mm (may be substantially shorter if the tail has been shed, since the tail comprises up to half of this length); body grey or grey-brown (males), or with a coppery or golden-brown back and darker flanks (females), may have a thin stripe running down the back; widespread across Britain, in open woodland, heaths, grassland, and human-disturbed habitats such as urban wasteland and gardens.

– Body over 450 mm long; end of tail not blunt snakes 4

X.2

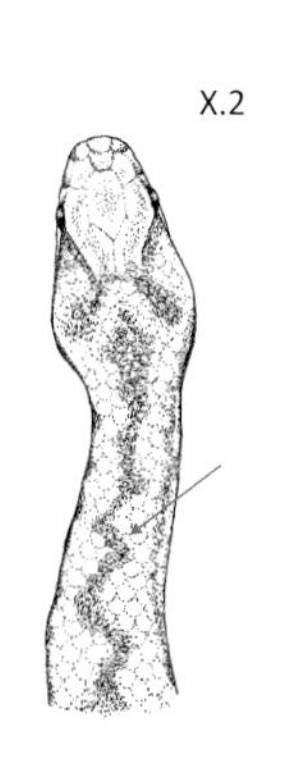

4 (3) Body dark or silvery/reddish brown with a darker zig-zag pattern along the length of the body (X.2) *Vipera berus*

- *V. berus* – adder – L:700 mm; body typically greyish with black markings (males), or brownish with darker brown markings (females); throughout Britain, but less common in central England, in open woodland, grassland, heathland and moorland.

The rarer *Coronella austriaca* (smooth snake – L:550 mm) also keys out here. These are greyish or brownish, with an arrangement of indistinct pairs of dots (which may be joined up) along the back and a darker heart-shaped mark on the head (X.3). They are found in heathland in the south and south-west of England.

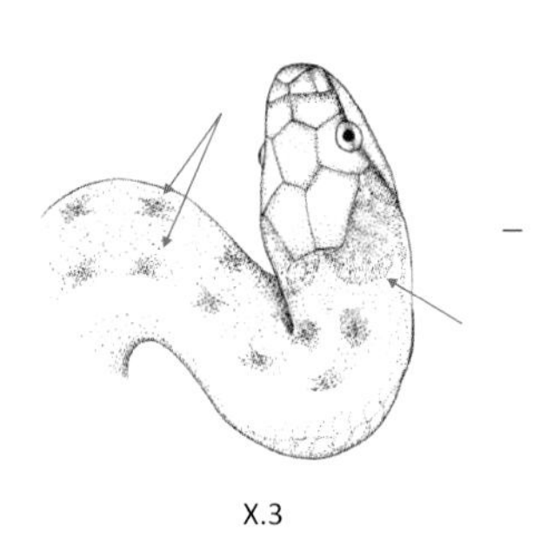

X.3

– Body greenish with darker markings restricted to the neck and flanks *Natrix helvetica*

- *N. helvetica* – (barred) grass snake – L:1,000 mm; body olive-green, brownish green, or greyish green, dark vertical marks along the flanks and a distinct yellow/orange collar bordered at the back with black; more common in England and Wales, rarer towards southern Scotland, in open woodland, hedgerows, heathland, and human-disturbed sites such as road or railway verges and gardens.

Several other species of snakes may also be found occasionally. *Pantherophis guttatus* (corn snake – L:1,500 mm) is a popular pet snake which, as an escape, may be found in gardens and near buildings. It typically has a light reddish-brown body with darkly outlined reddish patches along the back and sides. *Zamenis longissimus* (Aesculapian snake – L:1,750 mm) has established a colony near to the Welsh Mountain Zoo in north Wales. It has a uniform grey, brown or olive body with a yellow underside.

## Key Y Small mammals

You will be very lucky to find a small mammal under a log or stone that stays long enough for you to get a clear view of it for identification purposes. However, even with a quick glimpse, it may be possible to see enough to get an idea of the species seen. There are relatively few species that might be encountered and these can be separated on the basis of the shape of the head and ears, the shape of the body and the length of the tail. Measurements of lengths (L) for adult animals are of the head and body (not including the tail). Where an animal does not fully match the description, consult Harris & Yalden (2008), Strachan (2010), Couzens *et al.* (2021), or the Mammal Society website (see Chapter 6), which also includes images and distribution maps. See Section 3.6 for further information on small mammals.

1 Back and sides of body covered with spines *Erinaceus europaeus*

- *E. europaeus* (hedgehog) – L:150–300 mm (tail – 10–20 mm); body speckled brown and cream, with many spines on the back and sides, head furry and pointed; common in woods, grassland and farmland.

– Body covered in hair (no spines on back or sides of body) 2

2 (1) Animal small (under 100 mm) and relatively compact 3

– Animal long (over 150 mm) and slender Carnivora (Mustelidae) 5

Y.1

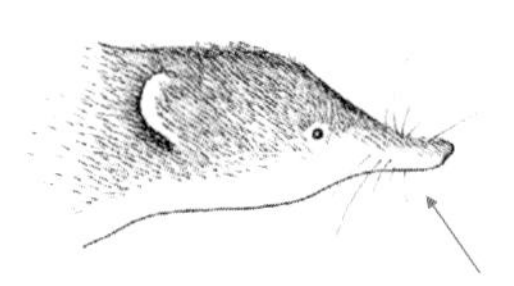

3 (2) Muzzle long and sharply pointed (Y.1) Eulipotyphla (Soricidae – shrews)

There are three species of shrews in the UK, which can be separated on the basis of their size and colour:

- *Neomys fodiens* – water shrew – L:67–96 mm (tail – 45–77 mm); body black above, and greyish white or yellowish underneath; throughout Britain but more localised in Scotland, usually in habitats close to water including the banks of rivers and ponds, but occasionally away from water in woodlands, scrub and grasslands.
- *Sorex araneus* – common shrew – L:48–80 mm (tail – 24–44 mm); body dark brown with paler sides and a whitish belly; widespread throughout Britain, in woodland, hedgerows, scrub and grassland.
- *S. minutus* – pygmy shrew – L:40–60 mm (tail – 32–46 mm – proportionally longer than in *S. araneus*); body greyish brown; widespread throughout Britain

and Ireland, in woodland, grassland and urban areas including gardens.

– Muzzle blunt — Rodentia (rodents) 4

Y.2

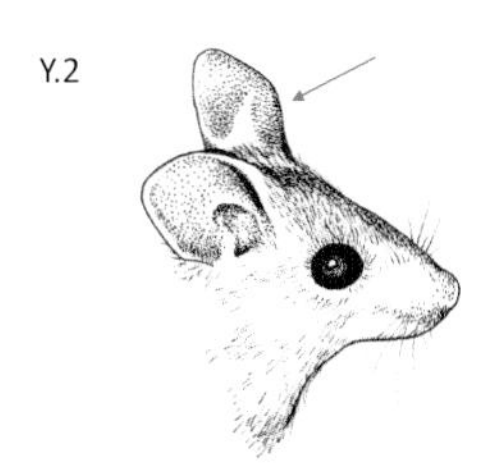

4 (3) Ears large and prominent (highly visible above fur – Y.2); tail long (about the same length as the head and body combined) — Muridae (mice)

There are two species of mice that may be seen under refuges:

- *Apodemus sylvaticus* – wood mouse – L:81–103 mm (tail – 71–95 mm); body sandy or reddish brown above, and greyish white underneath, eyes protrude, as do ears, tail long; widespread and common throughout Britain, in woodland, grassland, heathland, moorland and urban sites including gardens.
- *A. flavicollis* – yellow-necked mouse – L:95–120 mm (tail – 77–118 mm); body brown above, and white underneath, with a band of yellowish fur across the neck; restricted to southern England and Wales, in woodland and urban sites including gardens.

The slightly smaller house mouse (*Mus musculus* – L:70–90 mm, tail – 50–100 mm) is more usually found in buildings and can be distinguished by being greyer and having smaller hind feet (resulting in a less bouncing gait) than *Apodemus* species.

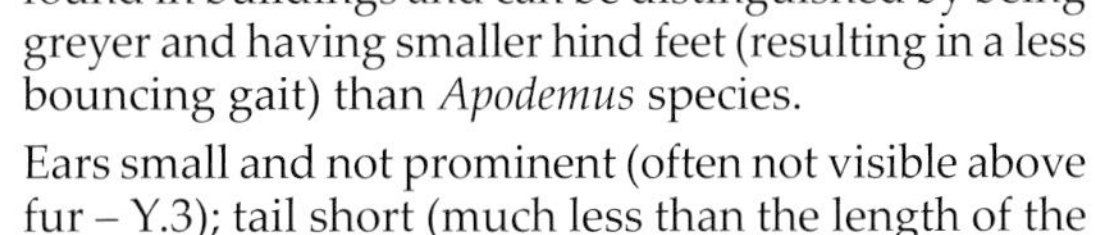

Y.3

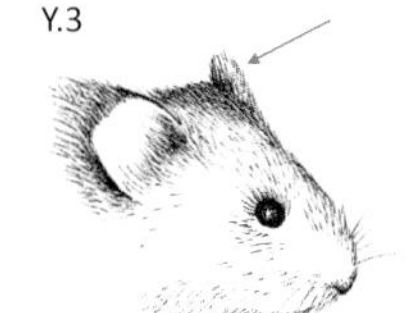

– Ears small and not prominent (often not visible above fur – Y.3); tail short (much less than the length of the head and body combined) — Cricetidae (voles)

There are two species of voles that may be seen under refuges:

- *Microtus agrestis* – (short-tailed) field vole – L:90–115 mm (tail – 10–50 mm); body grey-brown above, and creamy grey underneath, eyes less prominent than in *Apodemus* species, ears also less prominent and obviously furry, tail short (about 30% of the length of the head and body combined); widespread and common throughout Britain, in woodland, grassland, heathland, moorland and urban sites including gardens.
- *Myodes glareolus* – bank vole – L:90–110 mm (tail – 30–70 mm); body reddish brown above, and creamy grey underneath, small eyes and small ears hidden under fur, tail short (about 50% of the length of the head and body combined); widespread across Britain, less frequent in northern Scotland, in woodland, grassland, heathland and urban sites including gardens.

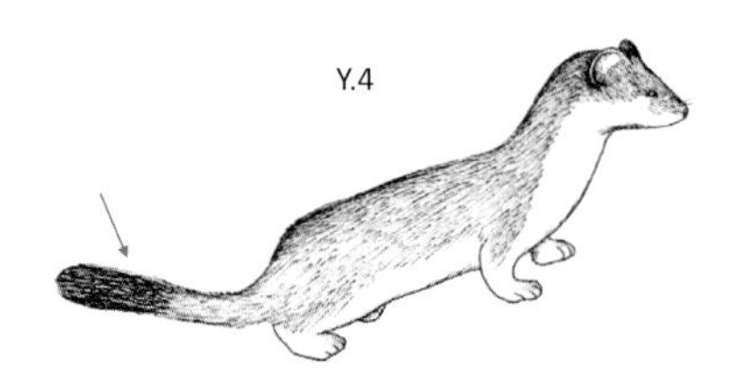


5 (2) With black tip to the tail (Y.4) *Mustela erminea*

- *M. erminea* – stoat – L:242–312 mm (tail – 95–140 mm); body ginger to reddish brown above, and cream or white underneath (whole animal may be all white or more substantially white in winter), tail with black tip; widespread and common throughout Britain, in woodland, grassland, heathland, wetland and urban sites including gardens.

– Without black tip to the tail *Mustela nivalis*

- *M. nivalis* – (least) weasel – L:173–217 mm (tail – 34–52 mm); body ginger to reddish brown above and cream underneath; widespread and common throughout Britain, in woodland, pasture, moors, marshes and urban sites.

# 5 Studying animals under logs and stones

Investigating the animals that can be found under logs and stones can both be highly satisfying and add to the knowledge that we have of the species and communities involved. Surprisingly little is known about many of the species and communities of animals that can be found beneath refuges, and even a small research project can be novel and informative. This chapter covers a range of types of projects and then details some of the techniques that can be used to implement them. The chapter also covers those factors that should be considered when planning research projects – including aspects of health and safety, ethical and legal issues, and time management. The collection, analysis and interpretation of information is also discussed, as is writing up and presenting the findings. Wheater *et al.* (2020) cover these elements in much more detail, albeit for wider ecological field research, while Wheater & Cook (2015) concentrate on studying invertebrates. Many studies of the ecology of individual species are quite old and may be restricted to only very common organisms or groups and hence there is considerable scope in extending these either by looking at additional elements, or by comparing the findings of research on one species with those from similar or contrasting species. There is a wide range of possible research avenues, which can be divided into those projects examining: the distribution of organisms from a geographical perspective; behavioural aspects of animals; interactions between organisms; interactions between animals and their environments; and the composition of the community found under logs and stones.

## 5.1 Types of research projects

There are two major types of research project: observational surveys and experiments. The former involve recording data from unmanipulated situations, such as how many animals of each species are encountered under logs from different species of trees. While these surveys mean that the results reflect the real-life situation, the results may be complicated by the variation in the environment (e.g. the age and size of the logs may vary confounding any analysis of the impact of tree species). Experiments involve manipulation of the situation, for example by placing similar-sized pieces of wood from different tree species that are all at the same stage of decomposition. Note that the important thing here is to vary the experimental refuges in only one way (e.g. tree species), if at all possible. By doing this, experiments can be designed to reduce the factor(s) being manipulated to only those that are of direct interest to the research project. However, because of this tightly controlled manipulation, they do not represent a real-life situation. So, while surveys can give an indication of the aspects of the environment that may be important in determining the presence of animal species in natural situations, experiments may enable confirmation of a direct effect (in the absence of other confounding factors) in a controlled situation.

### Geographical projects

Simple assessments of what is present and absent may help our understanding of the distribution, rarity and spread of particular groups of animals. Locally obtained information can feed into recording schemes for specific groups of animals and, where these have been under-recorded in the past, may help to plug certain geographical gaps (e.g. Figs 5.1, 5.2). For

**Fig. 5.1** Distribution map showing national distribution – gaps may indicate a lack of sampling coverage

**Fig. 5.2** Distribution map showing uneven (western) distribution

**Fig. 5.3** *Arthurdendyus triangulatus* – New Zealand flatworm (photo: HDJ)

some groups, there is even a reasonable chance of finding new species. For example, several species of flatworms, woodlice and millipedes new to the UK have been discovered in recent years. Many, but not all, of these are associated with imported plants and have been a particular feature of hot-houses and botanic gardens. Some of the organisations associated with the recording schemes that are listed in Chapter 6 will be very grateful for reliable records of species (including good quality photographs of specimens) together with their localities and details of the habitat in which they were found. Surveys to examine the spread of species over a wide area may provide information about the speed of establishment of introduced species. For instance, the ground beetle *Leistus rufomarginatus* has been spreading from its south-east England origins in the 1940s and 1950s, and is now found in the north of England (Northumbria to the east and Cumbria on the west) and lowland Scotland. Such studies may be of particular interest where an introduced species is potentially damaging to native organisms, as in the case of the New Zealand flatworm (*Arthurdendyus triangulatus* – Fig. 5.3), which predates earthworms. There are also some unanswered distributional questions regarding polymorphic species (those occurring in different forms, e.g. of different colours) and the factors that determine which morphs are dominant under which environmental conditions. For example, the occurrence of black-legged and red-legged forms of the ground beetle *Pterostichus madidus* (Figs 5.4, 5.5) appear to have some link to both climatic conditions and daily activity rhythms. The black-legged form appears to be more diurnal and is found less often in woodlands and more often in sites with lower minimum temperatures and drier conditions than the red-legged morph (Terrell-Nield, 1990).

**Fig. 5.4** *Pterostichus madidus* (common blackclock) – black-legged form

**Fig. 5.5** *Pterostichus madidus* (common blackclock) – red-legged form

Sometimes it will be the distributions of similar species that are of interest. The common centipede *Lithobius variegatus*, for instance, seems to have a very southerly and westerly distribution, whereas *L. forficatus* is abundant in the west but is also found in the east. It would be interesting to research how the distributions of these species fit together, i.e. whether in areas in which they overlap they are found together (or not) in particular habitat types.

Studies on the biogeography of refuge systems may be used to test the Theory of Island Biogeography (see Chapter 2). On real islands, animal communities are hard to monitor, so examination of communities under logs and stones may provide a model to explain some of the dynamics in island animal communities. These predictions might be tested by examining aspects of community structure, such as colonisation rates of newly deposited structures, the number of predators, herbivores and decomposers and the overall number of species present under objects of different sizes and at differing distances from one another. This could involve calculation of species–area curves (number of species associated with refuges of differing areas). Such projects could also include assessments of how (if at all) habitat diversity and/or habitat structure influence any biogeographical relationships found.

**Behavioural projects**

There is much scope for research into feeding behaviour, diurnal or seasonal activity levels, or dispersal rates and mechanisms of animals under logs and stones. Such projects can involve simple observations of animal behaviour. For example, the feeding behaviour of animals taken (temporarily) from the field can be observed by introducing prey species to predators and recording how they respond to prey, both from a distance and close at hand, and then how they catch and manipulate the prey, and subsequently consume it. Activity such as movements of animals between refuges may be monitored. Animals that are active at the surface can be caught in pitfall traps (see later Section 5.3). Movements into and out of the habitat may be assessed by enclosing some individual logs and stones with barriers and leaving others undisturbed (see the later section on barrier traps). Seasonal activity levels may be inferred from collecting or trapping at different times of the year, while diurnal activity can be examined using a simple actograph (e.g. Wheater, 1987). Human disturbance (including noise in urban environments) may affect the behaviour of invertebrates (e.g. Morley *et al.*,

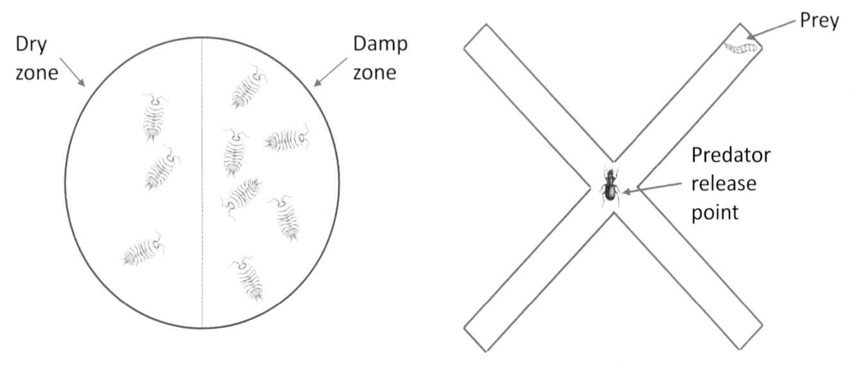


**Fig. 5.6** Circular choice chamber

**Fig. 5.7** X-shaped choice chamber

2014). Activity under different levels of noise or other disturbances (e.g. light levels) can be measured using actographs. In addition, choice chambers or X-shaped chambers can be used to assess animals' preferences for different conditions or prey. These involve placing animals in a circular arena or an X-shaped chamber which has been prepared with different conditions such as temperature or humidity (or prey animals) at different positions within the chamber (Figs 5.6, 5.7). Where several animals are placed within a chamber, the number in each zone can be recorded after a set amount of time (e.g. 5 minutes) or, when a single animal is being observed, the time that the animal spends in each area of the chamber over a set time period (e.g. 10 minutes) can be recorded to ascertain their preferences. Morris (1999) describes experiments to study orientation behaviour in woodlice, while Dixie *et al.* (2015) give details of experiments to observe behavioural and physiological changes in woodlice under differing microclimatic conditions.

### Projects examining interactions between organisms

There are many different ways in which organisms interact with each other, including mating, predation and aggregation to prevent desiccation. Projects investigating such interactions may include observations of mating behaviour (e.g. in captive animals). Choice chambers can be used to examine predatory behaviour, or to identify the environmental conditions that stimulate certain species (e.g. woodlice) to aggregate (see the previous section on Behavioural projects). Other research could include assessments of which species of plant possible pest species such as slugs and snails may avoid, and how predators such as spiders, ground beetles and earwigs may impact on aphid pests in gardens. The ways in which species with similar ecological functions interact is also an interesting area of investigation. For example, studies have shown that woodlice and earthworms may act synergistically in promoting the decomposition of some types of leaf litter, but do not do so in others (Zimmer *et al.*, 2005).

### Projects examining interactions between organisms and their environments

Impacts of environmental factors on the communities found under logs and stones may include how community structure changes with various attributes of the refuge (microclimate, size, depth buried into the soil, roughness of the underside and, in the case of logs, the species of wood and state of decay). The characteristics of the surrounding microhabitats can also affect the cryptozoa, where different species and communities may be found with differing types and depths of leaf litter, amounts of moss and ground vegetation cover, and the distance from other, similar refuges.

The effect of the degree to which a refuge is buried can be examined by collecting and identifying animals from a series of objects that differ in the depth of soil with which their lower surface is in contact. The complexity of the surfaces underneath a refuge can provide more niches and influence the ease of movement for smaller organisms. Surface features could be recorded and related to the animals using them. The dimensions of features such as cracks, depressions or holes might be listed on an appropriate ranked scale. For instance, cracks might be scored on the basis of width (1 = <0.2 mm, 2 = 0.2–0.5 mm, 3 = >0.5 mm) and length (1 = <1 mm, 2 = 1–5 mm, 3 = >5 mm). Perhaps the abundance, diversity and size range of the animals can be related to the architectural complexity of the habitat in this way. More simply, the complexity of the undersurface of the object could be assessed by measuring a number of axes with a ruler (straight line distance) and again with a map wheel (surface distance), giving a measure of the dimensions including any undulations. The ratio between these would give an estimate of surface complexity which could be related to community structure (Fig. 5.8): where the surface is perfectly flat, the two measurements will be the same and the ratio will be 1; indices greater than 1 indicate greater surface complexity. Experimental situations could be created by laying down refuges of the same overall size with differing degrees of roughness on the undersurface – perhaps using pieces of flat wood that have been scored deeply to differing extents. When using cut timber in this

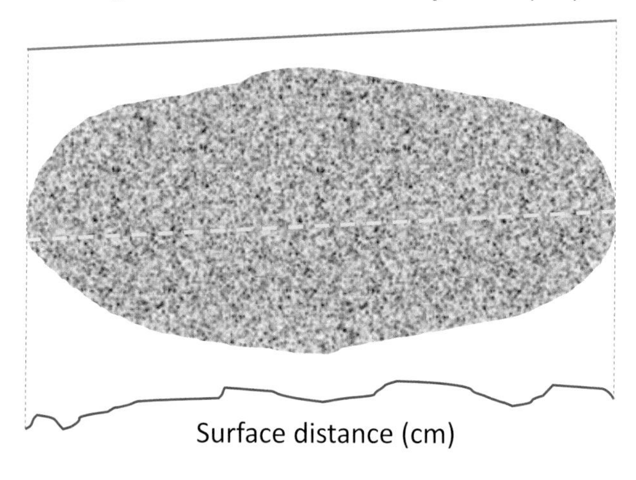


$$\text{Surface complexity} = \frac{\text{surface distance}}{\text{straight line distance}}$$

**Fig. 5.8** Measuring stone complexity

way, take care to avoid wood that has been treated in case this impacts on the invertebrates being studied.

The influence of the type of substrate on the invertebrates that make up the community structure may be revealed by comparisons between similar logs or stones resting on or in different substrates – for example, examining whether the presence of mor humus (see Section 2.2) is associated with a lack of animals other than earthworms. Differences in community structure may also be associated with the dominant vegetation types in the area. Comparing logs and stones in neighbouring deciduous or coniferous woodlands, or in woodland and grassland, may give some information about the importance of surrounding vegetation for the communities under examination. Butterfield & Malvido (1992) found that the presence of more than one species of tree in a woodland was associated with an increase in surface-active invertebrates. It may be that this is also true of animals under logs and stones. Grasslands may vary according to whether they are grazed or ungrazed. Different grazing animals (e.g. cows and horses) may produce differing conditions. Other habitats, such as gardens (Owen, 1990, 2010) and even refuse tips (Darlington, 1969), are also worthy of study.

The most diverse and rich communities of snails are found on calcareous soils, since snails need calcium to form their shells. Investigation of communities associated with different rock types may suggest other such relationships. Artificial stones of differing materials such as paving stones could be used as experimental refuges to identify the communities found under stones of differing chemical constituency. Soil surveys may also elicit differences between the cryptozoa of soils with different amounts of various nutrients, pH, and organic and water contents. Note that garden soil test kits (available from garden centres) may help with the measurement of some of these variables, although such kits tend to be rather low resolution and are only really suitable when relative values are required.

Much remains to be discovered about the influence of microclimate on the community of animals under logs and stones. The temperature can be measured in various ways. Simple but often relatively inaccurate probes may be obtained from garden centres. Soil thermometers (either alcohol-filled glass thermometers encased in metal tubes, or dial thermometers mounted on metal spikes) are reasonably accurate but may be difficult to position and read in situ (Figs 5.9–5.11). Various digital thermometers are available that have small probes that are remote from the meter and can be inserted precisely into a very small cavity.

Humidity is more difficult to measure. Accurate meters are expensive and the probes may require replacement if they are damaged by being forced under hard objects. Soil moisture content can be measured with simple spike meters from garden centres, but these usually provide only a crude

**Fig. 5.9** Alcohol-filled soil thermometer

**Fig. 5.10** Dial soil thermometer

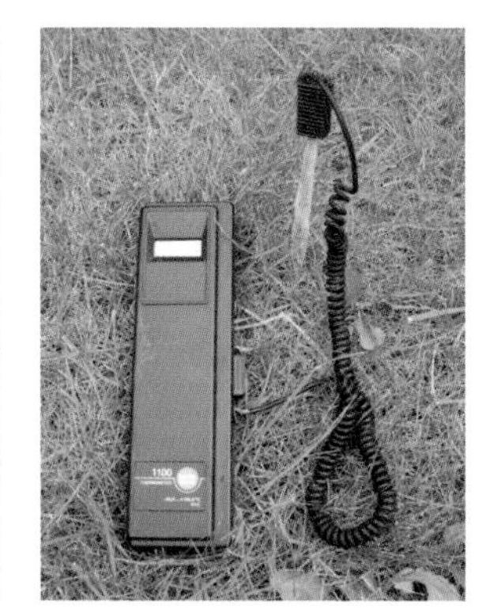

**Fig. 5.11** Digital thermometer with probe

readout (for example, on a scale of 1 (dry) to 5 or 10 (wet)) and therefore should be used only for comparative studies unless they have been calibrated against soils of known moisture content. More accurate soil moisture measurements can be done by taking soil samples of about 50 g. The soil is weighed, and then heated at 105 °C until it dries to constant mass, as indicated by three constant weighings following further heating. The percentage moisture content may be calculated as:

$$\frac{[(\text{Initial mass}) - (\text{Final mass})] \times 100}{\text{Initial mass}}$$

Unwin & Corbet (1991) discuss more accurate methods of measuring microclimate, especially in small spaces. Breckland Scientific, NHBS, RS Components and Watkins and Doncaster supply a range of temperature and humidity probes, as well as equipment to monitor soil nutrients and pH – see Chapter 6 for further details.

**Community structure projects**

Recording the community of animals under logs and stones involves identifying and establishing the abundance of each species (or group of species) found. The development of new communities can be recorded by regular monitoring of logs or stones introduced into a habitat. This can show not only the order in which species move in, but also whether certain groups of animals (e.g., predators, herbivores, decomposers, or generalist or specialist feeders) colonise first. To interpret these findings, it is useful to explore the potential source of colonisers, i.e. whether they come from the soil fauna, or from other logs and stones nearby, and how far they have travelled. As refuges become more established as part of a habitat, there may be successional changes in the animal community. For animals under stones, development of the communities may reflect interactions between species and changes in the soil underlying the refuge. In the case of logs, changes in community structure may be influenced by different stages of the decomposition process. Such impacts can be examined either by monitoring several sites over time or by simultaneously comparing logs at different stages of decay. It is important to use pieces of wood of the same tree species and of similar sizes. Decomposition proceeds differently in pieces of wood of different sizes and species, and further studies might examine the communities present under logs alike in age but differing in size. The chemical composition of wood may also affect the decomposition rate and, therefore, the community structure. This possibility could be explored by examining animal communities under wood from coniferous compared to deciduous tree species, or from different species of deciduous trees in the same habitat.

## 5.2 Planning research

In planning any research project it is essential to be clear about the aims and objectives of the proposed study. The research aims should express what it is intended that the study will achieve, while the objectives indicate the steps necessary to achieve the aims. You must understand the limitations of any techniques and practise them before gathering data, to ensure that the implementation at the beginning of the study does not suffer from a lack of familiarity with the process compared to the end of the study. Time management and a recognition of any logistical issues will be important to ensure that what is planned is possible (including the availability of any equipment). Where multiple pieces of equipment are being used, they should be as similar as possible to each other and preferably calibrated against each other to ensure that there are no built-in biases. A knowledge of the limits of any equipment used is also important: you should only measure variables within the known accuracy and precision of any monitoring device. Any potential health and safety, ethical and legal issues must be explored and a mitigation plan developed and implemented. It is always useful to carry out a pilot study in advance of full implementation of any research programme to ensure that most (if not all) issues have been understood and addressed. Fig. 5.12 summarises the process of planning and implementing a research project.

### Health and safety

Health and safety is an important consideration in any research project, not least because legally each individual is responsible for the safety of themselves and those around them. Even where a study appears quite straightforward, a full risk assessment should be undertaken, for example covering potential risks caused by the terrain, techniques being used and changes in the weather (e.g. Barrow, 2004). Wheater *et al.* (2020) provide an overview of addressing health and safety issues and of risk assessment; see also USHA/UCEA (2011) and Daniels & Laverlee (2014). Be particularly aware of the risks of working in more hazardous environments and avoid working alone whenever possible. Note that although there are few venomous species to be found under logs and stones in the UK (there are some – adders and

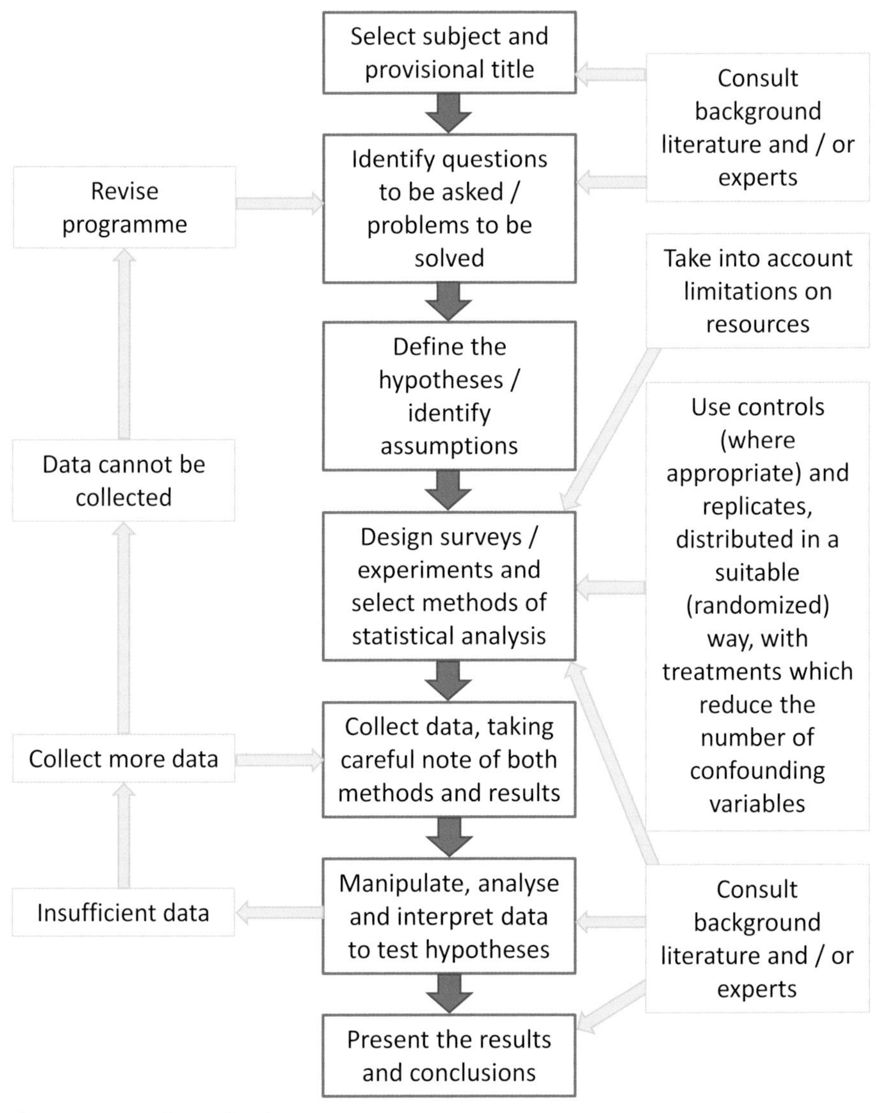


**Fig. 5.12** Project design flowchart

scorpions come to mind!), this may be a significant risk in other parts of the world. The use of any chemicals hazardous to health also needs to be considered carefully, including not only how they are handled, but also transportation, potential impacts on the environment and safe disposal after use. The use of power tools to cut logs (e.g. to create grooved surfaces, or to produce pieces of wood of a particular size) will also pose hazards and may require training and/or possession of a licence (e.g. when chainsaw use is necessary). All potential risks should be both understood and mitigated.

**Legal and ethical issues**

There are few legal issues when working with terrestrial invertebrates in the UK. This is not the case with some vertebrates, and catching and handling some species may be covered by legislation. Consult Beebee (2013) for information on amphibians and reptiles, and Gurnell & Flowerdew (2019) regarding catching and handling small mammals. When working in the field, and particularly if accessing private land, you should always ask for permission from the landowners or site managers, as well as any other interested parties (e.g. relevant statutory bodies if the sites have some sort of protection as a Site of Special Scientific Interest or as a National Nature Reserve). Note that even public land has an owner (such as a local authority) who should be contacted for permission. It is worth noting that much useful research could take place in domestic gardens. Care should always be taken to behave professionally and with due regard to the Countryside Code (see Chapter 6). Take into account any ethical considerations associated with your research project. These include keeping disturbance to a minimum, not removing material from a site unless absolutely necessary, avoiding killing animals if at all possible, returning any samples to the same site they were taken from following identification and/or behavioural experiments, and replacing any refuges that you turn over as close to the original position as possible. See Reed & Jennings (2007) and Parris *et al.* (2010) for more discussion on ethics in ecological fieldwork. Note that the Joint Committee for the Conservation of British Insects has published a code of conduct for collecting insects and other invertebrates (see Chapter 6).

## 5.3 Techniques

Many of the techniques required to collect and handle animals under logs and stone are fairly simple and straightforward. However, care does need to be taken to ensure that any technique used is standardised across a research project to ensure that the results are not biased in some way by the type of technique employed or its implementation. Wheater & Cook (2015) and Wheater *et al.* (2020) give further details of a wide range of techniques that are applicable to studying animals under logs and stones.

## Collecting animals

Notwithstanding the ethical consideration to avoid removing animals from a site if at all possible, there are several situations where it is necessary to collect organisms. Catching animals for identification can result in on-site release once they have been identified (the preferred scenario), although many species will require more detailed examination, requiring removal (possibly temporarily) from a site. In addition, behavioural experiments that are not implemented in the field will also require animals to be collected. Similarly, any study involving rearing/culturing animals will also necessitate the collection of specimens. For instance, different colour morphs of the same species could be reared to see whether different environmental conditions influence the development of different colour patterns (see Curl, 2016 for an example using woodlice). Whatever the justification – and this needs to be thought about carefully – you should take as few animals as possible, and where feasible return them to the site from which they were taken. The next section describes some of the common capture methods and aspects that should be considered when using such techniques.

### *Collecting by hand*

Collecting animals living under logs and stones by hand requires a quick eye and hand, rather than sophisticated equipment. Some species are small and may be well camouflaged against the background soil, wood or stone. Although many species are slow moving (like slugs and snails) or freeze when first disturbed (like some ground beetles), others move very quickly once they get going (including springtails and some centipedes). In addition, some, such as lithobiomorph centipedes, disappear rapidly into cracks and holes as soon as they are disturbed. It can be particularly tricky to gather all the animals disturbed by lifting a log or stone where there are large groups of animals exposed. Note that when you turn a log or stone over, some animals will be on the substrate that is revealed, but others may be on the underside of the refuge itself. Where it is important to collect all the animals from under a log or stone, it is worth considering using a large, deep-sided tray into which the refuge can be placed so that animals do not readily escape. One of the most important aspects of collecting from this habitat is familiarity; you need to be able to react quickly once the log or stone is lifted. This will come with experience. Before beginning to collect data for any research project, it is a good idea to practise capturing animals from similar situations to ensure that all samples collected reflect the actual situation rather than your (or your co-worker's) ability.

In some investigations, all the collecting can be done by active sampling in the field, for example when looking at the communities of animals found under stones of different sizes. There are various ways of picking up the animals. Large creatures such as beetles and centipedes can be picked up by hand, but some move quickly and may possibly bite (or sting). For these, a soft pair of stork-billed forceps is useful to hold the specimen without damaging it (Fig. 5.13). Small animals, such as springtails, mites and small spiders, are all too easily squashed and are best collected with a pooter (Figs 5.14, 5.15). Although pooters are available commercially, they are easily made from a small plastic container sealed with a lid or a bung with two rubber tubes leading from it. The collector sucks on one tube while holding the end of the other one close to the animal to be collected and the animal is then sucked into the collecting vessel. It is essential to cover the inner end of the sucking tube with gauze to avoid getting a mouthful of animals! A much simpler pooter can be made from two lengths of polythene tubing with one fitting inside

**Fig. 5.13** Millipede held in stork-billed forceps

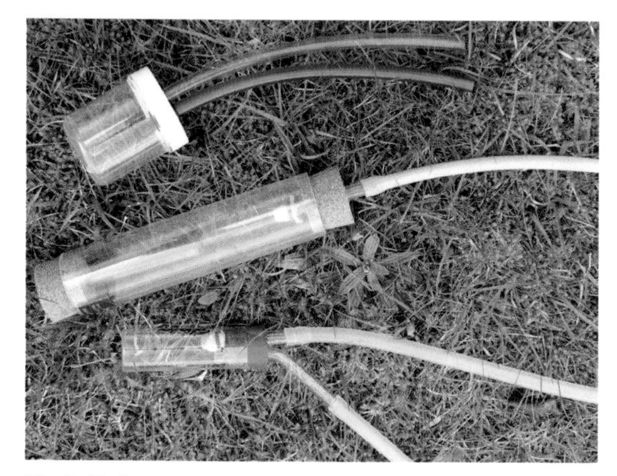

Fig. 5.14 Pooters

Fig. 5.15 Top of pooter

the other and covered in gauze. Trapped animals may then be transferred to a collecting tube. The best designs work if the tube being sucked is wider than the one held over the animal. This creates more suction and has a better chance of dislodging more animals.

For collecting animals in the field it is best to use sturdy plastic tubes or plastic bottles, which will not harm you if you fall over with them in your pocket. Remember to keep predators in separate containers if they are being transported alive, and to label everything clearly. It is important to note when and where animals were caught, especially if the investigation involves looking at animals from individual logs or stones. It is preferable to make notes in a book or on recording sheets as well as in, or on, the tubes containing the animals. Some animals may spoil labels; slugs may eat them or render them illegible, and millipedes, when preserved in alcohol, may stain the liquid dark red-brown, making the labels difficult to read. Furthermore, labels written on the outside of the tubes with marker pens may rub off if the preserving solution, such as alcohol, is spilt from another container. Using a soft pencil to write on paper labels placed in a plastic bag within which the specimen tube is placed, can be useful in avoiding the loss of information. Sealable polythene bags are often useful collecting containers, particularly for collecting the whole assemblage under one stone, as long as the bag is large enough to hold a handful of soil or leaf litter to keep the animals fresh and separated from each other.

### *Collecting by pitfall sampling*

Surface-active communities may be collected using pitfall traps; plastic cups or jam jars, buried in the ground so that the lip is level with the surface, will collect animals moving along the surface (Figs 5.16, 5.17). Highly active species are more likely to encounter the trap; therefore, such techniques (similar to hand collecting from under logs and stones) do not catch individuals of different species in strict proportion to their abundance. Pitfall traps only catch active individuals and so the catch can really be deemed to refer to the active trapability density of animals. Any comparisons must therefore be made with caution. However, they are a very efficient method of catching ground-active invertebrates, and should enable you to identify the dominant types of animals in each habitat and draw conclusions about their habitat requirements.

Although trapping is not practical under the refuge, it can be a useful technique next to a log or stone and, when used in association with barrier traps, can be employed to examine what is moving into and out of the refuge along the ground. Barrier traps can be used to surround a log or stone in order to ascertain which animals may be migrating from the refuge. The barriers can be made of thin (about 3 mm thick) wood, plastic or metal and should be smooth, about 150 mm high, and long enough so that four pieces will surround a log or stone. If pitfall traps are placed at the corners of these, animals that make contact with the barriers will tend to be funnelled towards the pitfall traps (Fig. 5.18). A double barrier with traps at the corners of the inner and outer sets will allow the separate capture of animals moving away from and towards the log or stone.

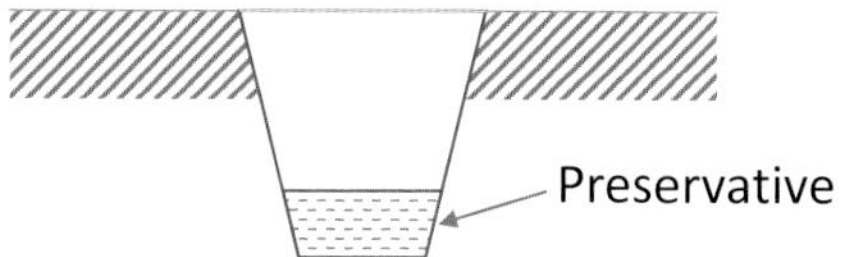


**Fig. 5.16** Pitfall trap

**Fig. 5.17** Pitfall trap in situ

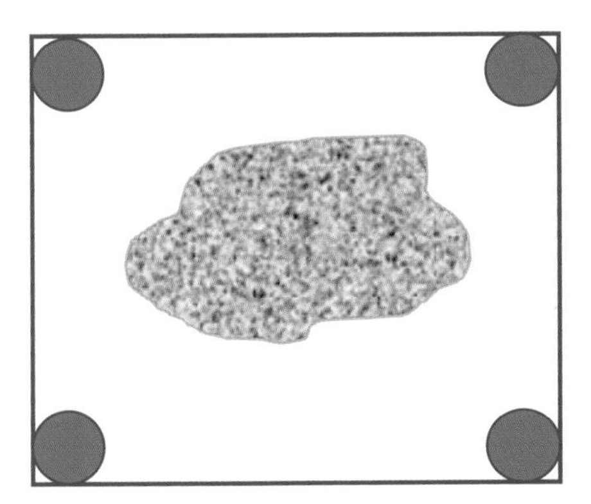

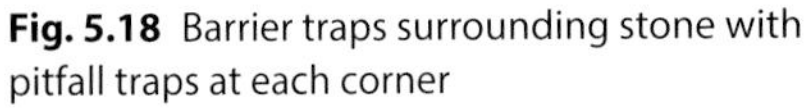

**Fig. 5.18** Barrier traps surrounding stone with pitfall traps at each corner

**Fig. 5.19** Pitfall trap with rain cover

Pitfall traps come in a variety of shapes and sizes. The easiest to deal with are plastic cups (not polystyrene) for vending machines. Bulb planters, from gardening shops, make a hole just the right size for a plastic cup and greatly ease the back ache of digging the holes. If two cups are used in each hole, one inside the other, the top one can be removed to be emptied and refilled without the hole collapsing. Glass jam jars, biscuit tins or plastic guttering make equally good traps but are not so easy to dig into the ground. The larger the diameter of the trap the more animals it will catch, and the deeper the trap the better the retention. The lip of the trap should be flush with the ground and a rain shield (a tile or square of hardboard raised above the trap by balancing on a couple of stones, or mounting it on a 150 mm nail) can be used to prevent the traps from overflowing with rain and also reduce the likelihood that small mammals will fall in (Fig. 5.19). If the inadvertent capture of vertebrates is a problem at the study site, a cover of chicken wire may prevent larger (unwanted animals) from falling in but allow small invertebrates through. Note that larger invertebrates may also be precluded by this adaptation.

If the traps are set dry they need to be emptied frequently to prevent animals from crawling out, and there is a high risk that a top predator will eat everything else that falls in. Dry traps are useful when animals are required for behavioural experiments or for rearing and culturing. When surveys are short term, with traps collected daily, live trapping will mean that animals can be released unharmed back into the study site. Alternatively, a small amount of killing and preserving fluid can be put into each trap to stop in-trap predation and to prevent animals from climbing out. Various solutions can be used for this purpose. The simplest is water with a few drops of washing-up liquid. This detergent is an important ingredient as it reduces the surface tension so that small animals are wetted and sink to the bottom. If pitfall traps containing water are left for more than a few days, especially during summer, specimens deteriorate badly.

Ethanediol (most commonly available as antifreeze), or preferably propylene glycol, which is less toxic to those handling it, as a 25–80% solution in water will allow the traps to be left for up to a week. These are preservatives but not fixatives, so it is important that animals are transferred into alcohol or another fixative as soon as possible after the traps have been collected. Antifreeze has another disadvantage; it is coloured, usually blue or brown, and will stain some paler specimens.

If traps are to be left for a long period, alcohol (70–90% methanol) can be a useful solution. However, alcohol evaporates rapidly and may be of limited use in the summer. Some researchers add a little glycerol (up to 5%) to the alcohol to prevent samples from drying out completely, but the viscosity that prevents drying can be a problem when examining the animals during identification.

Both substances – ethanediol and alcohol – are potentially harmful and should be treated with care. They should not be ingested or inhaled and contact with the skin should be avoided. Rubber gloves should be worn when handling them, and you should understand how to dispose of unwanted and contaminated fluids safely and legally. Control of substances hazardous to health (COSHH) data sheets for such chemicals are available online from chemical supply companies and the Health and Safety Executive (HSE – see Chapter 6). Increasingly, specimens are collected for DNA analysis, often to help to confirm phylogenetic relationships between species. Where DNA analysis is planned or may be used in the future, it is important to use non-denatured ethanol, or another suitable preservative that does not degrade DNA (see Lear *et al.*, 2018 for further details).

**non-denatured ethanol** Non-denatured (technical grade) ethanol is of high purity, without additives, and is suitable as a preservative for specimens intended to be used in DNA analysis. Denatured ethanol, on the other hand, has chemicals (denaturing agents) added to it that make it undrinkable, and hence is not pure enough a preservative for specimens that will be used in DNA analysis.

The presence of solutions in the traps may attract or repel certain species of animal. This can be taken a step further by baiting the traps with cheese, meat or something sweet in a dry trap. This technique is interesting to try, but is not appropriate if the aim is simply to catch animals that are naturally occurring in the vicinity of the refuge. Brown & Matthews (2016), and Hohbein & Conway (2018) review the ways in which pitfall traps have been used in a wide range of studies on invertebrates.

### *Extracting animals from soil and leaf litter*

Studying the invertebrate communities found within the soil and leaf litter layers under and near to logs and stones will require the animals to be separated from the substrate. The simplest method is to pass a soil or litter sample through a succession of soil sieves of differing sizes – from coarse to fine mesh – examining each subsample for obvious signs of living animals (Fig. 5.20). This is time consuming and may be tricky to use with smaller, more delicate species; however, even a fairly coarse gardening sieve can be useful in separating out larger animals and pieces of material. Small animals can be more readily captured if the finer fraction is spread onto a white sheet or tray. Alternatively, animals can be separated from the substrate (soil and leaf litter) by flotation techniques. Here, the sample is placed in a solution with a specific gravity of 1.2 or more (use a hygrometer to check this). The most commonly used solution for flotation is aqueous magnesium sulphate, although other solutions including water with the addition of table salt or sugar have also been used. The sample is gently stirred into the solution and left for 2 minutes. Small animals will tend to float to the surface, while most mineral soil particles will sink.

Animals can also be extracted from leaf litter and soil using light or heat, or a combination of the two. The soil invertebrate community may be examined using extraction techniques such as Tullgren funnels where soil or leaf litter samples are placed near a warm light source (Figs 5.21–5.24).

**Fig. 5.20** Soil sieves of various mesh sizes

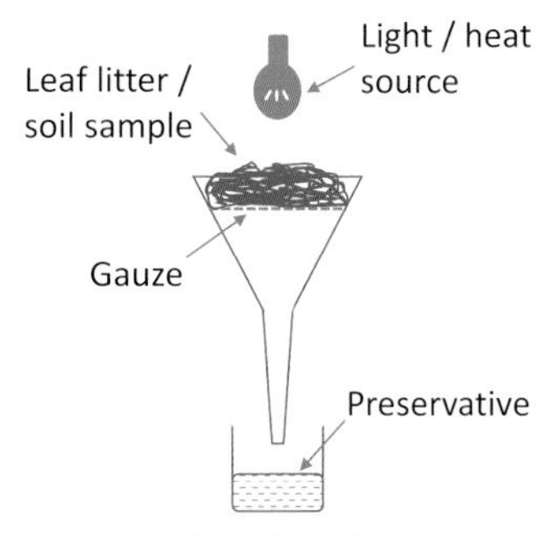


**Fig. 5.21** Tullgren funnel

**Fig. 5.22** Commercial Tullgren funnels

**Fig. 5.23** Home-made Tullgren funnels

**Fig. 5.24** Portable Tullgren funnel

Since most soil animals prefer damp, dark environments they move away from the light (i.e. are negatively phototactic), and the bulb may also dry out the substrate. If the soil or litter is placed so that the animals fall into preservative or water when escaping, they may be retained for more detailed examination. Kempson bowl extractors act in a similar way but create more of a temperature gradient between the heat of the lamp above the sample and the cooled preservative below it (Fig. 5.25). Funnel or bowl extractors need to be left for a couple of days in order to extract most of the animals. Winkler

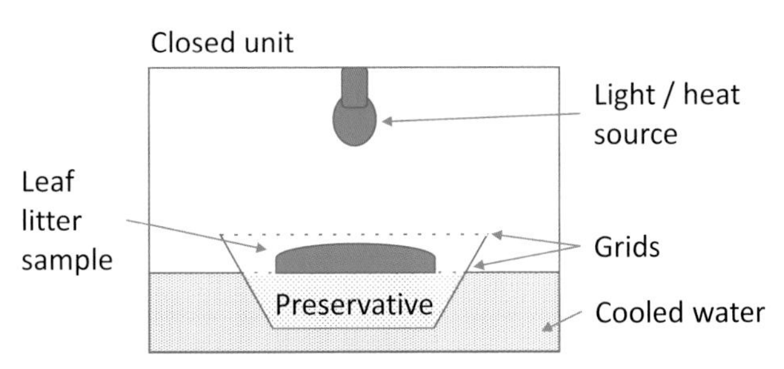


**Fig. 5.25** Kempson bowl extractor

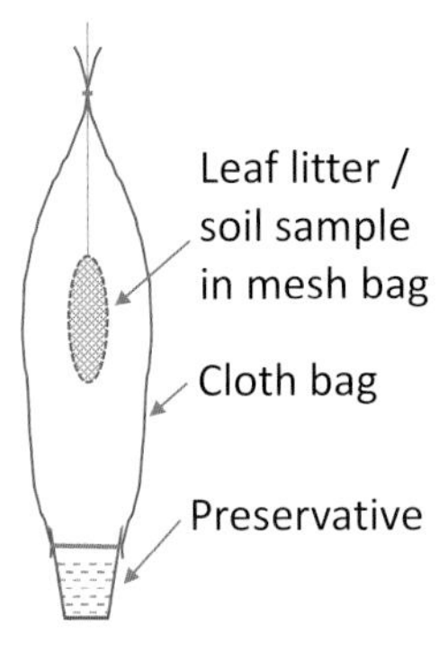


**Fig. 5.26** Winkler sampler

**Fig. 5.27** Winkler sampler in situ

samplers operate on the same principle, although here there is no light source (Figs 5.26, 5.27). The sample is placed within a mesh bag, which is suspended within a light-proof cloth bag, which itself is hung in a sheltered area (i.e. away from possible rainfall). The sample is left to slowly dry out, during which time (may be several days) the animals migrate away from the sample and fall into a collecting jar at the base of the cloth bag. The same principle can be used on a tabletop by shining a bright light on one side of a heap of soil. Most of the animals can be picked out if the soil is simply spread out on a plastic sheet, white tray or table. A bench lamp placed at one side will encourage animals to move to the cooler, darker side. This method can be extended by using a tray-light extractor (Fig. 5.28). This is simply a slightly inclined tray with a soil or leaf litter sample at the raised end and some preservative at the lower end. If a light is shone on the sample and the rest of the tray is covered with a light-proof material such as black polythene, animals will tend to move away from the light and down towards the preservative. Using a hot bulb will speed this process up since the animals will move away from heat as well as light. The trick with all of the methods using lights is to find a cool room in which to set it all up, and

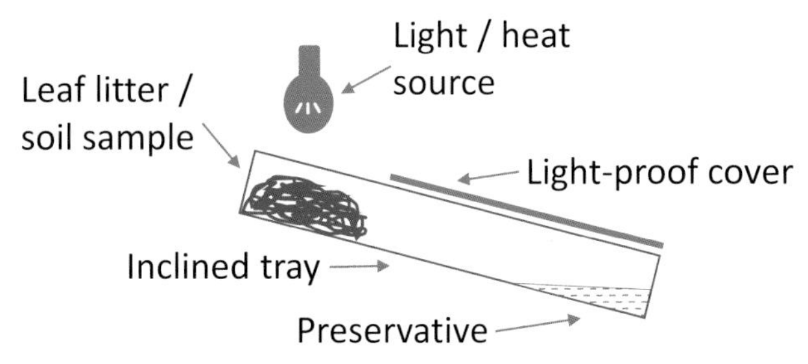


**Fig. 5.28** Tray-light extractor

to get the lamp at the right height so that the animals are not roasted before they can move away but are warm enough to feel the need to move downwards. This can be quite tricky in practice. It is important that once you have decided on a method, the same procedure is used throughout your study. For example, different bulbs may provide different drying capacities and so, for consistency, samples should all be treated with the same type of bulb.

### *Marking animals*

Some studies require animals to be marked individually in order to determine, for example, how frequently they move between logs and stones, or to estimate the size of the population. Snails are easily marked and numbered either using waterproof paint, by gluing numbers onto their shells, or by using a small craft drill to engrave the shells (taking care not to press too hard and go through the shell). Animals such as beetles can be marked using a coding system in which different colours and positions on the back of the animal give a large number of different combinations (Fig. 5.29).

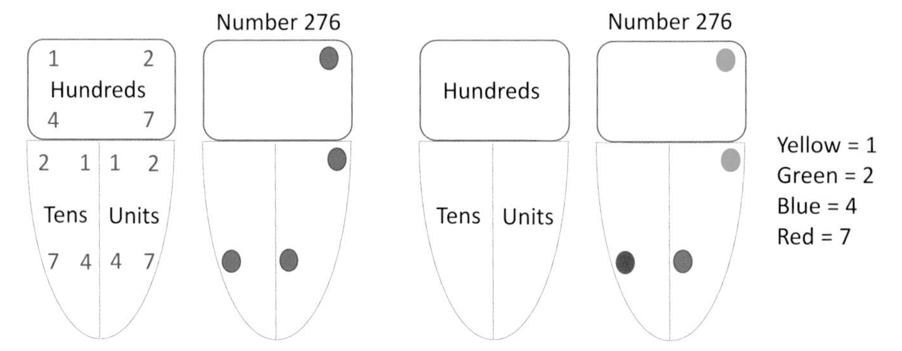


**Fig. 5.29** Marking individual animals: the left pair of animals indicate the use of position to mark large numbers with a single colour; the right pair show how multiple colours can be similarly utilised

A flexible, waterproof, fast-drying paint such as cellulose dope, acrylic paint or nail varnish is ideal. It is possible to get pens containing suitable paints which make marking animals easier and less messy.

There are several methods of estimating population sizes using marked animals. In most of the methods animals are caught and marked, and then released to mix with the population. On a subsequent occasion, perhaps the following day, animals are again caught and any recaptures are noted. Techniques for estimating the size of the population range from the very simple, such as the Lincoln Index (Petersen Method), to the more complex (e.g. see Blower *et al.*, 1981). All such methods rely on thorough mixing of marked animals with the unmarked population to no detriment of either. Therefore, marking should not impact on this, either by influencing the activity of marked animals or by altering their survival (e.g. by making them more or less attractive to predators). Wheater & Cook (2015) give a worked example and further details of other useful methods.

Using the Lincoln Index (Petersen Method) of population estimation

The population estimate = $\frac{M_1 \times M_2}{n_2}$

Where:
$M_1$ = the number of animals captured (and marked) on sampling occasion 1
$M_2$ = the number of animals captured on sampling occasion 2
$n_2$ = the number of animals marked on sampling occasion 1 and recaptured on sampling occasion 2

### *Killing and preserving invertebrate animals*

Many groups of cryptozoa are unfortunately much easier to identify using dead and preserved specimens. Once the commonest species have been learnt in this way it is often possible for some identification to be done in the field without killing the animals. Immersion in methylated spirits will usually both kill and preserve animals, especially soft-bodied ones. Specimens preserved in liquid should be stored in their vials submerged in alcohol in glass bottles or jars with tight-fitting lids (Fig. 5.30). Alcohol is really the best substance to use but it evaporates quickly and needs to be checked and topped up fairly frequently; some animals such as spiders are impossible to identify once they have dried up. Some people prefer to add 5% glycerol to the alcohol as a preserving fluid since this does not dry up as readily. However, it is very viscous and this may cause problems for identification purposes by sticking parts of a specimen together. If this

**Fig. 5.30** Animals stored in preserving fluid in labelled jars

happens, adding more alcohol may ease the situation. A range of alcohols (including methanol, ethanol and isopropanol) have been used as both killing and preserving fluids, although some alcohols (such as ethanol) are difficult to get without a licence. Each bottle or jar should be well labelled.

**Fig. 5.31** Specimen (*Eurynebria complanata*) mounted on card

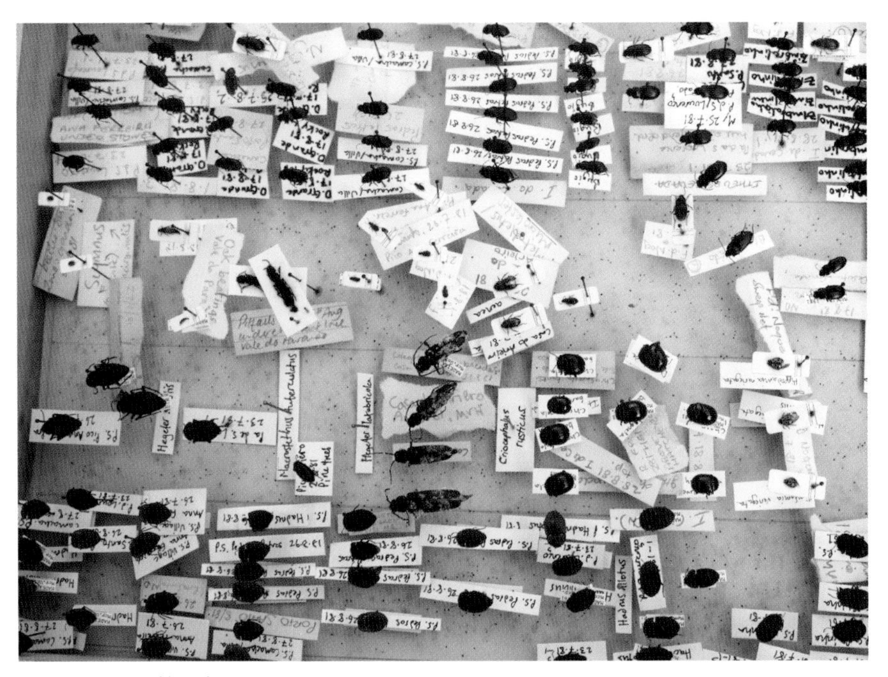

**Fig. 5.32** Pinned beetle specimens

Remember to use pencil or indelible ink for labels in alcohol since many types of ink will fade in alcohol.

Hard-bodied insects, such as beetles, may be killed in a closed jar containing blotting or filter paper which has been moistened with a few drops of ethyl acetate; this process may take some time to actually kill the specimens, especially if the killing jar is being used in the field where the chemical vapour can escape if opened frequently to add more specimens. Once the specimen has stopped moving, it should be kept in the jar for at least another 15 minutes to make sure it will not recover. They may then be mounted, either by gluing the feet and antennae to small pieces of thin card or acetate with water-soluble glue and piercing the mount with a pin, or, if the specimen is larger, pushing a thin entomological pin through the abdomen (Figs 5.31, 5.32). In either case, labels should be fixed onto the pin below the specimen, giving the date and place of capture, together with other information, such as who caught it, its specific name and the person who identified it. Note that ethyl acetate is an irritant and is toxic to humans and must be used with caution (it will also dissolve some plastic containers and their lids and so should be used in glass jars with metal or cork stoppers).

### *Keeping animals alive*

It may be important to keep animals alive in order to use them in behavioural experiments, or to rear juveniles to adults so as to be able to confirm their identification, or to breed them. Transporting such animals back from the field will be better for them if they are supplied with some 'comfort' in the form of vegetation within the specimen jar to provide cover and help to maintain appropriate humidity levels (Fig. 5.33). Many animals can be kept and even cultured if care is taken to keep them in appropriate environmental conditions and if they are supplied with suitable food. Most herbivores or detritivores (including woodlice, millipedes and springtails) can be kept at cool temperatures in the dark as long as they are given fresh food from time to time. Such animals should be kept in a plastic lunchbox with a good thick layer of leaf mould. The lid of the box should have several small holes cut into it, or one larger hole covered with fine mesh to allow the passage of air and prevent too much build-up of condensation. Although culture chambers need to be kept moist, care needs to be taken not to add too much water. The animals need to be supplied with fresh leaves occasionally, and they can also be given vegetables such as slices of potato or cucumber, preferably organic

**Fig. 5.33** Specimen jar containing animals and 'comfort'

produce that will not have been treated with pesticides. Some of the small predators can also be kept in this way, eating tiny soil invertebrates. Larger predators, such as large carabid beetles, large spiders and centipedes, need to be fed more actively and will often only take live prey. Substitutes for naturally available food are vestigial-wing *Drosophila* (fruit flies – available from exotic pet shops or biological suppliers), maggots (available in most fishing tackle shops) or mealworms (available from pet shops). Scavengers may feed on fish food such as freeze-dried tubifex worms which is available in small cubes from pet shops. In all cases, removing uneaten food and avoiding over-watering the substrates will reduce the chances of problematic fungal infestations.

Very small animals, such as mites, can be cultured in small containers. However, in order to examine these, first prepare a small container about 50 mm diameter and 70 mm deep and pour in plaster of Paris mixed with water to a depth of 5–10 mm. When set, the base is hard, and makes a good substrate since the animals can easily be seen walking on top of it. When the atmosphere in the container starts to dry out, a drop or two of water can be added which soon soaks into the plaster of Paris. Many common soil mite species can be fed on yeast, so the occasional addition of a few granules of dried baking yeast is sufficient to support them. Some species are parthenogenetic – females reproduce without mating – and may reproduce quite happily in captivity. *Damaeus* is a good genus with which to start experimenting with culturing; they are larger than many other mites and can be fascinating to watch. Millipedes can also be housed in a similar set-up, in a larger container since, using plaster of Paris, it is relatively easy to keep the container moist without overwatering it (Fig. 5.34). Here a layer of leaf litter can be placed on top of the plaster of Paris and replaced when necessary.

There are many insect larvae that are not possible to identify to species and so rearing these to adulthood will enable retrospective identification. Rearing through successive stages of moulting and metamorphosis may require more targeted conditions. Wong (1972) collated much of the early literature on culture methods, while Lutz *et al.* (1937) and Mahoney (1966) describe methods appropriate to a wide range of invertebrates, and Burakowski (1993) examines methods appropriate for soil beetles. The Amateur Entomologists' Society website features a number of care sheets for a range of invertebrates (see Chapter 6).

**Fig. 5.34** Terrarium with leaf litter on a base of plaster of Paris

## 5.4 Project design

It is important to be very clear about the aims of any research project before designing the study. Certain principles should always be addressed in planning your project design. Where the study involves experiments, it is good practice to use a control – that is, a baseline treatment. For example, in an experiment where the aim was to see whether woodlice aggregate under different types of shelter, animals were released into an arena which had two shelters that were identical except for the amount of light that they blocked out. To ensure that there was no intrinsic bias towards one shelter rather than the other in the absence of a difference in shading, a control was run where the shelters had the same degree of light-blocking ability. See Devigne *et al.* (2011) for further details.

Choosing an appropriate sample size is also important to gather sufficient data to identify any impacts, while avoiding gathering so much that there is a danger of environmental damage (e.g. turning over too many logs) or taking so much time that the project runs the risk of not being completed. The rule of thumb is that where the system is not very variable (e.g. where roughly the same number of animals are found under similar types of refuge) then sample sizes of 10–20 refuges may be sufficient. However, if the system is very variable (e.g. where even very similar refuges have very different numbers of animals underneath them), then much larger samples (>30 refuges) should be taken. Various online calculators can help to estimate the sample size required for particular situations (see Chapter 6). As well as determining how many replicates should be used, it is important to ensure that these are appropriately selected. Studying the number of animals under logs from deciduous woods and coniferous woods provides one example of how carefully replicates should be selected. If a number of logs in one coniferous woodland are sampled and compared with a number of logs from a single deciduous woodland, then any differences found may reflect not only the type of woodland but also other differences between just two sites (such as altitude, tree density, soil type, etc.). This is a form of pseudoreplication, where replicates are not independent of each other. It is far better to sample from several woodlands of each type.

It is an important statistical requirement that any samples (or treatments within an experiment) are independent of each other. If a study of logs of different sizes resulted in all large ones being in a clump and all small ones being scattered further away, any differences found might not relate to the size of the log but could be related to the location (perhaps being damper, or with more organic material, or in the shade). It is better to survey logs scattered randomly throughout a woodland. This could even be planned by dividing the woodland into sections and selecting logs using random numbers within the sections. This involves using pairs of random numbers as coordinates in each section and selecting the nearest log to that position (alternating first a large log and then a small one). Random numbers can be generated in Microsoft Excel and the free program FCStats (see marginal note) allows the generation of random coordinates for any size of area. In a similar way, randomly allocating treatments in experiments is also best practice. For instance, if animals are placed in a choice chamber (Fig. 5.6) to examine preferences for dry and damp substrates, the chamber should not always be placed with the damp side to the right in case there is anything else in that direction that might influence the results (e.g. a slight wind, more shade, etc.).

**FCStats statistics program**

This program runs in Microsoft Excel and is designed for small datasets obtained from ecological field research projects. This program covers both descriptive and inferential statistics together with some specialist ecological techniques and includes data from the examples found in Wheater & Cook (2000, 2015) – see Chapter 6.

## Analysis

Most research projects generate a body of data that can be difficult to interpret without some analysis. This may be to explore and describe a target variable (descriptive statistics),

**Table 5.1** Types of data

| Type | Explanation |
|---|---|
| *Categories* (categorical/nominal data) | Attributes are grouped under particular categories (e.g. male or female; green, brown or grey). Categories do not overlap and have no order to them (i.e. one category is not larger or stronger than any of the others). |
| *Ranks or scores* (ranked or ordinal data) | Data can be ordered in a logical way but there is no clearly defined spacing between ranks (e.g. small, medium, large; colourless, pale coloured, dark coloured). |
| *Measurements** (measurement data) | Data can be ordered in a logical way and there is a measurable degree of magnitude between them (e.g. lengths of logs, depth in the soil, numbers of animals of a particular species). |

* Some texts describe two different types of measured data: ratio data and interval data. The former are measured on a scale with an absolute zero value (as in counts of the number of animals – it would be nonsense to say there were fewer than zero animals under a log), while the latter scales do not have absolute zero values (i.e. measurements below zero are possible, as in temperature measured on the Celsius scale). In practice, these distinctions are not important for most types of data analysis.

or to more deeply investigate whether there are any factors that may lead to differences, or be related to, or associated with the target variable (inferential statistics). A detailed examination of these techniques is beyond the scope of this book and when planning a project you should consult a more detailed guide such as Wheater & Cook (2000, 2015). Here we give a summary of some of the basic techniques together with the types of data (see Table 5.1) and variables that are involved. Note that converting from one type of data to another is possible, but this only works when converting from measurements to ranks to categories and not the other way around. So, the colour of the elytra of an insect species could be measured using fairly sophisticated techniques such as spectrophotometry to obtain measurement data on how iridescent individual beetles are (measurement data), or this could be summarised on a scale of 1–5 from highly iridescent to not (ranked data), or the animals could be placed into categories based on the background colour of predominantly blue or brown (categorical data). A simple spreadsheet-based program (FCStats) is available to analyse smaller datasets using many of these techniques (see Chapter 6).

### Descriptive statistics

These can be used both to explore data (examining data as an interim step in progressing to more interpretive analysis) and to display data characteristics (both numerically and graphically) for presentation. These techniques include calculating the average value (mean or median) of measured or ranked data and an estimation of how variable the data are (standard deviation or variance, or interquartile range). The average value is a useful summary of the data (for example, so you can say that stones sampled were an average of 205 mm in length rather than list the lengths of all stones measured). The variation in the data helps to understand how useful a particular average value is as a summary of the data. If all the stones were about 200 mm in length, then there is not much variation (the standard deviation would have a low value) and the mean would be a useful summary of the data. However, if the stones are highly variable, with some being very small while others are huge, this would give a large value for the standard deviation and the mean would not be a very useful summary of the data. Other measures of variation include the range, standard error and confidence intervals/limits. Averages and measures of variation are also used in inferential statistics.

Other useful numerical summaries of ecological data include estimations of population sizes and densities, assessments of the distribution of organisms (i.e. whether they are clumped, or regularly or randomly distributed), and measurements of community richness (or diversity). See Wheater & Cook (2015) for further details. There are also more sophisticated techniques that enable the community structure to be explored and analysed including ordination, classification and pattern analysis (see Wheater *et al.*, 2020 for further details).

Graphical techniques can be used both to explore data and to present research findings. Depending on the type of data, different graphical approaches can be used (Table 5.2). Microsoft Excel enables the creation of many of these types of graph, as does FCStats (see Chapter 6).

**Table 5.2** Range of graphical techniques for displaying different types of data

| Type of graph | Common usage |
|---|---|
| Bar/point graphs with means and error bars | Displays the mean of a measured variable either as a point (Fig. 5.35a) or the top of a bar (Fig. 5.35b) together with a measure of the variation (usually the standard error or confidence intervals) |
| Box and whisker chart | Displays the median together with the interquartile range and full range for values of ranked variables (Fig. 5.36) |
| Scattergraph | Displays the relationship between two measured or ranked variables (Fig. 5.37a) – where a causal relationship is proposed between two measured variables, a regression line can be drawn onto the scattergraph (Fig. 5.37b) |
| Frequency table/histogram | Displays frequencies of a single measured variable for a single sample (Fig. 5.38a,b) |
| Pie chart | Displays frequencies (including percentages) of several categories of a single sample (Fig. 5.39) |
| Clumped and stacked bar charts | Displays frequencies (including percentages) of several categories of several samples with categories either clustered together (Fig. 5.40a) or stacked on top of each other (Fig. 5.40b) – data are in Fig. 5.40c |

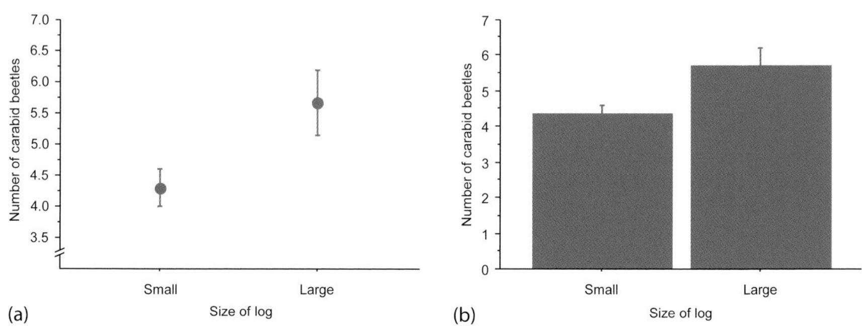


**Fig. 5.35** Point and bar graphs of the number of carabid beetles under different-sized logs: (a) point graph displaying the mean number (with standard error bars); (b) bar graph displaying means (with standard error bars)

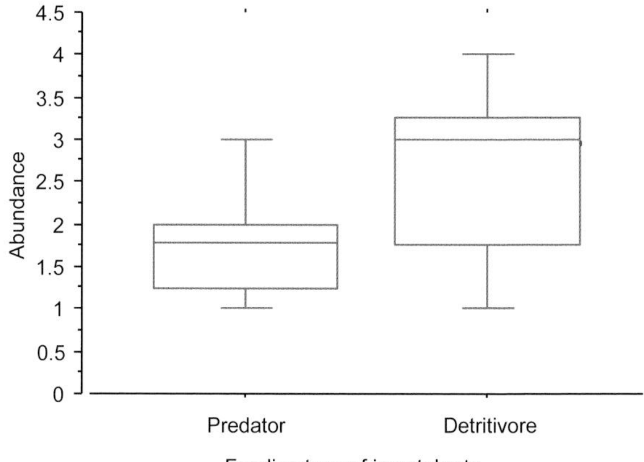


**Fig. 5.36** Box and whisker plot of the ranked abundance of different invertebrate feeding types under logs (showing medians and interquartile ranges)

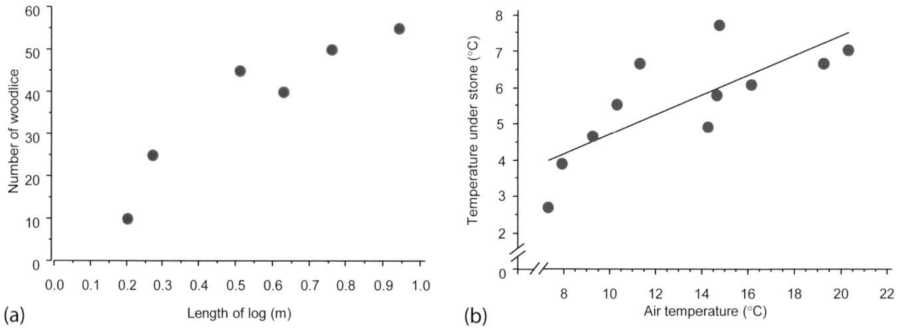


**Fig. 5.37** Scatterplots of the relationship between length of log and the number of woodlice found beneath them – (a) scatterplot; and the relationship between the air temperature and the temperature under stones – (b) scatterplot with regression line

| Number of woodlice | Frequency |
|---|---|
| 2 | 1 |
| 3 | 1 |
| 4 | 2 |
| 5 | 4 |
| 6 | 2 |
| 7 | 7 |
| 8 | 5 |
| 9 | 5 |
| 10 | 2 |
| 11 | 2 |
| 12 | 2 |
| 13 | 1 |
| 15 | 2 |
| 17 | 2 |
| 20 | 1 |
| 28 | 1 |

(a)

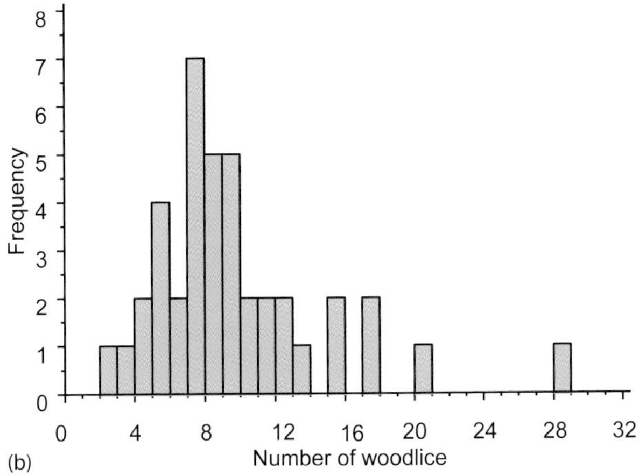

(b)

**Fig. 5.38** Frequencies of the occurrence of woodlice under refuges: (a) frequency table; (b) frequency histogram

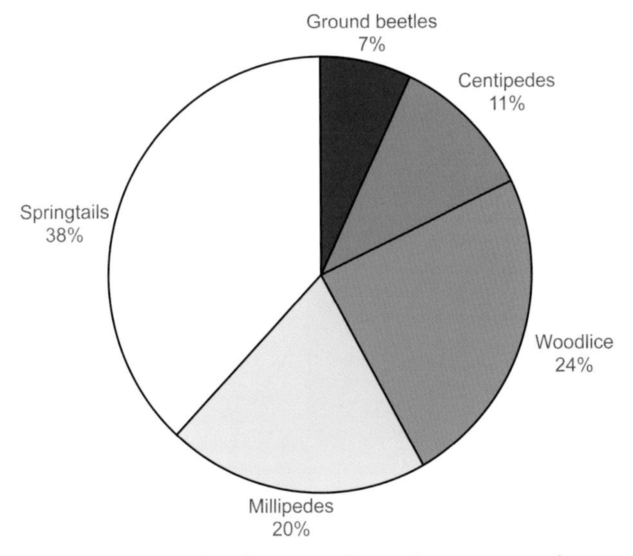


**Fig. 5.39** Pie chart of the frequency of invertebrate groups under logs

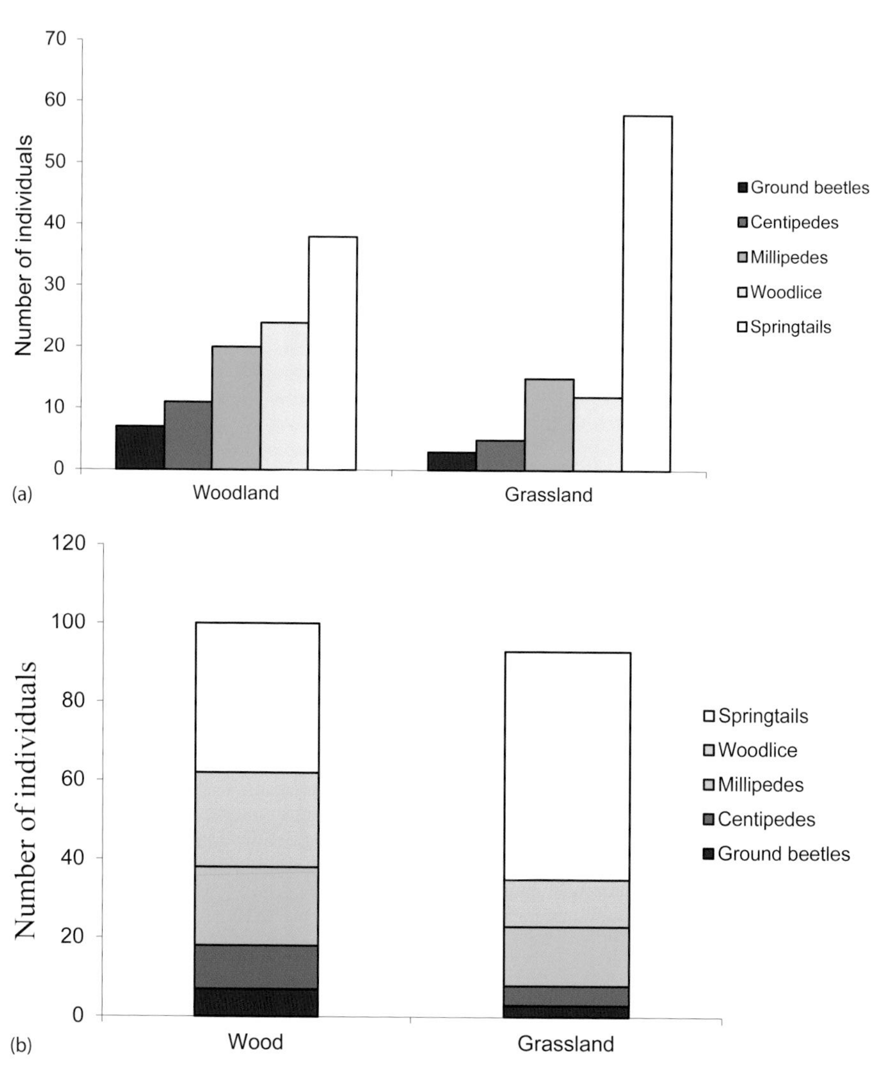


| Invertebrate | Wood | Grassland |
|---|---|---|
| Ground beetles | 7 | 3 |
| Centipedes | 11 | 5 |
| Millipedes | 20 | 15 |
| Woodlice | 24 | 12 |
| Springtails | 38 | 58 |

(c)

**Fig. 5.40** Frequency data for invertebrate groups in different habitats: (a) clustered bar chart; (b) stacked bar chart; (c) frequency table

**Inferential statistics**

It will often be necessary to apply statistical techniques to test whether a difference between two or more samples is real, or whether there is a relationship between one or more variables, or whether two or more frequency distributions are associated in some way. Although various textbooks will help, this is an area where expert advice can contribute much to the planning, as well as the analysis, of the work. A range of techniques exist that depend upon the question being asked and the types of data that have been gathered. Knowing in advance the type of analysis that will be used is important so that the project can be properly planned to gather an appropriate amount of data from suitable data types.

A traditional starting point of statistical analyses is to specify the null hypothesis. A null hypothesis predicts that samples do not differ significantly (or that variables are not related, or that frequency distributions are not associated). Each analysis produces a test statistic and uses the sample size to assess whether this could be a chance effect or is likely to be statistically significant. It does this by calculating the probability that the effect could statistically be obtained by chance (i.e. that the null hypothesis should be accepted) and by convention probabilities lower than 1 in 20 (<5% or, when measured on a 0–1 scale, <0.05) are deemed statistically significant (i.e. that the null hypothesis should be rejected). Fig. 5.41 shows the sequence of stages within these types of statistical analyses.

For most analyses, there are different tests depending on the types of data collected (with measurement data generally enabling better tests to be used than do ordinal or categorical data). Where data can be matched in some way, this will remove some of the natural variation within the system being studied. For example, if the number of woodlice under logs were being compared in the summer and winter, the logs could be tagged and each one surveyed both in the summer and winter. Since these data are paired (each log has two measurements: one in summer and a second in winter) then differences between the logs themselves can be taken into account using a special type of difference test – a paired test. There are other tests that facilitate situations when comparisons are between more than two samples (e.g. in comparisons between the number of woodlice under logs from three different species of trees: oak vs beech vs ash).

It is important to record the appropriate values resulting from the test when presenting any results (usually the test

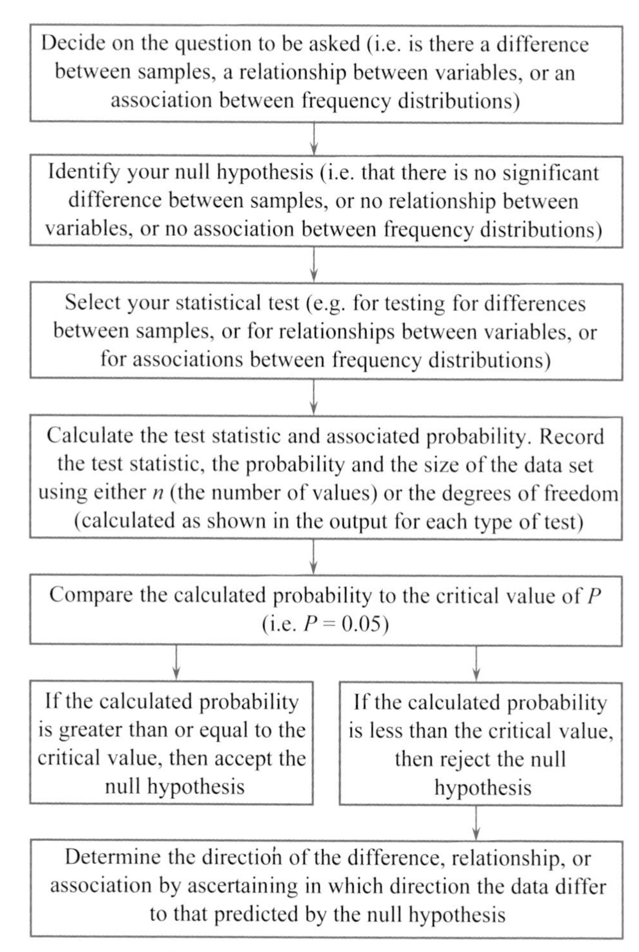


**Fig. 5.41** Flow chart for using statistical tests

statistic, some measure of the sample size, and the probability). Where statistically significant results are obtained, aspects of the dataset can be calculated to aid interpretation, i.e. to see where these differences occur.

## 5.5 Writing up and presenting the results

Writing up and presenting the results of a research project is important in order to ensure that the results and conclusions are accessible to those who are interested or would find the results useful. A really thorough, critical investigation that has established new information of general interest may be worth publishing in a scientific journal or presenting at a scientific conference. Several scientific journals may publish short articles on these kinds of investigations. Such journals include the *British Journal of Entomology and Natural History* and the *Entomologist's Monthly Magazine*, or various local publications such as *The Naturalist* (a journal that concentrates on work of relevance to Yorkshire). Various scientific organisations and societies have newsletters and bulletins that also accept papers of interest (e.g. the *Bulletin of the British Myriapod and Isopod Group* and both the *Newsletter* and the *Bulletin of the British Arachnological Society*). *The Journal of Biological Education* deals with articles of an educational nature. Examine recent editions of these journals to see the sort of thing they publish, and then write a paper on similar lines, keeping it as short and concise as possible while still presenting enough information to establish the conclusions. Advice from an expert is helpful at this stage.

Many scientific societies also run periodic conferences and workshops and it may be possible to present your research either as a poster or oral presentation. Posters are usually A1 (or sometimes A0) in size and need to be highly visual. They should have as little text as possible and give a simple message. Oral presentations also need to focus on a simple message and should be short and concise (depending on the amount of allocated time). Both types of presentation should use clear images and a limited amount of text or words to give the message. In oral presentations, a brief introduction explaining what will be covered and why, should be followed by a (slightly) more detailed description of what was done, what was found and what it means, finishing with a brief conclusion stating what was important about the study.

Publishing information on the internet can also be useful. Interested individuals often have websites, as do local wildlife groups where records and other information of interest to like-minded people can be posted. In addition, specialist Facebook pages and other social media such as Instagram and Twitter feeds can be used to let followers know of anything of interest that you have found. Social media can also help you to reach out to find advice on identification, collecting techniques, culture methods and help with interpreting results.

It is an unbreakable convention of scientific publications and presentations that results are reported with scrupulous honesty. It is therefore essential to keep detailed and accurate records throughout the investigation, and to distinguish in the write-up between certainty and probability, and between deduction and speculation. Wheater *et al.* (2020) give more details on writing for, and presenting to, different audiences.

Much worthwhile work remains to be done on even the most basic aspects of the natural history of these groups of animals and their relationships with their environment. Experiments and surveys that are not too complicated, based on sound ideas supported by the wider literature, and that have been carefully planned, should produce results that are capable of being appropriately analysed and interpreted. Remember that so-called negative results (not finding a definitive answer) may still be interesting and useful in advancing our knowledge of the fascinating animals that inhabit the relatively hidden world of the cryptosphere.

# 6 Useful addresses and links

A wide range of organisations provide information for (and often support) the study of the animals covered by this book. These have been organised to cover general links followed by those supporting groups of animals (in order of the keys in this book).

## 6.1 Equipment

A number of organisations supply monitoring equipment and traps for both field and laboratory work, including:

**Breckland Scientific** – for field and laboratory equipment
Antom Court, Tollgate Drive, Stafford ST16 3AF. Tel.: 01785 227227 https://www.brecklandscientific.co.uk

**NHBS Ltd** – for books and field and laboratory equipment
1–6 The Stables, Ford Road, Totnes, Devon TQ9 5LE. Tel.: 01803 865913. https://www.nhbs.com

**RS Components Ltd** – for monitoring equipment
Birchington Road, Corby, Northamptonshire, NN17 9RS. Tel.: 03457 201201. https://uk.rs-online.com/web/

**Watkins & Doncaster** – for field and laboratory equipment
PO Box 114, Leominster, HR6 6BS. Tel.: 0333 800 3133. https://www.watdon.co.uk

## 6.2 General links

**Classification and taxonomy** – several websites show the classification of species

- Catalogue of Life – https://www.catalogueoflife.org
- Tree of Life Project – http://tolweb.org/tree/

**Distribution maps and databases**

- Biological Records Centre – https://www.brc.ac.uk
- National Biodiversity Network Atlas (NBNatlas) – https://nbnatlas.org

**Fieldwork information**

- Countryside Code – https://assets.publishing.service.gov.uk/government/uploads/system/uploads/attachment_data/file/897289/countryside-code-leaflet.pdf
- HSE Health and safety executive (chemical safety data sheets) – https://www.hse.gov.uk/coshh/basics/datasheets.htm

**Statistical resources**

- FCStats – https://bcs.wiley.com/he-bcs/Books?action=resource&bcsId=11891&itemId=1119413222&resourceId=47144
- Sample size calculators (note that some idea of the data to be analysed is required, either from other similar studies or from a pilot study) – http://www.stat.ubc.ca/~rollin/stats/ssize/index.html; http://abs.gov.au/websitedbs/D3310114.nsf/home/Sample+Size+Calculator

## 6.3 General invertebrate resources

**Amateur Entomologists' Society (AES)** – offers help with identification and insect care sheets (including caring for insect larvae)

PO Box 8774, London, SW7 5ZG.
General website – https://www.amentsoc.org
Care sheets – https://www.amentsoc.org/insects/caresheets/
Joint Committee for the Conservation of British Insects (code of conduct for collecting insects and other invertebrates) – https://www.amentsoc.org/publications/online/collecting-code.html

**British Entomological and Natural History Society (BENHS)** – provides information for those interested in insects

The Pelham-Clinton Building, Dinton Pastures Country Park, Davis Street, Hurst, Reading, Berkshire RG10 0TH. https://www.benhs.org.uk

**Buglife** – charity concerned with the conservation of insects

G.06 Allia Future Business Centre, London Road, Peterborough PE2 8AN. https://www.buglife.org.uk

**Field Studies Council (FSC)** – publishes their own identification guides (e.g. the AIDGAP series) and some for other organisations including the Linnean Society and the Royal Entomological Society, as well as some species distribution atlases

Preston Montford, Montford Bridge, Shrewsbury, Shropshire SY4 1HW. Tel.: 01743 852100. https://www.field-studies-council.org

**Royal Entomological Society (RES)** – provides information for those interested in insects including publishing handbooks for the identification of a range of insect groups

General website – https://www.royensoc.co.uk

Handbooks for the Identification of British Insects (covers Hymenoptera – ants, etc., Diptera – true flies, Coleoptera – beetles, and larvae) – https://www.royensoc.co.uk/publications/handbooks

pdf versions of out of print handbooks – https://www.royensoc.co.uk/out-print-handbooks

## 6.4 Annelida (leeches, potworms and earthworms)

**Earthworm Society of Britain (ESB)** – provides information, including on identification and distribution of earthworms

c/o Field Studies Council Head Office, Preston Montford, Shrewsbury, Shropshire SY4 1HW. https://www.earthwormsoc.org.uk

**NBNatlas** – earthworm distribution data

https://registry.nbnatlas.org/public/show/dp88

## 6.5 Mollusca (slugs and snails)

**Conchological Society of Great Britain and Ireland** – provides information, including a species checklist and on identification of a range of mollusc species

https://conchsoc.net

**Malacological Society of London** – supports research and education on molluscs

http://malacsoc.org.uk

**MolluscIreland (mollusc pages)** – provides information on the slugs and snails of Ireland

http://www.habitas.org.uk/molluscireland/index.html

**National Museum Wales (slug pages)** – provides information on British and Irish slugs

https://naturalhistory.museumwales.ac.uk/slugs/Home.php

## 6.6 Arachnida (Acari: mites, Opiliones: harvestmen, Scorpiones: scorpions, Pseudoscorpiones: pseudoscorpions, and Aranea: spiders)

**Biodiversity Research Centre of the University of British Columbia** – provides information on mites and other microarthropods

https://www.zoology.ubc.ca/~srivast/mites/group.html

**Bristol University Tick ID** – provides information on the identification of ticks

http://www.bristoluniversitytickid.uk

**British Arachnological Society (BAS)** – provides information, including on identification of a range of arachnid species

General website – https://www.britishspiders.org.uk
Harvestmen webpages – https://srs.britishspiders.org.uk/portal.php/p/Harvestmen
Pseudoscorpion Recorders Group – https://nbn.org.uk/biological-recording-scheme/pseudoscorpion-recorders-group/
Spiders and harvestmen species index including distribution maps) – http://srs.britishspiders.org.uk/portal.php/p/A-Z+Species+Index
Spider checklist – https://www.britishspiders.org.uk/spider-checklist-2019
Spider webpages – https://www.britishspiders.org.uk/araneae

**Gerald Legg's website** – provides a wealth of information about pseudoscorpions

https://www.chelifer.com/?page_id=61

**Soil Biodiversity UK** – provides information on mites and other soil animals

Website – http://soilbiodiversityuk.myspecies.info
Facebook Public Group – https://www.facebook.com/groups/438740999565613

## 6.7 Crustacea (Amphipoda: landhoppers, and Isopoda: woodlice) and Myriapoda (Diplopoda: millipedes, Chilopoda: centipedes, Symphyla: symphylids, and Pauropoda: pauropods)

**British Myriapod and Isopod Group (BMIG)** – provides information, including on the identification, of a range of species of woodlice and myriapods

General website – https://www.bmig.org.uk
Centipede checklist – http://www.bmig.org.uk/checklist/centipede-checklist
Millipede checklist – https://www.bmig.org.uk/checklist/millipede-checklist
Pauropods – https://www.bmig.org.uk/page/pauropod-checklist
Symphylids – https://www.bmig.org.uk/page/symphylan-checklist
Woodlouse (and terrestrial amphipod) checklist – http://www.bmig.org.uk/checklist/woodlice-waterlice-checklist

## 6.8 Collembola (springtails)

**Checklist of the Collembola of the world** (Bellinger P.F., Christiansen K.A. & Janssens F. 1996–2021) – provides a checklist and identification keys

https://www.collembola.org

**Springtails website** (Philippe Garcelon) – provides information on identification

https://collemboles.fr/en/classification-dna-barcode/family-determination.html

**UK Collembola taxonomy and ecology website** – pages originally created by Steve Hopkin and maintained by Peter Shaw, providing information on UK species, including species lists and distribution maps

https://urweb.roehampton.ac.uk/collembola/

## 6.9 Dermaptera (earwigs), Dictyoptera (cockroaches) and Orthoptera (crickets, grasshoppers, etc.)

**Cockroach Species File Online** – website which provides an up-to-date taxonomic catalogue of world cockroaches

http://cockroach.speciesfile.org/HomePage/Cockroach/HomePage.aspx

**Orthoptera & Allied Insects website** – provides information, including on the identification, recording, and distribution of species

C/o Biological Records Centre, Centre for Ecology and Hydrology, Wallingford OX10 8BB. Tel.: 01491 692564
General website – https://orthoptera.org.uk
Dermaptera (earwigs) – https://www.orthoptera.org.uk/taxonomy/dermaptera
Dictyoptera (cockroaches) – https://www.orthoptera.org.uk/taxonomy/dictyoptera
Orthoptera (grasshoppers, crickets, etc.) – https://orthoptera.org.uk/taxonomy/orthoptera

## 6.10 Hymenoptera (Formicidae: ants)

**AntWiki** – provides information on UK species of ants

https://www.antwiki.org/wiki/United_Kingdom_of_Great_Britain_and_Northern_Ireland

**Bees, Wasps & Ants Recording Society (BWARS): Formicidae webpages** – provides information, including on the identification and distribution of species of ants

https://www.bwars.com/category/taxonomic-hierarchy/ant/formicidae

## 6.11 Coleoptera (beetles including ground beetles and rove beetles)

**Dung Beetle UK Mapping Project** – provides information, including identification keys to *Aphodius* and *Onthophagus*

https://dungbeetlemap.wordpress.com/finding-and-recording-dung-beetles/identification/

**Mike Hackston's Insect Keys (keys for the identification of British beetles – Coleoptera)** – provides identification keys for a wide range of British beetle families

https://sites.google.com/view/mikes-insect-keys/mikes-insect-keys/keys-for-the-identification-of-british-beetles-coleoptera
Keys for the identification of British Carabidae, including separate keys for different tribes – https://sites.google.com/view/mikes-insect-keys/mikes-insect-keys/keys-for-the-identification-of-british-beetles-coleoptera/keys-for-the-identification-of-british-carabidae
Keys for the identification of British Staphylinidae, including separate keys for different subfamilies – https://sites.google.com/view/mikes-insect-keys/mikes-insect-keys/

keys-for-the-identification-of-british-beetles-coleoptera/keys-for-the-identification-of-british-staphylinidae

**NBN Atlas – beetle distribution data**

https://registry.nbnatlas.org/datasets#filters=contains%3Acoleoptera

**Silphidae Recording Scheme** – provides information, including identification keys to Silphidae

https://www.coleoptera.org.uk/sites/www.coleoptera.org.uk/files/Silphidae%20Key_2019%20version.pdf

**UK Beetles website** – provides information, including on identification, of beetle species

General website – https://www.ukbeetles.co.uk
Carabidae webpages – https://www.ukbeetles.co.uk/carabidae
Key to the British Geotrupidae – https://www.ukbeetles.co.uk/geotrupidae-key
Staphylinidae webpages – https://www.ukbeetles.co.uk/staphylinidae

**UK Beetle Recording website** – provides information, including on identification and distribution of UK beetles

General website – https://www.coleoptera.org.uk
Carabidae webpages – https://www.coleoptera.org.uk/family/carabidae
Staphylinidae webpages – https://www.coleoptera.org.uk/family/staphylinidae

**John Walters and Mark Telfer** – provide identification guides to British beetles, mainly Carabidae, but also Geotrupidae (dor beetles)

http://johnwalters.co.uk/publications/guide-to-british-beetles.php

## 6.12 Vertebrates (amphibians, reptiles and small mammals)

**Amphibian and Reptile Conservation (ARC)** – provides information, including on identification and recording species

744 Christchurch Road, Boscombe, Bournemouth, Dorset BH7 6BZ. Tel.: 01202 391319. https://www.arc-trust.org

**Amphibian and Reptile Groups of the UK (ARG UK)** – umbrella organisation for local amphibian and reptile groups
https://www.arguk.org

**Froglife** – provides information, including on the identification of British species
1 Loxley, Werrington, Peterborough PE4 5BW.
General website – https://www.froglife.org
Amphibians – https://www.froglife.org/info-advice/amphibians-and-reptiles/amphibians/
Reptiles – https://www.froglife.org/info-advice/amphibians-and-reptiles/reptiles/

**Mammal Society** – provides information, including on identification, recording and distribution of British species
Black Horse Cottage, 33 Milton Abbas, Blandford Forum, Dorset DT11 0BL. Tel.: 02380 010981. https://www.mammal.org.uk/species-hub/full-species-hub/discover-mammals/

# 7 References and further reading

## 7.1 Finding books and journals

Many journal articles are now available on the internet, either complete or in abstract form. It is worth searching there first. Where references have a 'doi' (digital object identifier) number this can be copied into a computer browser to locate the relevant article online.

It is often possible to make arrangements to see or borrow books or journal articles by visiting the library of a local university, or by asking your local public library to borrow the work (or a photocopy of it) for you via the British Library Document Supply Centre (see https://www.bl.uk/articles). This may take some time, and it is important that your librarian, or online request, has a reference that is correct in every detail. References are acceptable in the form given here, namely the author's name, the date of publication, followed by (for a book) the title and publisher or (for a journal article) the title of the article, the journal title, the volume number, and the first and last pages of the article.

## 7.2 Reference list


Alexander, K.N.A. (2002) *The Invertebrates of Living and Decaying Timber in Britain & Ireland: A Provisional Annotated Checklist*. English Nature Research Reports ENRR467. Peterborough: English Nature. http://publications.naturalengland.org.uk/publication/132027

Alexander, K.N.A. (2008) Tree biology and saproxylic Coleoptera: issues of definitions and conservation language. *Revue d'Écologie (Terre Vie)* 63: 9–13. https://core.ac.uk/download/pdf/33521294.pdf

Anderson, R. & Rowson, B. (2020) *Annotated List of the Non-marine Mollusca of Britain and Ireland*. https://conchsoc.org/special_publications

Baker A.S. (2002) Chapter 15 Mites and Ticks. In: Hawksworth, D.L. (ed.) *The Changing Wildlife of Great Britain and Ireland*. The Systematics Association, Special Volume 62. Florida: CRC Press. pp. 230–239.

Barber, A.D. (2008) *Key to the Identification of British Centipedes*. AIDGAP OP130. Shrewsbury: Field Studies Council.

Barber, A.D. (2009) *Centipedes: Keys and Notes for the Identification of the Species*. Linnean Society Synopses of the British Fauna (New Series) 58. Shropshire: Field Studies Council.

Barber, A.D. (2022) *Atlas of the Centipedes of Britain and Ireland*. Shrewsbury: Field Studies Council.

Barber, A.D., Blower, J.G. & Scheller, U. (1992) Pauropoda, the smallest myriapods. *Bulletin of the British Myriapod Group* 8: 13–24. https://www.bmig.org.uk/sites/www.bmig.org.uk/files/bulletin_bmg/BullBMG08-1992.pdf

Barnard, P.C. (2011) *The Royal Entomological Society Book of British Insects*. Chichester: Wiley-Blackwell. https://doi.org/10.1002/9781444344981

Barrett, K.E.J. (1979) *Provisional Atlas of the Insects of the British Isles: Part 5, Hymenoptera: Formicidae, Ants*. Huntingdon: Biological Records Centre. https://www.brc.ac.uk/biblio/provisional-atlas-insects-british-isles-part-5-hymenoptera-formicidae-ants

Barrow, C. (2004) 10. Risk assessment and crisis management. In: Winser, S. (ed.) *The Expedition Handbook*. London: The Royal Geographical Society, Profile Books. pp. 106–116. https://www.rgs.org/CMSPages/GetFile.aspx?nodeguid=fdb2060e-3c47-4aac-96e5-fc17bbf67d95&lang=en-GB

Bee, L., Oxford, G. & Smith, H. (2017) *Britain's Spiders: A Field Guide*. WildGuides. New Jersey: Princeton University Press. https://doi.org/10.1515/9781400885060

Beebee, T. (2013) *Amphibians and Reptiles*. Naturalists' Handbooks 31. Exeter: Pelagic Publishing.

Beebee, T. & Griffiths, R. (2000) *Amphibians and Reptiles*. New Naturalist 87. London: HarperCollins.

Bellmann, H. (1988) *A Field Guide to Grasshoppers and Crickets of Britain and Northern Europe*. London: Collins.
Benton, T. (2012) *Grasshoppers and Crickets*. New Naturalist 120. London: HarperCollins.
Benton, T. (2017) *Solitary Bees*. Naturalists' Handbooks 33. Exeter: Pelagic Publishing.
Benton, T.G. (1991) The life history of *Euscorpius flavicaudis* (Scorpiones, Chactidae). *Journal of Arachnology* 19(2): 105–110. https://www.jstor.org/stable/3705658
Berry, R.J. (2009) *Islands*. New Naturalist 109. London: HarperCollins.
Bevan, D. (1987) *Forest Insects: A Guide to Insects Feeding on Trees in Britain*. Forestry Commission Handbook 1. London: HMSO.
Blower, J.G. (1985) *Millipedes*. The Linnean Society Synopses of the British Fauna 35. London: E.J. Brill/Dr W. Backhuys.
Blower, J.G., Cook, L.M. & Bishop, J.A. (1981) *Estimating the Size of Animal Populations*. London: George Allen & Unwin.
Boddy, L. (2021) *Fungi and Trees: Their Complex Relationships*. Gloucestershire: Arboricultural Association.
Bolton, B. & Collingwood, C.A. (1975) *Hymenoptera: Formicidae*. Royal Entomological Society Handbooks for the Identification of British Insects Vol. 6, Pt 3(c). London: Royal Entomological Society of London. https://www.royensoc.co.uk/wp-content/uploads/2021/12/Vol06_Part03c.pdf
Brian, M.V. (1977) *Ants*. New Naturalist 59. London: Collins.
Brock, P.D. (2019) *A Comprehensive Guide to Insects of Britain and Ireland*. Berkshire: Pisces Publications.
Brown, G.R. & Matthews, I.M. (2016) A review of extensive variation in the design of pitfall traps and a proposal for a standard pitfall trap design for monitoring ground-active arthropod biodiversity. *Ecology and Evolution* 6: 3953–3964. https://doi.org/10.1002/ece3.2176
Brown, V.K. (1990) *Grasshoppers*. Naturalists' Handbooks 2. Slough: The Richmond Publishing Co. Ltd.
Burakowski, B. (1993) Laboratory methods for rearing soil beetles (Coleoptera). *Memorabilia Zoologica* 46: 1–66.
Butterfield, J. & Malvido, J.B. (1992) Effect of mixed-species tree planting on the distribution of soil invertebrates. In Cannell, M.G.R., Malcolm, D.C. & Robertson, P.A. (eds) *The Ecology of Mixed-Species Stands of Trees*. Special Publication Series of The British Ecological Society 11. Oxford: Blackwell Scientific Publications.
Cameron, R.A.D. (2008) *Land Snails in the British Isles*. AIDGAP OP79. Shrewsbury: Field Studies Council.
Cameron, R. (2016) *Slugs and Snails*. New Naturalist 133. London: HarperCollins.
Chu, H.F. & Cutkomp, L.K. (1992) *How to Know the Immature Insects*. 2nd edition. New York: McGraw-Hill Education.
Coleman, D.C., Callaham, M.A. & Crossley, D.A. (2017) *Fundamentals of Soil Ecology*. Academic Press, New York.
Corbet, G.B. (1989) *Finding and Identifying Mammals: A Guide to the Mammals of Britain and Ireland*. 2nd edition. London: British Museum (Natural History).
Couzens, D., Swash, A., Still, R. & Dunn, J. (2021) *Britain's Mammals: A Field Guide to the Mammals of Britain and Ireland*. WildGuides. New Jersey: Princeton University Press.
Crowson, R.A. (1956) *Coleoptera: Introduction and Keys to Families*. Handbooks for the Identification of British Insects Vol. 4, Pt 1. London: The Royal Entomological Society. https://www.royensoc.co.uk/wp-content/uploads/2021/12/Vol04_Part01.pdf
Curl, J. (2016) 'Border collies' and other colour variations in *Porcellio scaber*. *British Myriapod and Isopod Group Newsletter* 33: 5–6.
Curry, J.P. (1994) *Grassland Invertebrates: Ecology, Influence on Soil Fertility and Effects on Plant Growth*. London: Chapman & Hall.
Dallimore, T. & Shaw, P. (2013) *Illustrated Keys to the Families of British Springtails (Collembola)*. AIDGAP OP157. Shrewsbury: Field Studies Council.
Daniels, L.D. & Lavallee, S. (2014) Better safe than sorry: planning for safe and successful fieldwork. *Bulletin of the Ecological Society of America* 95: 264–273. https://doi.org/10.1890/0012-9623-95.3.264
Darlington, A. (1969) *The Ecology of Refuse Tips*. London: Heinemann.
Darwin, C.R. (1881) *The Formation of Vegetable Mould Through the Action of Worms, with Observations on their Habits*. London: John Murray. https://doi.org/10.5962/bhl.title.107559
Davidson, M.B. (2019) The British harvestman (Opiliones) fauna: 50 years of biodiversity change, and an annotated checklist. *Arachnology* 18(3): 213–222. https://doi.org/10.13156/arac.2019.18.3.213

Davis, B.N.K., Walker, N., Ball, D.F. & Fitter, A.H. (1992) *The Soil*. New Naturalist 77. London: HarperCollins.
Delaney, M.J. (1954) *Thysanura and Diplura*. Royal Entomological Society Handbooks for the Identification of British Insects Vol. 1, Pt 2. London: Royal Entomological Society. http://www.uop.edu.pk/ocontents/Thysanura%20and%20Diplura.pdf
Devigne, C., Broly, P. & Deneubourg, J.L. (2011) Individual preferences and social interactions determine the aggregation of woodlice. *PLoS ONE* 6(2): e17389. https://doi.org/10.1371/journal.pone.0017389
Dibb, J.R. (1948) *Field Book of Beetles*. Hull: A. Brown & Sons.
Dixie, B., White, H. & Hassall, M. (2015) Effects of microclimate on behavioural and life history traits of terrestrial isopods: implications for responses to climate change. *ZooKeys* 515: 145–157. https://doi.org/10.3897/zookeys.515.9399
Duff, A.G. (2018) *Checklist of Beetles of the British Isles*. Iver: Pemberley Books.
Eberhard, W.G. (2020) *Spider Webs: Behaviour, Function and Evolution*. Chicago: University of Chicago. https://doi.org/10.7208/chicago/9780226534749.001.0001
Edwards, C.A. (1959). A revision of the British Symphyla. *Proceedings of the Zoological Society of London* 132: 403–439. https://doi.org/10.1111/j.1469-7998.1959.tb05529.x
Eiseman, C. & Charney, N. (2010) *Tracks & Sign of Insects & Other Invertebrates: A Guide to North American Species*. Pennsylvania: Stackpole Books.
Eisenbeis, G. & Wichard, W. (1987) *Atlas on the Biology of Soil Arthropods*. Springer, Berlin. https://doi.org/10.1007/978-3-642-72634-7
Elliott, J.M. & Dobson, M. (2015) *Freshwater Leeches of Britain and Ireland: Keys to the Hirudinea and a Review of their Ecology*. FBA SP69. Ambleside: Freshwater Biological Association.
Evans, G.O., Sheals, J.G. & MacFarlane, D. (1961) *The Terrestrial Acari of the British Isles. An Introduction to their Morphology, Biology and Classification. Vol. 1. Introduction and Biology*. London: British Museum (Natural History).
Evans, M. & Edmondson, R. (2007) *A Photographic Guide to the Grasshoppers and Crickets of Britain and Ireland*. Norfolk: WildGuides.
Falk, S. & Lewington, R. (2018) *Field Guide to Bees of Great Britain and Ireland*. London: Bloomsbury Publishing.
Fjellberg, A. (1998) *The Collembola of Fennoscandinavia and Denmark, Part I, Poduromorpha*. Fauna Entomologica Scandinavica, Volume 35. Leiden: Brill.
Fjellberg, A. (2007) *The Collembola of Fennoscandinavia and Denmark, Part II, Entomobryomorpha and Symphypleona*. Fauna Entomologica Scandinavica, Volume 42. Leiden: Brill. https://doi.org/10.1163/ej.9789004157705.i-265
Foelix, R. (2010) *Biology of Spiders*. 3rd edition. Oxford: Oxford University Press.
Ford, R.L.E. (1963) *Practical Entomology*. London: Frederick Warne & Co. Ltd.
Forsythe, T.G. (2000) *Ground Beetles*. Naturalists' Handbooks 8. 2nd edition. Slough: The Richmond Publishing Co. Ltd.
Freckman, D. (ed.) (2012) *Nematodes in Soil Ecosystems*. Texas: University of Texas Press.
Fry, R. & Lonsdale, D. (eds) (1991) *Habitat Conservation for Insects – a Neglected Green Issue*. Volume 21. Middlesex: The Amateur Entomologists' Society.
Galli, L., Capurro, M., Colasanto, E., Molyneux, T., Murray, A., Torti, C. & Zinni, M. (2019) A synopsis of the ecology of Protura (Arthropoda: Hexapoda). *Revue suisse de Zoologie* 126(2): 155–164. https://zenodo.org/record/3463443/files/155_164_Galli_etal.pdf?download=1
Geoffroy, J-J. (1981). Modalités de la coexistence de deux Diplopodes *Cylindroiulus punctatus* (Leach) et *Cylindroiulus nitidus* (Verhoeff) dans un écosystème forestier du Bassin Parisien. *Acta Oecologia Oecologia Generalis* 2: 357–372.
Gibson, R. (1994) *British Nemerteans: Keys and Notes for Identification of the Species*. Linnean Society Synopses of the British Fauna 24. Cambridge: Cambridge University Press.
Gilbert, O.L. (1989) *The Ecology of Urban Habitats*. London: Chapman & Hall. https://doi.org/10.1007/978-94-009-0821-5
Gormon, M. (1979) *Island Ecology*. Outline Studies in Ecology. London: Chapman & Hall. https://doi.org/10.1007/978-94-009-5800-5
Greenslade, P.J.M. (1963) Daily rhythms of locomotor activity in some Carabidae (Coleoptera). *Entomologia Experimentalis et Applicata* 6: 171–180. https://doi.org/10.1111/j.1570-7458.1963.tb00615.x
Gregory, S. (2009) *Woodlice and Waterlice (Isopoda: Oniscidae & Asellota) in Britain and Ireland*. Shrewsbury: Field Studies Council.
Gregory, S.J. (2016) On the terrestrial landhopper *Arcitalitrus dorrieni* (Hunt, 1925) (Amphipoda: Talitridae): identification and

current distribution. *Bulletin of the British Myriapod and Isopod Group* 29: 2–13. https://www.bmig.org.uk/sites/www.bmig.org.uk/files/bulletin/BullBMIG29-p02-13_Gregory-Arcitalitrus.pdf

Gurnell, J. & Flowerdew, J.R. (2019) *Live Trapping of Small Mammals: A Practical Guide*. 5th edition. Occasional Publications 3. London: Mammal Society.

Guthrie, M. (1989) *Animals of the Surface Film*. Naturalists' Handbooks 12. Slough: The Richmond Publishing Co. Ltd.

Haes, E.C.M. & Harding, P.T. (1997) *Atlas of Grasshoppers, Crickets and Allied Insects in Britain and Ireland*. ITE Research Publication 11. London: The Stationary Office. http://nora.nerc.ac.uk/id/eprint/7358/1/Grasshoppers.pdf

Hammond, P.M., Marshall, J.E., Cox, M.L., Jessop, L., Garner, B.H. & Barclay, M.V.L. (2019) *British Coleoptera Larvae: A Guide to the Families and Major Subfamilies*. Royal Entomological Society Handbooks for the Identification of British Insects Vol. 4, Pt 1a. Shrewsbury: Field Studies Council.

Harde, K.W. & Hammond, P.M. (2000) *A Field Guide in Colour to Beetles*. Leicester: Silverdale Books.

Harding, P.T. & Sutton, S.L. (1985) *Woodlice in Britain and Ireland: Distribution and Habitat*. Huntingdon: Institute of Terrestrial Ecology.

Harris, S. & Yalden, D.W. (eds) (2008) *Mammals of the British Isles: Handbook*. London: The Mammal Society.

Harvey, P., Davidson, M., Dawson, I., Fowles, A., Hitchcock, G., Lee, P., Merrett, P., Russell-Smith, A. & Smith, H. (2017) *A Review of the Scarce and Threatened Spiders (Araneae) of Great Britain*. Species Status 22. NRW Evidence Report 11. Bangor: Natural Resources Wales.

Harvey, P.R., Nellist, D.R. & Telfer, M.G. (2002a) *Provisional Atlas of British Spiders (Arachnida, Araneae)*. Vol. 1. Huntingdon: Biological Records Centre. http://nora.nerc.ac.uk/id/eprint/8094/1/Spidersv1.pdf

Harvey, P.R., Nellist, D.R. & Telfer, M.G. (2002b) *Provisional Atlas of British Spiders (Arachnida, Araneae)*. Vol. 2. Huntingdon: Biological Records Centre. http://nora.nerc.ac.uk/id/eprint/8096/1/Spidersv2.pdf

Hillyard, P.D. (2005) *Harvestmen*. Linnean Society Synopses of the British Fauna 4. 3rd edition. London: Linnean Society.

Hincks, W.D. (1956) *Dermaptera and Orthoptera*. Royal Entomological Society Handbooks for the Identification of British Insects. Vol. 1, Pt 5. London: Royal Entomological Society. https://www.royensoc.co.uk/wp-content/uploads/2021/12/Vol01_Part05.pdf

Hodge, P.J. & Jones, R.A. (1995) *New British Beetles: Species not in Joy's Practical Handbook*. Hurst: British Entomological and Natural History Society.

Hohbein, R.R. & Conway, C.J. (2018) Pitfall traps: a review of methods for estimating arthropod abundance. *Wildlife Society Bulletin* 42: 597–606. https://doi.org/10.1002/wsb.928

Hopkin, S.P. (1991) *A Key to the Woodlice of Britain and Ireland*. AIDGAP OP204. Shrewsbury: Field Studies Council.

Hopkin, S.P. (1997) *Biology of the Springtails*. Oxford: Oxford University Press.

Hopkin, S.P. (2007) *A Key to the Collembola (springtails) of Britain and Ireland*. AIDGAP OP111. Shrewsbury: Field Studies Council.

Hopkin, S.P. & Read, H.J. (1992) *Biology of Millipedes*. Oxford: Oxford University Press.

Hopkin, S.P. & Roberts, A.W. (1988) Symphyla – The least studied of the most interesting soil animals. *Bulletin of the British Myriapod Group* 5: 28–34. https://www.bmig.org.uk/sites/www.bmig.org.uk/files/bulletin_bmg/BullBMG05-1988.pdf

Humphrey, J. & Bailey, S. (2012) *Managing Deadwood in Forests and Woodlands*. Practice Guide. Cheshire: Forestry Commission. https://cdn.forestresearch.gov.uk/2012/04/fcpg020.pdf

Inns, H. (2018) *Britain's Reptiles and Amphibians: A Guide to the Reptiles and Amphibians of Great Britain, Ireland and the Channel Islands*. Hampshire: WildGuides. https://doi.org/10.2307/j.ctvs32rjq

Iorio, E. & Labroche, A. (2015) Les chilopodes (Chilopoda) de la moitié nord de la France: toutes les bases pour débuter l'étude de ce groupe et identifier facilement les espèces. Les chilopodes de la moitié nord de la France. *Invertébrés Armoricains* 13: 1–108.

Jessop, L. (1986) *Dung Beetles and Chafers – Coleoptera: Scarabaeoidea*. Royal Entomological Society Handbooks for the Identification of British Insects Vol. 5, Pt 11. London: Royal Entomological Society. https://www.royensoc.co.uk/wp-content/uploads/2021/12/Vol05_Part11.pdf

Jones, D. (1983) *The Country Life Guide to Spiders of Britain and Northern Europe*. Middlesex: Hamlyn.

Jones, H.D. (1998) The African and European land planarian faunas, with an identification guide for field workers in Europe. *Pedobiologia* 42: 477–489.

Jones, H.D. (2005) Identification: British land flatworms. *British Wildlife* 16: 189–194.
Jones, P.E. (1980) *Provisional Atlas of the Arachnida of the British Isles Part 1: Pseudoscorpiones*. European Invertebrate Survey, Biological Records Centre. Huntingdon: ITE. http://nora.nerc.ac.uk/id/eprint/7030/1/Arachnida_1_A1b.pdf
Jones, R. (2018) *Beetles*. New Naturalist 136. London: HarperCollins.
Joy, N.H. (1932) *A Practical Handbook of British Beetles*. 2 volumes. London: Witherby (E.W. Classey reprint, 1976). https://www.pisces-conservation.com/softjoy.html
Kennedy, F. (2002) *The Identification of Soils for Forest Management*. Field guide. Edinburgh: Forestry Commission. https://www.forestresearch.gov.uk/publications/the-identification-of-soils-for-forest-management/
Kerney, M. (1999) *Atlas of the Land and Freshwater Molluscs of Britain and Ireland*. Colchester: Harley Books.
Kerney, M.P. & Cameron, R.A.D. (1979) *A Field Guide to the Land Snails of Britain and North-west Europe*. London: Collins.
Kirby, K.J. & Drake, C.M. (1993) (eds) *Dead Wood Matters: The Ecology and Conservation of Saproxylic Invertebrates in Britain*. Proceedings of a British Ecological Society Meeting, Dunham Massey Park, 24 April 1992. Peterborough: English Nature. http://publications.naturalengland.org.uk/publication/2260356
Kirby, P. (1992) *Habitat Management for Invertebrates: A Practical Handbook*. Sandy, Bedfordshire: RSPB.
Krediet, A. (2020) *Earthworms: Photographic Guide to the Species of Northwestern Europe*. 's-Graveland, Netherlands: Jeugdbondsuitgeverij.
Lane, S.A. (2016) *A Review of the Beetles of Great Britain: The Staphylinidae: Tachyporinae Beetles*. Natural England Commissioned Report NECR265. Species Status 38. Peterborough: Natural England. http://publications.naturalengland.org.uk/publication/5694765406617600
Lane, S.A. & Mann, D.J. (2016) *A Review of the Status of the Beetles of Great Britain: The Stag Beetles, Dor Beetles, Dung Beetles, Chafers and their Allies – Lucanidae, Geotrupidae, Trogidae and Scarabaeidae*. Natural England Commissioned Report NECR224. Species Status 31. Peterborough: Natural England. http://publications.naturalengland.org.uk/publication/5488450394914816
Lavery, A. (2019) A revised checklist of the spiders of Great Britain and Ireland. *Arachnology* 18(3): 196–212. https://doi.org/10.13156/arac.2019.18.3.196
Lear, G., Dickie, I., Bank, J., Boyer, S., Buckley, H.L., Buckley, T.R., Cruickshank, R., Dopheide, A., Handley, K.M., Hermans, S. & Kamke, J. (2018) Methods for the extraction, storage, amplification and sequencing of DNA from environmental samples. *New Zealand Journal of Ecology* 42: 10–50A. https://doi.org/10.20417/nzjecol.42.9
Leather, S.R. (2004) *Insect Sampling in Forest Ecosystems*. Chichester: Wiley-Blackwell. https://doi.org/10.1002/9780470750513
Leather, S.R. & Barbour, D.A. (1999). *Animals on Cherry Trees*. Naturalists' Handbook 27. Slough: The Richmond Publishing Co. Ltd.
Lebas, C., Galkowski, C., Blatrix, R. & Wegnez, P. (2019) *Ants of Britain and Europe: A Photographic Guide*. London: Bloomsbury Wildlife.
Lee, P. (2015) *A Review of the Millipedes (Diplopoda), Centipedes (Chilopoda) and Woodlice (Isopoda) of Great Britain*. Natural England Commissioned Report NECR186. Species Status 23. Peterborough: Natural England. http://publications.naturalengland.org.uk/publication/4924476719366144
Lee, P. & Harding, P. (2006) *Atlas of the Millipedes (Diplopoda) of Britain and Ireland*. Pensoft Series Faunistica 59. Sofia-Moscow: Pensoft Publishers.
Legg, G. (2019) Changes in the British pseudoscorpion fauna over the last 50 years. *Arachnology* 18(3): 189–195. https://doi.org/10.13156/arac.2019.18.3.189
Legg, G. & Farr-Cox, F. (2016) *Illustrated Key to the British False Scorpions (Pseudoscorpions)*. AIDGAP OP173. Shrewsbury: Field Studies Council, Shrewsbury.
Legg, G. & Jones, R.E. (1988) *Pseudoscorpions*. Linnean Society Synopses of the British Fauna (New Series) 40. Leiden: E.J. Brill/Dr W. Backhuys.
Lewis, J.G.E. (1974). *The Biology of Centipedes*. Cambridge: Cambridge University Press.
Locket, G.H. & Millidge, A.F. (1951) *British Spiders*. Vol. 1. London: Ray Society.
Locket, G.H. & Millidge, A.F. (1953) *British Spiders*. Vol. 2. London: Ray Society.
Locket, G.H., Millidge, A.F. & Merrett, P. (1974) *British Spiders*. Vol. 3. London: Ray Society.

Lott, D.A. (2009) *The Staphylinidae (Rove Beetles) of Britain and Ireland: Scaphidiinae, Piestinae, Oxytelinae*. Royal Entomological Society Handbooks for the Identification of British Insects Vol. 12, Pt 5. Shrewsbury: Field Studies Council.

Lott, D.A. & Anderson, R. (2011) *The Staphylinidae (Rove Beetles) of Britain & Ireland: Oxyporinae, Steninae, Euaesthetinae, Pseudopsinae, Paederinae, Staphylininae*. Royal Entomological Society Handbooks for the Identification of British Insects Vol. 12, Pts 7 & 8. Shrewsbury: Field Studies Council.

Luff, M.L. (1998) *Provisional Atlas of the Ground Beetles (Coleoptera, Carabidae) of Britain*. Huntingdon: Biological Records Centre. http://nora.nerc.ac.uk/id/eprint/8106/1/Ground_Beetles.pdf

Luff, M.L. (2007) *The Carabidae (Ground Beetles of Britain and Ireland)*. Royal Entomological Society Handbook Vol. 4, Pt 2. 2nd edition. Shrewsbury: Field Studies Council.

Lutz, F.E., Welch, P.S., Galtsoff, P.S. & Needham, J.G. (1937) *Culture Methods for Invertebrate Animals*. New York: Constable & Company.

Mahoney, R. (1966) *Laboratory Techniques in Zoology*. London: Butterworth.

Marsh, R.J. (2016) *A Provisional Atlas of the Coleoptera of Yorkshire (Vice-Counties 61–65): Part 5 – Staphylinidae – Groups other than Aleocharinae*. York: The Yorkshire Naturalists' Union.

Marshall, J.A. & Haes, E.C.M. (1988) *Grasshoppers and Allied Insects of Great Britain and Ireland*. Colchester: Harley Books.

Marshall, J. & Ovenden, D. (1999) *Guide to Grasshoppers and Allied Insects*. OP54. Shrewsbury: Field Studies Council.

Marshall, S.A. (2012) *Flies: The Natural History and Diversity of Diptera*. New York: Firefly Books.

Marshall, S.A. (2017) *Insects: Their Natural History and Diversity*. 2nd edition. New York: Firefly Books.

Marshall, S.A. (2018) *Beetles: The Natural History of Coleoptera*. New York: Firefly Books.

Mekonen, S., Petros, I. & Hailemariam, M. (2017) The role of nematodes in the processes of soil ecology and their use as bioindicators. *Agriculture and Biology Journal of North America* 8(4): 132–140.

Moore, J., Gibson, R. & Jones, H.D. (2001) Terrestrial nemerteans thirty years on. *Hydrobiologia* 456(1–3): 1–6. https://doi.org/10.1023/A:1013052728257

Morley, E.L., Jones, G. & Radford, A.N. (2014) The importance of invertebrates when considering the impacts of anthropogenic noise. *Proceedings of the Royal Society B* 281: 1–8. https://doi.org/10.1098/rspb.2013.2683

Morris, M.C. (1999) Using woodlice (Isopoda, Oniscoidea) to demonstrate orientation behaviour. *Journal of Biological Education* 33: 215–216. https://doi.org/10.1080/00219266.1999.9655669

Morris, M.G. (1997–2012) *Weevils. Coleoptera: Curculionidae*. Royal Entomological Society Handbooks for the Identification of British Insects Vol. 5, Pts 17a–d. London: Royal Entomological Society.

Murchie, A.K. & Gordon, A.W. (2013) The impact of the 'New Zealand flatworm', *Arthurdendyus triangulatus*, on earthworm populations in the field. *Biological Invasions* 15(3): 569–586. https://doi.org/10.1007/s10530-012-0309-7

Naggs, F., Preece, R.C., Anderson, R., Peiris, A., Taylor, H. & White, T.S. (2014) *An Illustrated Guide to the Land Snails of the British Isles*. Joint Conchological Society and Malacological Society Publication. Exeter: SRP Ltd.

Nielsen, U.N. (2019) *Soil Fauna Assemblages: Global to Local Scales*. Cambridge: Cambridge University Press. https://doi.org/10.1017/9781108123518

O'Connor, F.B. (1967) *8. The Enchytraeidae*. pp. 213–258. In Burgess A. & Raw F. (eds) *Soil Biology*. London: Academic Press.

Oldroyd, H. (1970) *Diptera: Introduction and Key to Families*. Royal Entomological Society Handbooks for the Identification of British Insects. Vol. 9, Pt 1. London: Royal Entomological Society. https://www.royensoc.co.uk/wp-content/uploads/2022/01/Vol09_Part01.pdf

Oldroyd, H. (1973) *Collecting, Preserving and Studying Insects*. 2nd edition. London: Hutchinson Scientific and Technical.

Oliver, P.G. & Meechan C.J. (1993) *Woodlice*. Linnean Society Synopses of the British Fauna (New Series) 49. Field Studies Council, Shrewsbury.

Owen, J. (1990) *The Ecology of an Urban Garden: The First Fifteen Years*. Cambridge: Cambridge University Press.

Owen, J. (2010) *Wildlife of a Garden: A Thirty-Year Study*. London: Royal Horticultural Society.

Parris, K.M., McCall, S.C., McCarthy, M.A., Minteer, B.A., Steele, K., Bekessy, S. & Medvecky, F. (2010) Assessing ethical trade-offs in ecological field studies. *Journal of Applied Ecology* 47: 227–234. https://doi.org/10.1111/j.1365-2664.2009.01755.x

Pass, G. & Szucsich, N.U. (2011) 100 years of research on the Protura: many secrets still retained. *Soil Organisms* 8(3): 309–334.

Plant, C.W. (1997) *A Key to Adults of British Lacewings and their Allies*. AIDGAP OP245. Shrewsbury: Field Studies Council.

Prŷs-Jones, O.E. & Corbet, S.A. (2014) *Bumblebees*. Naturalists' Handbooks 6. 3rd edition. Exeter: Pelagic Publishing.

PTES (undated) *Build a Log Pile for Stag Beetles*. London: People's Trust for Endangered Species. https://ptes.org/get-involved/wildlife-action/help-stag-beetles/build-a-log-pile-for-stag-beetles/

Read, H.J. (2020) *Veteran Trees: A Guide to Good Management*. Peterborough: English Nature. http://publications.naturalengland.org.uk/publication/75035

Reed, B.T. & Jennings, M. (2007) Promoting consideration of the ethical aspects of animal use and implementation of the 3Rs. *Proceedings of the 6th World Congress on Alternatives & Animal Use in the Life Sciences, August 21–25, 2007, Tokyo, Japan. AATEX* 14, Special Issue: 131–135.

Rhyder, J. (2021) *Track and Sign: A Guide to the Field Signs of Mammals and Birds of the UK*. Cheltenham: The History Press.

Richards, P. (2010) *Guide to Harvestmen of the British Isles*. FSC OP140. Shrewsbury: Field Studies Council.

Richards, P. & Burkmar, R. (2017) *UK Harvestmen* (Version 1.0) (Knowledge-base) (for FSC Identikit). Shrewsbury: Field Studies Council. https://harvestmen.fscbiodiversity.uk

Roberts, M.J. (1993) *The Spiders of Great Britain and Ireland*. Compact Edition (2 vols). Colchester: Harley Books.

Roberts, M.J. (1995) *Spiders of Britain and Northern Europe*. London: HarperCollins.

Roberts, P., Froglife & Ovenden D. (2016) *Guide to the Reptiles and Amphibians of Britain and Ireland*. OP55. Shrewsbury: Field Studies Council.

Rowson, B., Turner, J., Anderson, R. & Symondson, B. (2014) *Slugs of Britain and Ireland*. AIDGAP OP160. Shrewsbury: Field Studies Council.

Roy, H.E., Brown, P.M.J., Comont, R.F., Poland, R.L. & Sloggett, J.J. (2013) *Ladybirds*. Naturalists' Handbooks 10. 2nd edition. Exeter: Pelagic Publishing.

Savory, T. (1971) *Biology of the Cryptozoa*. Herts: Merrow.

Scheller, U. (2008) A reclassification of the Pauropoda (Myriapoda). *International Journal of Myriapodology* 1(1): 1–38. https://brill.com/downloadpdf/journals/ijm/1/1/article-p1.xml

Schmelz, R. & Collado, R. (2010) A guide to European terrestrial and freshwater species of Enchytraeidae (Oligochaeta). *Soil Organisms* 82(1): 1–176.

Schmelz, R.M. & Collado, R. (2015) Checklist of taxa of Enchytraeidae (Oligochaeta): an update. *Soil Organisms* 87(2): 149–153.

Schmidt-Rhaesa, A. (1997) *Nematomorpha*. New York: Gustav Fischer.

Shepherd, M. & Crotty, F. (2020) A Key to the Soil Mites of Britain and Ireland: Test Version 9. https://drive.google.com/file/d/1Hvg3FdQbvtQaanIb83SONKuRsQipm4h4/view?usp=sharing

Sheppard, D., Jones, D. & Eggleton, P. (2014) *Earthworms in England: Distribution, Abundance and Habitats*. NECR145. Peterborough: Natural England. http://publications.naturalengland.org.uk/file/5824256822738944

Sherlock, E. (2018) *Key to the Earthworms of the UK and Ireland*. AIDGAP OP180. 2nd edition. Shrewsbury: Field Studies Council.

Sims, R.W. & Gerard, B.M. (1999) *Earthworms*. Linnean Society Synopses of the British Fauna (New Series) 31 (revised). Shrewsbury: Field Studies Council.

Skinner, G.J. & Jarman, A. (2024) *Ants*. 2nd edition. Naturalists' Handbooks 24. Exeter: Pelagic Publishing.

Smith, K.G. (1989) *An Introduction to the Immature Stages of British Flies: Diptera Larvae, with Notes on Eggs, Puparia and Pupae*. Royal Entomological Society Handbooks for the Identification of British Insects Vol. 10, Pt 14. London: Royal Entomological Society. https://www.royensoc.co.uk/wp-content/uploads/2022/01/Vol10_Part14_MainText.pdf

Southwood, T.R.E. & Leston, D. (1959) *Land and Water Bugs of the British Isles*. London: Frederick Warne & Co. (Available on CD-ROM and as a facsimile copy produced by Pisces Conservation – https://www.pisces-conservation.com/softlwb.html.)

Speight, M.C.D. (1989) *Saproxylic Invertebrates and their Conservation*. Strasbourg: Council of Europe.

Spooner, B. & Roberts, P. (2005) *Fungi*. New Naturalist 96. London: HarperCollins.

Sterling, P., Parsons, M.S. & Lewington, R. (2018) *Field Guide to the Micro-Moths of Great*

*Britain and Ireland*. London: Bloomsbury Publishing.

Stokland, J.N., Siitonen, J. & Jonsson, B.G. (2012) *Biodiversity in Dead Wood*. Cambridge: Cambridge University Press. https://doi.org/10.1017/CBO9781139025843

Strachan, R. (2010) *Mammal Detective*. Linton: Whittet Books.

Stubbs, A.E. (1972) Wildlife conservation and dead wood. *A supplement to the Journal of the Devon Trust for Nature Conservation*, 1–18.

Sutton, P.G. (2015) *A Review of the Orthoptera (Grasshoppers and Crickets) and Allied Species of Great Britain. Orthoptera, Dictyoptera, Dermaptera, Phasmida*. Natural England Commissioned Report NECR187. Species Status 21. Peterborough: Natural England. http://publications.naturalengland.org.uk/file/5727271299055616

Sutton, S.L. (1980) *Woodlice*. Oxford: Pergamon Press.

Swift, M.J., Heal, O.W. & Anderson, J.M. (1979) *Decomposition in Terrestrial Ecosystems*. Studies in Ecology 5. Oxford: Blackwell Scientific Publications.

Telfer, M.G. (2016) *A Review of the Beetles of Great Britain: Ground Beetles (Carabidae)*. Natural England Commissioned Report NECR189. Species Status 25. Peterborough: Natural England. http://publications.naturalengland.org.uk/publication/6270849377107968

Terrell-Nield, C. (1990) Distribution of leg-colour morphs of *Pterostichus madidus* (F.) in relation to climate. In: Stork, N.E. (ed.) *The Role of Ground Beetles in Ecological and Environmental Studies*. Andover: Intercept. pp. 39–51.

Thiele, H-U. (1977) *Carabid Beetles in their Environments: A Study on Habitat Selection by Adaptations in Physiology and Behaviour*. Berlin: Springer-Verlag.

Tilling, S.M. (2014) *A Key to the Major Groups of British Terrestrial Invertebrates*. 2nd edition. AIDGAP OP167. Shrewsbury: Field Studies Council.

Tottenham, C.E. (1954) *Coleoptera: Staphylinidae (Part). Section (a) Piestinae to Euaesthetinae*. Royal Entomological Society Handbooks for the Identification of British Insects Vol. 4, Pt 8a. London: The Royal Entomological Society. https://www.royensoc.co.uk/wp-content/uploads/2021/12/Vol04_Part08a.pdf

Tronquet, M. (2006) *Catalogue Iconographique des Coléoptères des Pyrénées-Orientales. Volume 1: Staphylinidae*. Perpignan: Association Roussillonnaise d'Entomologie.

Trudgill, S. (1989) *Soil Types: A Field Identification Guide*. AIDGAP OP196. Shrewsbury: The Field Studies Council.

Unwin, D.M. (1981) *A Key to the Families of British Diptera*. AIDGAP OP143. Shrewsbury: Field Studies Council.

Unwin, D.M. (1984) *A Key to the Families of British Beetles*. AIDGAP OP166. Shrewsbury: Field Studies Council.

Unwin, D.M. (2001) *A Key to the Families of British Bugs (Insecta, Hemiptera)*. AIDGAP OP269. Shrewsbury: Field Studies Council.

Unwin, D.M. & Corbet, S.A. (1991) *Insects, Plants and Microclimate*. Naturalists' Handbooks 15. Slough: The Richmond Publishing Co. Ltd.

USHA/UCEA (2011) *Guidance on Health and Safety in Fieldwork*. Brighton and London: The Universities Safety and Health Association (USHA) in association with the Universities and Colleges Employers Association (UCEA). https://www.stir.ac.uk/media/stirling/services/policy-and-planning/documents/guidance-on-health-and-safety-in-fieldwork.pdf

Vaucel, J., du Plouy, N.L.B., Courtois, A., Bragança, C. & Labadie, M. (2020) *Euscorpius flavicaudis* sting is not lethal but not harmless either: First record of neurological symptoms in child after sting. *Toxicologie Analytique et Clinique* 32(1): 85–88. https://doi.org/10.1016/j.toxac.2019.09.003

Wallwork, J.A. (1970) *Ecology of Soil Animals*. Berkshire: McGraw-Hill Publishing Co. Ltd.

Wallwork, J.A. (1976) *The Distribution and Diversity of Soil Fauna*. London: Academic Press.

Waring, P., Townsend, M. & Lewington, R. (2018) *Field Guide to the Moths of Great Britain and Ireland*. London: Bloomsbury Publishing.

Watkinson, S.C., Boddy, L. & Money, N. (2015) *The Fungi*. London: Academic Press.

Wheater, C.P. (1987) Measurement of the activity of mesofauna using an actograph-microcomputer system. *Pedobiologia* 31: 219–222.

Wheater, C.P. & Cook, P.A. (2000) *Using Statistics to Understand the Environment*. London: Routledge.

Wheater, C.P. & Cook, P.A. (2015) *Studying Invertebrates*. Naturalists' Handbooks 28. Exeter: Pelagic Publishing.

Wheater, C.P., Cook, P.A. & Bell, J.R. (2020) *Practical Field Ecology: A Project Guide*. 2nd edition. Chichester: Wiley-Blackwell.

Wijnhoven, H. (2009) *De Nederlandse Hooiwagens (Opiliones)*. Entomologische Tabellen 3, Nederlandse Entomologische Vereniging. Leiden: Museum Naturalis (English translation of the text available at: http://srs.britishspiders.org.uk/portal.php/p/Harvestman+Resources).

Wilmer, P. (1985) *Bees, Ants and Wasps: A Key to Genera of the British Aculeates*. AIDGAP OP7. Shrewsbury: Field Studies Council.

Wong, H.R. (1972) *Literature Guide to Methods for Rearing Insects and Mites*. Alberta: Canadian Forestry Service.

Yeo, P.F. & Corbet, S.A. (2015) *Solitary Wasps*. Naturalists' Handbooks 3. 2nd edition. Exeter: Pelagic Publishing.

Zimmer, M., Kautz, G. & Topp, W. (2005) Do woodlice and earthworms act synergistically in leaf litter decomposition? *Functional Ecology* 19: 7–16. https://doi.org/10.1111/j.0269-8463.2005.00926.x

# Index

References to figures and photographs appear in *italic* type; those in **bold** type refer to tables.